Dynamics of Passive Margins

Dynamics of Passive Margins

Edited by R. A. Scrutton

Geodynamics Series
Volume 6

American Geophysical Union
Washington, D.C.

Geological Society of America
Boulder, Colorado

1982

Final Report of Working Group 8, Dynamics of Passive
Margins, coordinated by A. L. Hales on behalf of the
Bureau of Inter-Union Commission on Geodynamics

American Geophysical Union, 2000 Florida Avenue, N.W.
   Washington, D.C.   20009

Geological Society of America, 3300 Penrose Place; P.O. Box 9140
   Boulder, Colorado   80301

Library of Congress Cataloging in Publication Data

Main entry under title:

Dynamics of passive margins.

   (Geodynamics series; v. 6)
   Includes bibliographies.
   1. Continental margins. 2. Geodynamics.
I. Scrutton, R. A. II. Series.
GC84.D96   551.4'1   81-20504
ISBN 0-87590-509-9   AACR2

Printed in the United States of America

CONTENTS

After a decade of intense and productive scientific cooperation between geologists, geophysicists and geochemists the International Geodynamics Program formally ended on July 31, 1980. The scientific accomplishments of the program are represented in more than seventy scientific reports and in this series of Final Report volumes.

The concept of the Geodynamics Program, as a natural successor to the Upper Mantle Project, developed during 1970 and 1971. The International Union of Geological Sciences (IUGS) and the International Union of Geodesy and Geophysics (IUGG) then sought support for the new program from the International Council of Scientific Unions (ICSU). As a result the Inter-Union Commission on Geodynamics was established by ICSU to manage the International Geodynamics Program.

The governing body of the Inter-Union Commission on Geodynamics was a Bureau of seven members, three appointed by IUGG, three by IUGS and one jointly by the two Unions. The President was appointed by ICSU and a Secretary-General by the Bureau from among its members. The scientific work of the Program was coordinated by the Commission, composed of the Chairmen of the Working Groups and the representatives of the national committees for the International Geodynamics Program. Both the Bureau and the Commission met annually, often in association with the Assembly of one of the Unions, or one of the constituent Associations of the Unions.

Initially the Secretariat of the Commission was in Paris with support from France through BRGM, and later in Vancouver with support from Canada through DEMR and NRC.

The scientific work of the Program was coordinated by ten Working Groups.

WG 1 Geodynamics of the Western Pacific-Indonesian Region

WG 2 Geodynamics of the Eastern Pacific Region, Caribbean and Scotia Arcs

WG 3 Geodynamics of the Alpine-Himalayan Region, West

WG 4 Geodynamics of Continental and Oceanic Rifts

WG 5 Properties and Processes in the Earth's Interior

WG 6 Geodynamics of the Alpine-Himalayan Region, East

WG 7 Geodynamics of Plate Interiors

WG 8 Geodynamics of Seismically Inactive Margins

WG 9 History and Interaction of Tectonic, Metamorphic and Magmatic Processes

WG 10 Global Syntheses and Paleoreconstruction

These Working Groups held discussion meetings and sponsored symposia. The papers given at the symposia were published in a series of Scientific Reports. The scientific studies were all organized and financed at the national level by national committees even when multinational programs were involved. It is to the national committees, and to those who participated in the studies organized by those committees, that the success of the Program must be attributed.

Financial support for the symposia and the meetings of the Commission was provided by subventions from IUGG, IUGS, UNESCO and ICSU.

Information on the activities of the Commission and its Working Groups is available in a series of 17 publications: Geodynamics Reports, 1-8, edited by F. Delany, published by BRGM; Geodynamics Highlights, 1-4, edited by F. Delany, published by BRGM; and Geodynamics International, 13-17, edited by R. D. Russell. Geodynamics International was published by World Data Center A for Solid Earth Geophysics, Boulder, Colorado 80308, USA. Copies of these publications, which contain lists of the Scientific Reports, may be obtained from WDC A. In some cases only microfiche copies are now available.

This volume is one of a series of Final Reports summarizing the work of the Commission. The Final Report volumes, organized by the Working Groups, represent in part a statement of what has been accomplished during the Program and in part an analysis of problems still to be solved. This volume from Working Group 8 (Chairman, M.H.P. Bott) was edited by R. Scrutton.

At the end of the Geodynamics Program it is clear that the kinematics of the major plate movements during the past 200 million

years is well understood, but there is much
less understanding of the dynamics of the
processes which cause these movements.

   Perhaps the best measure of the success
of the Program is the enthusiasm with
which the Unions and national committees
have joined in the establishment of a
successor program to be known as:
Dynamics and evolution of the lithosphere: The framework for earth resources
and the reduction of the hazards.

   To all of those who have contributed
their time so generously to the Geodynamics
Program we tender our thanks.

   C. L. Drake, President ICG, 1971-1975

   A. L. Hales, President ICG, 1975-1980

Members of Working Group 8:

   M.H.P. Bott
   H.A. Asmus
   C.C. Bon der Borch
   R.V. Dingle
   O. Eldholm
   C. Keen
   I.P. Kosminskaya
   H. Martin

   L. Montadert
   P.D. Rabinowitz
   R. Scrutton
   E. Seibold
   E.S.W. Simpson
   P. Vincent
   J.S. Watkins

PREFACE

The contributions to this volume have been organized into three parts reflecting the three pronged approach to the study of passive continental margins that has been made during the International Geodynamics Project. The first part contains six papers dealing with the data from some of the passive margins. These observational data are essential for testing of hypotheses with regard to the evolution of this type of margin. In general multi-channel seismic reflection data are regarded as providing essential information on structural conditions and evolution of continental margins. However it is probably equally important from the point of view of understanding the geodynamics involved that there should have been deep crustal studies of at least some of the margins. Recognizing that geophysical studies show only the structure as it is today recourse must be made to stratigraphic studies in order to obtain guides to the past development of the margins. The papers in Part One cover all these aspects of passive margin development.

The three papers in Part Two concentrate on one characteristic common to a number of margins as it appears around the world. These papers bridge the gap between observation and theory, on the one hand reviewing observations relevant to the particular characteristic while on the other examining evolutionary mechanisms that might explain that characteristic. A gravity anomaly indicator of the continent-ocean boundary may be a world-wide phenomenon whereas evaporite basins and sheared margins are more local characteristics. As pointed out in the Introduction, hypotheses with regard to the evolutionary mechanisms must be able to explain local as well as global features.

Part Three contains the papers on the mechanisms proposed for the evolution of passive continental margins. In the first instance the mechanisms can be subdivided into those based on stress and those based on thermal factors, but in practice the effects are inter-related, and in the last three papers aspects of this inter-relationship are tackled. Neugebauer and Spohn investigate the body forces resulting from a basalt-eclogite phase change at the Moho that is temperature controlled, and Steckler and Watts consider the effects of sediment loading on a passive margin that is undergoing thermal contraction; but the last word has been given to Sloss. He believes that not only are processes inter-related at passive margins themselves but also between passive margins and continent interiors. A successor programme to the Geodynamic Project may well find itself devoting some time to this possibility.

Roger Scrutton

PASSIVE CONTINENTAL MARGINS: A REVIEW OF OBSERVATIONS AND MECHANISMS

R.A. Scrutton

Grant Institute of Geology, West Mains Road, Edinburgh, Scotland

Introduction

In the study of the geodynamics of passive
continental margins greatest emphasis has been
placed on the evolution of the crystalline crust
and deeper rocks. Contributions from detailed
studies of the sedimentary succession have
perhaps played a smaller, but nonetheless
important role in determining the mechanisms of
evolution. As a consequence, this review concentrates on the crystalline rocks, only mention-
ing the important aspects of the sediments in
passing. It also concentrates on the rifted
passive margins at the expense of the sheared
ones in view of the fact that the state of the
art on the latter is less advanced, and they are,
in terms of length, less important. A few
comments on sheared margins will be made now,
however.

A good indication of the way our knowledge of
passive margins has advanced in the last decade
is given by the relatively new awareness of
sheared margins. Prior to the 1970s it was
thought that passive margins were, in effect,
rifted margins, despite the work of Wilson (1965)
relating fracture zones to margin offsets. In
the early 1970s the rectilinear pattern of
passive margins was explained in terms of
orthogonal rifted and sheared segments, making
it possible to understand abrupt changes in
margin direction, such as at the southern tip of
South Africa (Francheteau and Le Pichon, 1972).
The formation of marginal plateaus and of microcontinents was also easily explained by invoking
the presence of sheared segments of margin and
their associated fracture zones (Falvey, 1972).
By the time the Geodynamics Project reached its
end a distinction between rifted and sheared
margins could be drawn from most types of
observational data (Scrutton, this volume).

The understanding of thermal and dynamic
mechanisms at rifted margins has advanced at a
rapid rate, helped by an ever-increasing input
of new geophysical and geological data of all
sorts from university, government and industrial research groups. Methods of investigation
have improved most importantly in three ways -
the wider use of multichannel seismic reflection,
the introduction of ocean-bottom seismometers for
deep seismic investigations, and the drilling of
a number of boreholes in passive margins by
"Glomar Challenger". Without such improvements
it is certain the advances would not have been
made so quickly.

Observations at Passive Margins

It now appears that at rifted margins there are
two kinds of feature: the ubiquitous and the
local. Among the ubiquitous is, naturally, the
continent-ocean boundary; tensional faulting,
continental crustal thinning and subsidence of
the continental and oceanic domains are also
common to all margins. Among the more local
features are extensive volcanism, sediment starvation, marginal plateaus and major transverse
structures. Mechanisms for margin development
must explain this fundamental division.

With regard to tensional faulting, two exciting
discoveries have been made. Firstly, on the
mature margins much of the faulting in continental basement and the overlying sediments
apparently ceased in about mid-Cretaceous times
(Kent, 1977). Of all observations, this is the
most difficult to explain. It requires a worldwide mechanism, probably of a thermal nature (see
Dingle, this volume). Secondly, it seems that
listric faults are the common type of tensional
fault at rifted margins (but see Le Pichon et
al., in press). They have been observed at the
margin in the northern Bay of Biscay (de Charpel
et al., 1978), and in Afar (Morton and Black,
1975) and the land-locked Basin and Range
Province (Profett, 1977). By rotation of blocks
bounded by the faults extension of the brittle
upper crust can take place (up to a factor of
c. 2 has been estimated for the North Sea
(Christie and Sclater, 1980)).

Continental crustal thinning at rifted margins
has been known about since early gravity (Vening
Meinesz, 1941) and seismic refraction (Drake &
Nafe, 1968) measurements. More and more data
from the transition zones between normal continental and oceanic crusts have shown that the
zone varies considerably in width, from no more
than a few tens of kilometres to as much as a few

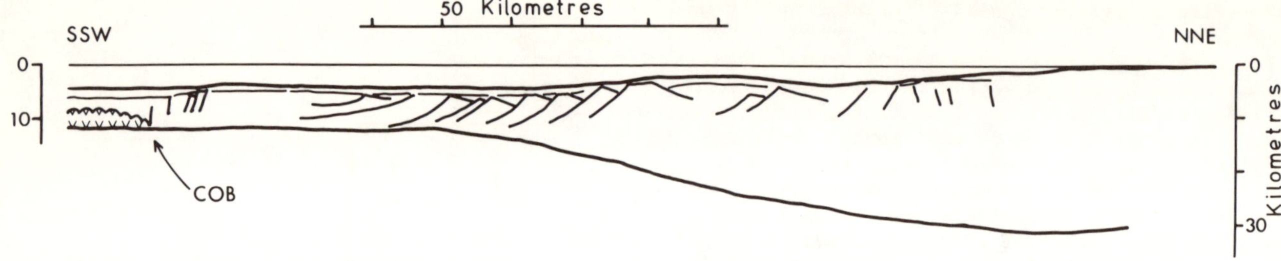

Fig. 1. Sketch of a crustal cross section in the northern Bay of Biscay extending from the Biscay abyssal plain to the Armorican shelf (redrawn from Guennoc, 1978). It illustrates drift-phase sediment starvation, listric faulting in rift-phase deposits and continental basement and drastic thinning of the lower continental crust. This was an area of rifting within a pre-existing basin where some crustal thinning may have already existed. COB: continent-ocean boundary.

hundred. Thinning may not take place uniformly across the zone. It is observed or inferred in several places to take place abruptly over a short distance with wider zones of gradual thinning to seaward and/or landward, for example beneath the marginal plateau of Goban Spur, southwest of Ireland (Scrutton, 1979). It is the zones of thin continental crust adjacent to the boundary that are attracting most interest at present. Keen (this volume) has suggested that there may be a minimum thickness of 15 km to which the crystalline continental crust will reduce. A thickness of only 6 km has been observed in the northern Bay of Biscay, however (de Charpel et al., 1978) (Fig. 1), which is little more than the upper, brittle part of the crust. Perhaps the thickness of the brittle layer, which varies from place to place, is the minimum to which the crust is reducible.

Knowing the location of the boundary between continental and oceanic crusts is crucial to identifying zones of greatly thinned continental crust. Observing the location in seismic data had been virtually impossible until deep penetration reflection systems were used in areas of limited drift-phase deposition. Thick sediments that mask the boundary in other areas had precluded the use of continuous reflection profiles whilst refraction techniques were not sensitive enough to define the feature. Nowadays, the change from faulted continental basement and rift valley deposits to volcanic oceanic basement can be seen under favourable conditions. It is still difficult to locate the boundary using seismic refraction, although use may be made of converted P-waves originating at the boundary, as has been done by Bott et al. (1976). The possibilities of there being thickened oceanic crust on one side of the boundary (Roots et al., 1979) and drastically thinned continental crust on the other hampers the use of crustal thickness changes as a guide to the location. The magnetic edge effect at the boundary between differently magnetised crusts remains a good method of detection despite possible limitations where the

boundary is deeply buried or where the nearby continental crust contains magnetization contrasts that could be confused with the boundary. During the Geodynamics Project a new indicator of the location of the boundary was suggested – a steep isostatic gravity gradient landward of a gravity high over a basement high in the oldest oceanic crust (Rabinowitz and La Brecque, 1977; Rabinowitz, this volume). This technique may work in some areas but there is little evidence as yet that the oceanic basement high is a widespread feature and it would be easy to confuse the genuine isostatic gradient with others due, say, to horsts in continental basement. A combination of seismic reflection and magnetic data is now probably the best tool for locating the continent-ocean boundary.

Substantial subsidence of rifted passive margins has also been known of for tens of years. Seismic refraction work carried out off the eastern United States in the 1950s revealed up to 10 km of Triassic and younger sediments (Drake et al., 1957). If these sediments accumulated by simply filling a pre-existing depression, the latter must have been (assuming local, Airy isostatic equilibrium) 7 km deep to begin with. This is unlikely, implying subsidence of the basin as sedimentation proceeded. An initial deep basin was completely disproved and extensive subsidence confirmed when deep drilling at rifted passive margins encountered largely shelf-type deposits. It has become apparent that limited, graben style subsidence occurs immediately prior to the onset of sea-floor spreading to be followed by more widespread subsidence, fault-controlled up to about mid-Cretaceous time but thereafter, flexural in nature. Even on rifted margins where there is little sediment to testify to the subsidence of continental crust, the planed off tops of fault blocks rotated during rifting suggest that these, now 3000 m or more deep in the Bay of Biscay (Montadert et al., 1977), were once near to sea level. A phase of erosion prior to subsidence is commonly observed in the sedimentary succession at rifted margins,

indicating an earlier period of crustal uplift or
doming.

The observation that oceanic crust subsides as
it increases in age away from the mid-ocean ridge
(Sclater et al., 1971) and that it is, except for
minor faulting, coupled to continental crust at
passive margins (but see Turcotte, this volume),
also leads to the conclusion that widespread
subsidence occurs in margin areas. If a ridge
crest subsides about 1-2 km since formation and
oceanic crust subsides about 2-3 km as it ages to
80 my, then the total of 3-5 km is the order of
margin subsidence in the absence of any other
effects.

Observing the less ubiquitous features of
rifted passive margins has proved to be as rewarding in terms of understanding margin geodynamics
as the study of the common features has been. By
analogy with continental rift zones and by geological and geophysical mapping of continental
shelves and adjacent landmasses it has been found
that dike intrusion and limited volcanism is
common in the continental crust at passive
margins. In a few places, however, the presence
of a hot spot trace leading away from a passive
margin (Morgan, 1972) and/or abundant volcanic
features on land has led to the conclusion that
rifted margins may experience excessive volcanism. Examples of this, in various stages of
development, are the Afar triangle, where volcanism has brought oceanic crust above sea level
(Barberi & Varet, 1978), the Greenland-Iceland-
Faeroes Ridge, and the Rio Grande Rise-Walvis
Ridge feature. These would qualify as the plume
areas of Kinsman (1975), who distinguishes plume
from inter-plume rifted margins (Fig. 2).

Another less ubiquitous feature is sediment
starvation of a rifted margin. Even as recently
as the beginning of the Geodynamics Project the
classical picture of a rifted passive margin
included a thick succession of sediments beneath
the shelf, slope and rise. Now a few areas have
been discovered where only 1-2 km of drift-phase
deposits occur. The slope and rise from Rockall
Trough to the Bay of Biscay in the N.E. Atlantic
is one area. Here, sediment source areas from
early Mesozoic times have fed the intracontinental basins of the shelf rather than the slope
and rise.

Major transverse structures at rifted margins
may be considered as local as opposed to ubiquitous features. In some cases these are aseismic
ridges, either of volcanic or tectonic origin.
The volcanic ones have already been mentioned and
may exist as continuous features or seamount
chains that meet the passive margin at a point
where volcanism may have been excessive. An
intriguing volcanic chain is the Cameroon line,
which crosses the African margin and is still
active. It apparently postdates the formation
of the margin, perhaps originating from the same
asthenospheric upwelling as the Benue Trough
(Fitton, 1980). The tectonic ridges are most
likely to arise from fracture zones, as the

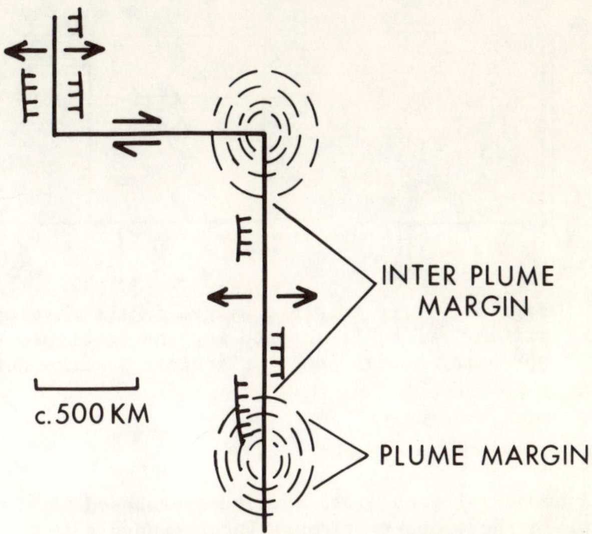

Fig. 2. The relationship of plume to inter-
plume rifted margins according to Kinsman
(1975). Hot-spot traces may develop from
the plume areas when the ocean opens.

Ninety East Ridge and Agulhas Ridge (du Plessis,
1977) have done.

Fracture zones are very important transverse
structures in their own right. Where they meet
passive margins, margin offsets occur, more often
than not too small to detect but reaching up to
1200 km long at the Falkland-Agulhas Fracture
Zone in the S. Atlantic (Scrutton, 1973).
Marginal plateaus and microcontinents can be
formed by an intersection of rifted and offset
margin segments, Falkland Plateau being an
excellent example of the former and Rockall
Plateau (related to the Gibbs Fracture Zone) of
the latter (Falvey, 1972) (Fig. 3). In some
places fracture zones impinging on the continent
are thought to line up with pre-existing tectonic
lineaments in the continental crust, perhaps one
of the best examples being the alignment of the
Romanche Fracture Zone with the eastern edge of
the West African craton. In other places, there
is an alignment of igneous events on the continent with the offshore fracture zone (e.g. Marsh,
1973), the events taking place at the same time
as the ocean opening, almost as though they are
exploiting a crustal weakness that was ancestral
to the fracture zone.

Finally, the distribution of salt-bearing
basins at rifted margins is seen to be related
to the presence of transverse structures (Rona,
1976, and this volume). A clear example of this
is the S. Atlantic basin between the Rio Grande-
Walvis Ridge feature and the equatorial fracture
zones. Palaeo-oceanographic reconstructions of
the South Atlantic (van Andel et al., 1977)
reveal that the transverse features would have
acted as barriers to water circulation in the
early stages of basin formation, thus allowing
evaporation and inducing salt deposition.

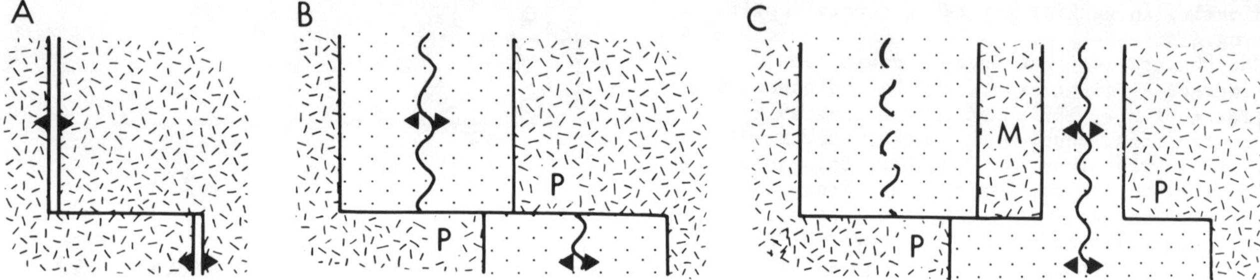

Fig. 3. Three possible stages in the development of a fracture zone at a passive margin. A,
rifting stage; B, sheared margins developed with P, potential marginal plateau areas; C, a
spreading centre jump can isolate a microcontinent M.

Mechanisms

A number of mechanisms have been proposed to explain these observations. Understandably they have concentrated on explaining the more ubiquitous features.

The Rifting Stage

The general idea of rifted passive margins evolving from intracontinental rift systems, with the processes operating at the rifts giving an indication of those operating in the early stages of formation of margins, is generally accepted. Doming of the continental crust often occurs with rifting and is almost certainly caused by thermal expansion of the lithosphere over and around a rising "diapir" of asthenosphere. Doming is most pronounced in areas of widespread igneous activity where the asthenosphere is forcibly intruding the lithosphere (Kinsman, 1975). Elsewhere, the asthenosphere probably only rises in response to stretching and thinning and rifting of the lithosphere and the release of confining pressure (Fig. 2). In these areas doming may be less pronounced but uplift of the rift flanks is likely because of heating and isostatic equilibration.

Geophysical studies in rift zones reveal that crustal thinning occurs in them, but in some it is in excess of that which can be produced purely by any extension that has occurred across the rifts. In these areas the Moho must have been raised by processes other than the stretching process. The processes proposed to account for extra thinning are a form of magmatic stoping or basification of the crust by rising lithosphere (van der Linden, 1977), and migration of the Moho as a part of the basalt-eclogite phase change under the influence of an increased geothermal gradient. Both processes have their problems: stoping of such large volumes of rock is disputed (Read, 1957), whilst the Moho need not necessarily be a phase change feature but a chemical boundary. More detailed work on the deep structure of rift zones and thinned continental crust at young rifted margins is required to solve this problem, as well as more definitive petrological models of the continental Moho.

That extension does take place in the continental crust at rift zones is evidenced by the presence of non-vertical or listric normal faults bounding subsided or rotated fault blocks. Doming of the crust can produce such faults (Cloos, 1939) but their existence in areas that have experienced little, if any, doming suggests that they can also be produced by tensional forces on a horizontal plate (Heiskanen & Vening Meinesz, 1958). By either method, the crust is stretched and thinned. In some areas, such as the Basin and Range Province, stretching of the continental crust can account for all the crustal thinning (Profett, 1977). However, fault block rotations of 20° were observed at the margin in the northern Bay of Biscay, giving rise to a crustal extension of only 10% (de Charpel et al., 1978). A 10% extension of an area would lead to a crustal thinning of no more than 10% assuming a cubical dilatation of 0. In the Biscay area the observed crustal thinning of over 50% cannot be explained by stretching alone and a mechanism, such as magmatic stoping must be invoked.

Much of what has just been said depends on the observed faults in fact being listric. Le Pichon et al. (in press) suggest that if what is observed is a complex zone of cross-cutting planar faults, the degree of stretching is compatible with the degree of crustal thinning.

During the rifting stage deposition of clastic material, volcanics and, perhaps, evaporites in graben act as loads on the lithosphere. Beaumont and Sweeney (1978) have shown that in the absence of thermal effects at depth a surface depression will be generated by the loading. Thus, potential subsidence of a rifted area exists prior to separation of the new passive margins. The extent to which thermally driven uplift offsets the loading or vice versa has not been studied.

The location of the ultimate split of the continents in the rift zone need not be central. Indeed, asymmetric splitting is advocated in a number of areas, e.g. Labrador Sea (Hinz and Schluter, 1979). A wide zone of rifted and already-thinned continental crust will exist on

one margin in such a case and this may well
behave differently in its subsequent evolution
from the margin with only a narrow zone of
attenuated continental crust. For instance, the
presence of new, hot upper mantle beneath the
wide attenuated continental crust (Fig. 4) will
give that area a greater potential for subsidence
on cooling. High geothermal gradients will be
experienced by sediments in these areas. From
purely isostatic consideration, the wide zone of
drastically thinned continental crust will find
equilibrium at depths nearer to those of oceanic
basement.

The Drifting Stage

Subsidence will take place following continental splitting as the continental lithosphere
cools and migrates away from the new sea-floor
spreading centre (Sleep, 1971 and this volume).
A depocentre is produced which will allow a sediment load to accumulate should source and trans-
portation conditions be favourable. It is now
known that the combined effect of thermal subsidence of continental and adjacent oceanic
lithosphere together with sediment loading can
explain the growth of a thick sedimentary
succession at a number of rifted passive margins
(Watts and Steckler, this volume).

Local sediment loading has been distinguished
from regional, flexural loading in such studies
and changes in the rigidity of the lithosphere as
it cools have been taken into account.

Seaward progradation of thick sedimentary
sequences, modified by gravity driven mass transport and bottom current erosion and redeposition
widens the load, especially at times of low sea-
level stands. The load encroaches on to oceanic
crust and masks the continent-ocean boundary.
Loading the oceanic lithosphere differs little
from loading the continental one in that the same
exponential decrease of subsidence rates with
time is observed. Different parameters for
plate thickness and rigidity are involved however.

Some mechanisms have been proposed for thinning
of the continental crust during the drifting
phase, but their effectiveness is difficult to
evaluate. The lower crustal creep hypothesis of
Bott (1971) depends on the lower crust having the
right rheological properties in the right thermal
regime to be effective. If the conditions are
right then the mechanism is a powerful means of
thinning the continental crust, leading to subsidence by isostatic equilibration. Increasing the
metamorphic grade of the continental crust in
rift zones and, thus, increasing its density,
leading to subsidence by isostatic equilibration
during later rifting and drifting has been proposed by Falvey (1972). This is dependent on the
metamorphic grade of the crust being low (green-
schist) initially, however, which contradicts
current ideas that the lower continental crust is
largely granulite facies rocks. Also, the possibility of retrogressive metamorphism taking place

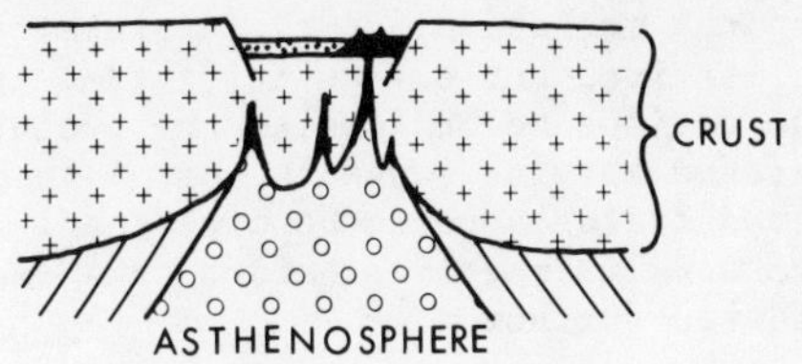

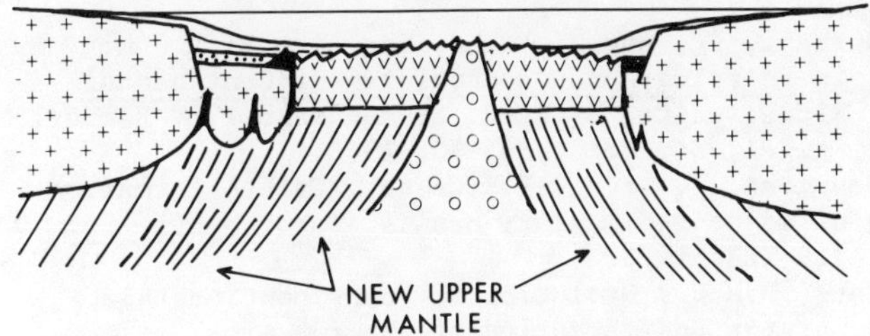

Fig. 4. Sketch to illustrate how different
opposing rifted margins can develop from
asymmetrical splitting in a rift zone. On
the left hand side of the lower diagram
highly attenuated continental crust is
underlain by "oceanic" upper mantle.
Vertical exaggeration approximately 2:1.

after the source of heat has   declined may
remove the density effect. Lastly, on the
assumption that the Moho represents one stage
in the basalt-eclogite phase transition, it is
thought that it will migrate upwards with time as
the geothermal gradient increases under a blanket
of sediments (Spohn and Neugebauer, 1978, and
this volume). This presumably occurs at an
advanced stage of drifting, since in the first
several tens of millions of years after break-up
the continental margin is cooling rather than
warming up. The unknown factors here are the
exact petrological nature of the Moho and the
detailed positions of the P, T boundaries of the
stability fields in the basalt-eclogite transition (see fig. 5.15 of Wyllie, 1971, and O'Hara
et al., 1971).

Conclusions

Our understanding of processes operating at
rifted passive margins has increased dramatically
over the past decade. Most success in explaining
subsidence of both oceanic and continental lithospheres has been achieved with the thermal
contraction-sediment loading model. Surface subsidence presumably also takes place to restore
the isostatic equilibrium of a thinned continental crust. It is the thinning mechanisms
that are perhaps least well understood at present.
Lack of detailed knowledge of mechanical and
petrological properties of the lower continental
crust and the way they change under differing
temperatures and pressures hinders evaluation of
most thinning mechanisms. In most cases it is
a matter of marginal change in the properties

making the mechanisms quite feasible.

Qualitative, but not quantitative, application of the various mechanisms to the evolution of sheared margins suggests that most of the observed features could be explained. Quantitative work is required before this can be considered further.

References

Barberi, F., and J. Varet, The Afar Rift Junction, in Petrology and Geochemistry of Continental Rifts, edited by E.-R. Neumann and I.B. Ramberg, Reidel, Dordrecht, 55-70, 1978.

Beaumont, C., and J.F. Sweeney, Graben formation of major sedimentary basins, Tectonophys. 50, T19-T23, 1978.

Bott, M.H.P., Evolution of young continental margins and formation of shelf basins, Tectonophys. 11, 319-327, 1971.

Bott, M.H.P., P.H. Neilson, and J. Sunderland, Converted P-waves originating at the continental margin between the Iceland-Faeroe Ridge and the Faeroe Block, Geophys. J.R. Astr. Soc. 44, 229-238, 1976.

Christie, P.A.F., and J.G. Sclater, An extensional origin for the Buchan and Witchground Graben in the North Sea, Nature, 283, 729-732, 1980.

Cloos, H., Hebung-Spaltung-Vulkanismus, Geol. Rundschau, 30, 405-527, 1939.

de Charpel, O., P. Guennoc, L. Montadert, and D.G. Roberts, Rifting, crustal attenuation and subsidence in the Bay of Biscay, Nature, 275, 706-711, 1978.

Drake, C.L., G.H. Sutton and M. Ewing, Continental margins and geosynclines, east coast of North America north of Cape Hatteras (abstract), Bull. geol. Soc. Amer. 68, 1718-1719, 1957.

Drake, C.L., and J.E. Nafe, The transition from ocean to continent from seismic refraction data, in Crust and Upper Mantle of the Pacific Area, edited by L. Knopoff, C.L. Drake and P.J. Hart, AGU Geophys. Mono. 12, 174-186, 1968.

du Plessis, A., Seafloor spreading south of the Agulhas fracture zone, Nature, 270, 719-721, 1977.

Falvey, D.A., The nature and origin of marginal plateaux and adjacent ocean basins off Northern Australia, Ph.D. thesis, Univ. New South Wales, 242 pp., 1972.

Fitton J.G., The Benue Trough and Cameroon Line - a migrating rift system in Africa, Earth Planet. Sci. Lett. 51, 132-138, 1980.

Francheteau, J., and X. Le Pichon, Marginal fracture zones as structural framework of continental margins in the South Atlantic Ocean, Bull. Amer. Assoc. Petrol. Geol. 56, 991-1007, 1972.

Guennoc, P., Structure et evolution geologique de la pente continentale d'un secteur de l' Atlantique nord-est: de la terrasse de Meriadzek a l'Eperon de Goban, Ph.D. thesis, Univ. de Bretagne Occidentale, 112 pp. + 56 figs., 1978.

Heiskanen, W.A., and F.A. Vening Meinesz, The Earth and its Gravity Field, McGraw-Hill, New York, 470 pp., 1958.

Hinz, K., and H.-U. Schluter, The North Atlantic - Results of Geophysical Investigations by the Federal Institute for Geosciences and Natural Resources on North Atlantic Continental Margins, Oil Gas - Europ. Mag. 3/79, 31-38, 1979.

Kent, P.E., The Mesozoic development of aseismic continental margins, J. geol. Soc. Lond. 134, 1-18, 1977.

Kinsman, D.J.J., Rift valley basins and sedimentary history of trailing continental margins, in Petroleum and Global Tectonics, edited by A.G. Fischer and S. Judson, Princeton Univ. Press, Princeton, 83-126, 1975.

Le Pichon, X., J. Angelier, and J.-C. Sibuet, Plate boundaries and extensional tectonics, Tectonophysics, in press.

Marsh, J.S., The origin of alkaline igneous rock lineaments in Africa and South America, Earth Planet. Sci. Lett. 18, 317-323, 1973.

Montadert, L., D.G. Roberts, G.A. Auffret, W. Bock, P.A. Du Peuble, E.A. Hailwood, W. Harrison, J. Kagami, D.N. Lumsden, C. Muller, D. Schnitker, R.W. Thompson, T.L. Thompson, and P.P. Timofeev, Rifting and subsidence on passive continental margins in the North-East Atlantic, Nature, 268, 305-309, 1977.

Morgan, W.J., Deep mantle convection plumes and plate motions, Bull. Amer. Assoc. Petrol. Geol. 56, 203-213, 1972.

Morton, W.H., and R. Black, Crustal attenuation in Afar, in Afar Depression in Ethiopia, edited by A. Pilgar and A. Rosler, Schweizerbart, Stuttgart, 55-65, 1975.

O'Hara, M.J., S.W. Richardson, and G. Wilson, Garnet-peridotite, stability and occurrence in crust and mantle, Contribs. Mineral. and Petrol. 32, 48-68, 1971.

Profett, J.M., Cenozoic geology of the Yerington district, Nevada, and implications for the nature and origin of Basin and Range faulting, Bull. geol. Soc. Amer. 88, 247-266, 1977.

Rabinowitz, P.D. and J.L. La Brecque, The isostatic gravity anomaly: key to the evolution of the ocean-continent boundary at passive continental margins, Earth Planet. Sci. Lett. 35, 145-150, 1977.

Read, H.H., The Granite Controversy, Arrowsmith, Bristol, 430 pp.

Rona, P.A., Salt deposits of the Atlantic, in Continental Margins of Atlantic Type, edited by F.F.M. de Almeida, Anais. Acad. Brasil. Ciencias, 48, Supplement, 265-274, 1976.

Roots, W.D., J.J. Veevers, and D.F. Clowes, Lithospheric model with thick oceanic crust at the continental boundary: a mechanism for shallow spreading ridges in young oceans, Earth Planet. Sci. Lett. 43, 417-433, 1979.

Sclater, J.G., R.N. Anderson, and M.L. Bell, Elevation of ridges and evolution of the central Eastern Pacific, J. geophys. Res. 76, 7888-7915, 1971.

Scrutton, R.A., Structure and evolution of the sea-floor south of South Africa, Earth. Planet. Sci. Lett. 19, 250-256, 1973.

Scrutton, R.A., Structure of the crust and upper mantle at Goban Spur, southwest of the British Isles - some implications for margin studies, Tectonophys. 59, 201-215, 1979.

Sleep, N., Thermal effects of the formation of Atlantic continental margins by continental break-up, Geophys. J. R. Astr. Soc. 24, 325-350, 1971.

Spohn, T., and H. Neugebauer, Metastable phase transition models and their bearing on the development of Atlantic-type geosynclines, Tectonophys. 50, 387-412, 1978.

van Andel, Tj.H., J. Thiede, J.G. Sclater, and W.W. Hay, Depositional history of the South Atlantic Ocean during the last 125 million years, J. Geol. 85, 651-698, 1977.

van der Linden, How much continent under the oceans? Mar. geophys. Res. 3, 209-224, 1977.

Vening Meinesz, F.A., Gravity over the continental edges, Koninkl. Ned. Akad. Wetenschop. Proc. 44, 1941.

Wyllie, P.J., The Dynamic Earth, Wiley, New York, 416 pp., 1971.

GEOPHYSICAL STUDIES ON OCEANIC MARGINS OF THE USSR

I.P. Kosminskaya

Institute of Physics of the Earth, USSR Academy of Sciences

Abstract. Seismic, gravity and magnetic
observations in the Arctic and North Atlantic
Oceans allow the identification of a number of
relict-continental and oceanic features. The
entire polar branch of the mid-oceanic ridge
system constitutes the oceanic group. Relict
features flanking this carry thick sedimentary
sequences in places, show varying degrees of isostatic compensation and a poor degree of correlation between magnetic field characteristics and
bottom topography. Five stages of development
are proposed for the Arctic Ocean: the onset of
crust-mantle processes causing a rise in the continental Moho, mantle intrusion and thinning of
continental crust and subsidence, growth of
volcanoes at the site of the mid-oceanic ridge,
rise of the ridge as a topographic feature, and
development of a flood basalt pile throughout the
deepwater basins.

Introduction

This report gives a brief review of results of
geophysical studies carried out on passive margins of the USSR during the period of the International Geodynamics Project. At the author's
request, the materials for the report were
supplied by the Sakhalin Complex Research
Institute (A.A. Popov), Institute of Volcanology
of the Far East Scientific Center (S.A. Redotov,
S.T. Balesta), Yuzhmorgeo (I.A. Garkalenko),
Institute of Geology of the Arctic (Yu. G.
Kiselev). Numerous publications were also used;
a list of the most important is appended.

The Arctic Ocean washes the shores of the
Soviet Union in the north and in the south-west
it joins with the Northern Atlantic; in the east
it merges with the Bering Sea, which is a marginal sea of the Pacific Ocean. Large stretches
of coasts on the Black, Caspian and Baltic Seas
also belong to the Soviet Union. Though the
margins cover large distances, the studies of
relationships between continental and oceanic
structures in the USSR reach far beyond their
boundaries. Soviet research vessels study the
ocean bottom of the Mediterranean, conduct
observations in the Atlantic, Pacific and Indian
Oceans and off the coasts of Antarctica.

During the period of the Geodynamics Project
a vast amount of material has been collected on
the problem studied by Working Group 8*. This
report briefly touches principally upon new
results of research obtained by Soviet scientists in the regions of the Arctic and Northern
Atlantic. Attention is paid to geophysical
zoning of the indicated regions and to relations
between crustal structure and tectonics. An
essential asset in the studies is that all the
marine areas adjoin land covered by highly
credible geological and geophysical maps.

The Arctic

Geological-geophysical surveys of coastal zones
and deepwater basins were carried out during the
whole period of the Geodynamics Project. The
surveys were made mainly from ice floes and were
therefore called ice geophysics. They showed
that the largest continental structures of the
territory of the USSR, i.e. the Russian platform,
the Urals, the Siberian lowland, the Siberian
platform and other smaller tectonic elements,
continue to the north on the shelf of the Eurasian part of the Arctic (Yegiazarov, 1977).
There are several shallow shelf seas: the
Barents Sea, the Kara Sea, the Laptev Sea, the
East Siberian and Chukotka Seas. The transition
from the shelf shallow zone to the deepwater
Eurasian and Amerasian basins is within the
continental slope. Scarps, terraces, steps and
other forms of deep fragmentation, destruction
and subsidence of continental margins complicate
bottom topography against a background of
general irregularity (Fig. 1).

The most important data with regard to both
methods and results were obtained in deepwater

---

* The Soviet Working Group 8 of the Geodynamics
Project concentrates its studies on the inter-
action between oceanic and continental structures.
The list of members is as follows: N.A. Beliaev-
sky, R.I. Demenitskaya, I.P. Kosminskaya (Chair-
man), Yu. G. Kiselev, I.P. Kuzin (Scientific
Secretary), S.L. Soloviev, S.A. Fedotov, Yu. M.
Sheinman.

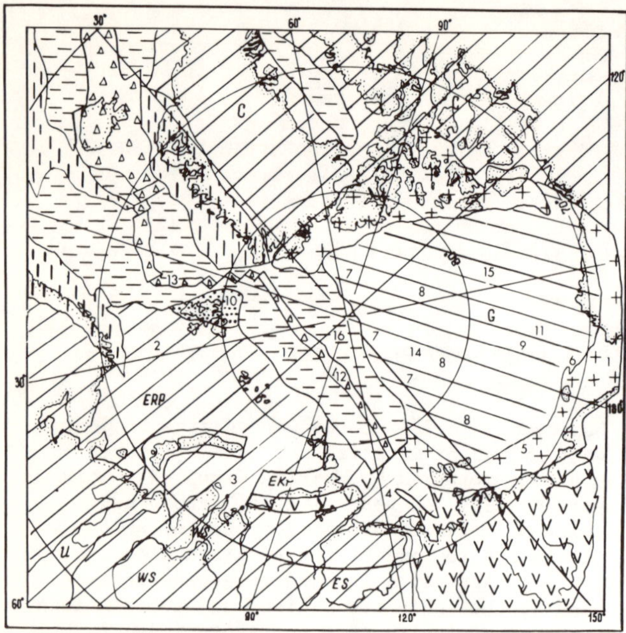

1 2 3 4 5 6 7 8

Fig. 1. key: 1 – Oceanic crust area; 2 –
Rift zone area; 3 – Area with transition type
of the crust; 4 – Area with continental crust,
ERP – East Russian platform, WS – West Siber-
ian plate, ES – East Siberian plate, C –
Canadian plate, U – Uralides, EKr – East
Karsk block; 5 – Atlantic orogenic belt;
6 – Arctic orogenic belt; 7 – Arctic–Atlantic
orogenic zone; 8 – Arctic–Pacific orogenic
zone.
Numbers on map: 1 – Bering Sea; 2 – Barents
Sea; 3 – Kara Sea; 4 – Laptev Sea; 5 – E.
Siberian Sea; 6 – Chukotka Sea; 7 – Lomonosov
Ridge; 8 – Mendeleiev Ridge; 9 – Chukotka
Dome; 10 – Yermak Plateau; 11 – Norvind
Plateau; 12 – Gakkel Ridge; 13 – Knipovich
Ridge; 14 – Makarov Ridge; 15 – Canadian
Basin; 16 – Eurasian Depression; 17 –
Nansen Depression.

basins bordering on the Arctic shelf. The bottom
of this micro-ocean differs considerably from the
pattern of the oceanic topography encountered in
the Atlantic and the Pacific oceans. Bottom
morphology of the Arctic basins has various positive structures which join the two continents,
America and Eurasia, a feature rarely observed in
the great oceans. In the Arctic we can easily
trace the structures and observe the changes in
them as they pass from the continental shelf to
deepwater oceanic basins.

The aseismic Lomonosov and Mendeleyev ridges,
the Chukotka dome, the Yermak and Norvind
plateaux are all classed as obvious relics of
continental structures by evidence of the

structure of their cover and basement. It should
be noted that seismic boundaries, recorded in the
sedimentary cover of the shelf, can be traced
continuously to deeply subsided ridges and
plateaux and oceanic basins.

It is an important fact that aseismic ridges
strike transversely to the Greenland and Eurasian
shelfs and their morphology is at considerable
variance with the active mid-oceanic ridges. In
the Arctic the only mid-oceanic ridge is the
Gakkel Ridge (Mid-Arctic); its strike is subparallel to the margin of the Barents Sea shelf.

According to the data of R.M. Demenitskaya and
Yu.G. Kiselev (1974), and Yu.C. Kiselev (1977), a
comparison of the bottom morphology of the shelf
and of the aforementioned uplifts gives evidence
that present-day sea-floor elevations in the deepwater part of the Arctic Ocean developed on
basements of both relict-continental and new
structures. They include in the first type the
Chukotka dome, the Norvind uplift, the Yermak
plateau, the Lomonosov and Mendeleyev ridges and
also the Faeroes-Icelandic and Icelandic-
Greenland thresholds, the Howgard threshold,
and the Voring plateau. The second type of
structures includes the entire polar branch of
the mid-oceanic ridges: Gakkel, Knipovich,
Mohns, and Kolbeinsey.

These authors state that all active ridges are
superimposed on oceanic basement which had
already formed by the time of their development
(Kiselev, 1977). In other words, the authors
reject the hypothesis of the opening of the
Arctic deepwater basins as the result of the
spreading process alone, i.e. moving apart of
continental blocks under the effect of growth
of the mid-oceanic ridge. They believe that
quite the opposite occurred; it started with
the collapse of continental margins and subsidence and transformation of the crust in parts
of margins; the crust was thinned almost to
oceanic thicknesses and on it later appear groups
of volcanoes piercing this thin crust and its
fairly thick sedimentary layer typical of the
Arctic basins.

Seismic Studies

The sedimentary layer covering deep basins and
shelf zones of the Arctic is studied in detail
by seismic methods. They reveal its stratification and divide it into three main intervals,
with the exception of the upper truly oceanic low
velocity pelagic sediments occasionally pierced
by plateau intrusions. The layer overlies the
basement of relict-continental structures. The
folded complex of the basement is dated as Middle
Riphean and Middle Paleozoic, in some places Late
Paleozoic, in age.

The first, lowest interval of the sedimentary
layer has a typically platform character well
studied on shelves. It was formed in the period
between the Middle Paleozoic and the Late Mesozoic. The composition of the interval, which

covers deep basins and uplifts, includes both
deepwater oceanic deposits and pre-oceanic, i.e.
protoshelf sediments. Seismic boundaries within
pre-oceanic formations are occasionally more
pronounced than the boundaries between the
oceanic and synoceanic rocks at shallower levels.

The second interval, traced from the shelf to
the basins, is almost horizontal with strongly
pronounced stratification and velocities of 2.5-
5.0 km/s. There is an obvious tendency to a
thickening of separate layers of the second
interval towards the centre of deepwater basins.
These are largely oceanic deposits.

The third interval, of synoceanic deposits, is
almost absent or very thin on the continental
slope, on shelves and horst-block uplifts. In
these regions the sedimentary layer is normally
represented by the lower part of the first and
by the second intervals.

A different pattern is observed in the regions
of intensive basin subsidence (the North Pole,
Toll, Makarov and Canadian basins). A steady
accumulation of substantial thicknesses of synoceanic deposits is observed towards the centre
of these basins; moreover, these deposits no
longer show definite signs of division into the
second and deeper intervals of the section, as
above the folded structural complexes.

The relationship of the sedimentary layer of
deep basins to the active Gakkel Ridge is peculiar. A complete section of the sedimentary
layer here is clearly defined only on the peri-
phery of the ridge. Towards the centre of the
basin, i.e. to the crest of the ridge, the layer
becomes much thinner. But even here, at the
approaches to the ridge, with continuous tracing
of reference seismic horizons from the edges of
the basin to its centre, a tendency is seen
towards a thickening of some of the layers,
which are already pierced and in some cases
essentially transformed by intrusive processes.

Potential Fields

The analysis of magnetic and gravity fields
over the deepwater regions of the Arctic basin
and their comparison with the fields for shelves
and continental margins show that magnetic and
gravity fields change considerably in the transition from the shelf and relict block structures
to the active mid-ocean ridges. Magnetic fields
on the shelves reflect mainly the morphology and
composition of the basement. For the Lomonosov
and Mendeleyev ridges the fields are similar to
those on shelves, i.e. there is no obvious correlation between bottom morphology and the field.
At the approaches to the mid-oceanic ridges,
however, this correlation becomes increasingly
pronounced, especially when moving south from
the Gakkel Ridge into the Northern Atlantic.

A similar pattern is true for the gravity field.
In the central part of the Arctic Ocean various
stages of isostatic compensation are observed.
The eastern part of the basin is characterised

by general recompensation, evidence of a later
stage of oceanisation of the region caused by
mantle upwelling and saturation of the crust by
heavy components. The present-day subsidence
here is still, however, uncompensated. This is
in contrast to the situation typical of the
Atlantic parts of the Arctic Ocean where the
field is in close correlation with topography.

Therefore, an evolution of geophysical fields
becomes obvious in the transition from deepwater
basins of the Arctic that are still forming to
fully formed basins and mid-oceanic ridges of
the Northern Atlantic, where both fields (gravity
and magnetic) are in good correlation with the
bottom topography of the ocean.

Development of the Arctic Basin and its Margins

Yu.G. Kiselev (1977) after many years of seismic studies and of other geophysicsl research in
the Arctic Basin, distinguishes several stages
in its development.

The first stage: crustal-mantle processes have
as yet no substantial effect on the transformation of the upper part of the (continental) crust
and on the sedimentary layer. This stage is
typical of depressions with ocean depths from
1800 to 3500 m (the North Pole, Toll, Severo-
mortsev depressions).

The second stage: a typical feature of this
stage is the process of transformation of not
only the uppermost crust, i.e. the basement, but
the folded sedimentary rocks and the lower levels
of the platform layer as well under the influence
of injection of mantle masses into the crust.
The ocean depths at this stage reach 3600-3900 m
(the Makarov and Canadian depressions).

The third and the fourth stages: at this
period appears the first evidence of the growth
of volcanoes (the mid-oceanic ridge) through the
sediment-filled depression, but as yet the volcanic ridge remains buried. The depths are now
3600-3800 m (the Labrador Sea). This is followed
by the extensive development of mantle diapirism
and the rise of the ridge with its clearly expressed topography. The depths vary from 3400 to
4300 m (the Eurasian basin, the Nansen and Amund-
sen depressions).

The last, fifth stage: a universal transformation of crustal structure takes place throughout
the deepwater basin; it is overlaid by a solid
basalt pile and the bottom topography alters
considerably (the Norwegian and Greenland Seas).

According to the author, the first and second
stages are reflected in the rise of the M discontinuity and in the reduction in thickness of
the layers of continental crust from the bottom
upwards. This process causes changes in gravity
and magnetic fields, which are unstable and exhibit features of preoceanic history, as observed,
in particular, in the Amerasian subbasin where
magnetic anomalies are at obvious variance with
the strike of submarine structures.

We observe the same variance on the Lomonosov

and Mendeleyev ridges and in the Toll, Makarov,
Canadian and Chukotka basins. Yu.G. Kiselev and
colleagues find the cause of the fields on these
different structures lies in the ancient basement
of the Hyperborean platform which is still intact
on them and which, in the past, was continuous
between these structures as a single unit.

During the third and fourth stages the active
mantle continues to rise into the crust, the
crust is further transformed, and basalt intrusions inject into the uppermost layers of the
crust. According to Yu.G. Kiselev, under the
synoceanic sediments there appears a second
oceanic layer, called in literature the acoustic
basement. This layer lies in remarkable unconformity with the subhorizontal deposition of the
sedimentary layer of the Arctic basins.

Intrusions of basalt masses and their appear-
ance immediately beneath the sea bottom causes
changes in the magnetic field; magnetic anomalies
are now found in good correlation with the features of bottom topography created by groups of
volcanoes (A.M. Karasik, 1973).

The fifth stage is described by Yu.G. Kiselev
as the period of full development of riftogenic
processes when basalt effusions cover large
territories reaching even beyond deepwater basins and affecting the development of passive
margins.

Northern Atlantic

Many of the features of the Arctic structures
have been observed in the Northern Atlantic as
well ("Nauka", 1977a, b). They are, in particular, the submerged transversal continental
structures, namely the Icelandic threshold,
the Faeroes-Scotland continental shelf, and
the peculiar structure of the Jan Mayen Ridge.
The Icelandic plateau has almost transformed
crust in the west and almost primary continental
crust in the east (Talwani, pers. comm., 1979).
The island of Iceland and its particularly thick
high velocity crust is being formed at the
"intersection" of a continental threshold
and the Reykjanes and Kolbeinsey rift ridges.
Nothing similar to this type of crust has been
as yet found either in the Arctic, or in any
other ocean.

Summary

All the cited material dealing with the structures of the Arctic and Northern Atlantic was
unfortunately obtained by means of different
techniques using various observation systems
and different approaches to the interpretation
of geophysical fields. This circumstance often
hinders and even obstructs comparison of many
physical parameters of the crust; to an even
greater extent it prevents justified generali-
sation of results. The conclusions offered
here were made mostly on the basis of analysis

of the major features of bottom topography, of
the sedimentary layer and of the general nature
of geophysical fields.

Data on the deep structure of the crust and
mantle are still very scarce. The progress in
the understanding of the structures of the Arctic and Northern Atlantic is, nevertheless, so
obvious and the results so essential, that they
may be used as the basis for the development of
all kinds of geological and geophysical studies
in these regions planned by any new global project to follow that on geodynamics.

It is now possible not only to draw maps of
the tectonics and history of formation of these
zones, but to establish the chronology of formation of structures of different types in the
process of disintegration of the protocontinent
of Laurasia to divide it into Eurasia and
America.

The authors studying the Arctic have arrived
at the conclusion that the process causing riftogenesis and the formation of the Mid-Atlantic
Ridge moves from south to north, varying under
different conditions as it encounters diverse
structures.

In the Arctic this process might be preceded
by the formation of deepwater basins and relics
of continents transformed by "mantle erosion".
In these regions the mid-oceanic ridges are
built up on the foundation of a thin oceanic
crust. In the Northern Atlantic riftogenesis
has developed on the foundation of both basins
(Norwegian, Greenland) and continental structures (Icelandic-Faeroes and Icelandic-Greenland
thresholds, the Rockall and Icelandic plateaux).
Oceanic crust formation cannot be considered
from this standpoint as the product of riftogenesis only, which is what is implied by the
basic concepts of plate tectonics hypothesis.

References

Demmitskaya, R.M., and Yu.G. Kiselev, To the
problem of continuation of structures in the
Eurasian basin of the Arctic Ocean, in Geophysical methods of reconnaissance in the
Arctic, Res. Inst. Geol. Arctic, 9, 94-96,
1974.

Karasik, A.M., Anomalous magnetic field of the
Eurasian sub-basin of the Arctic Ocean,
Doklady, 211, 86-89, 1973.

Kiselev, Yu.G., Structural evolution of sedimentary basins in the deepwater part of the
Arctic Ocean, Theses of papers, 5th Meeting
of Soviet Oceanographers, 1977.

"Nauka", Iceland and the Mid-Ocean Ridge, Deep
Structure, Seismicity, Geothermics, 1977a.

"Nauka", Iceland and the Mid-Ocean Ridge, Ocean
Bottom Structure, 1977b.

Yegiazarov, B.Kh. (Ed.), Memorandum to the
tectonic map of the Northern Polar Region
of the Earth, on scale 1:5000000, 200 pp,
1977.

STRUCTURE AND DEVELOPMENT OF THE POLAR MARGIN OF NORTH AMERICA

J.F. Sweeney

Earth Physics Branch, Department of Energy Mines and Resources, Ottawa, Canada K1A 0Y3

Abstract. Although the data are sparse, stratigraphic and paleontologic information combined with crustal seismic results indicate that the North American margin of the Canada Basin was created during the Early Cretaceous but not later than late Early Cretaceous time (about 120 Ma). The evidence suggests that the present polar margin resulted from rifting and separation of an adjacent continental mass. There may have been an initial rifting pulse at the outset of the Mesozoic but the present continental margin did not develop at this time. During the Early Paleozoic a continental block approached an earlier North American polar margin as intervening oceanic crust was eliminated between the encroaching landmasses until their suturing in mid-Paleozoic time. In the present, thick young sediment wedges appear to be the main source of magnetic anomaly lows and gravity anomaly highs along the polar shelf. Contemporary seismicity along the margin may be produced by stresses generated by the prograding sediment load in combination with a lowered threshold for seismic activity at the continental boundary.

Introduction

The polar continental shelf of North America lies adjacent to the Makarov and Canada basins of the Arctic Ocean (Figure 1). The easternmost part of this margin from the Alpha Ridge to the Lincoln Shelf remains largely unexplored. This discussion therefore focuses on the margin that faces the Canada Basin and emphasizes zones within Canada in which existing data are most closely spaced and interdisciplinary in nature.

Although there is now wide agreement that the Canada Basin is floored by oceanic crust and bordered by a rifted margin, the time and mode of origin of both the Canada and Makarov basins remains uncertain. An attempt is made here to resolve the uncertainty about the age of the Canada Basin by bracketing the most probable time of origin of the fringing continental boundary.

Because the data are few this review will address only the most fundamental questions regarding the Arctic margin, namely, its present gross structure and its geological history. The report presents geophysical data collected along the margin and uses this information to assess some major features of present margin structure and dynamics. Geological and crustal seismic data are then combined to reconstruct the probable sequence of Phanerozoic tectonic events that affected the region of the polar boundary both before and after its inception.

Physiography and Geophysical Character

The gross topography of the North American polar continental boundary is generally well defined except north of Axel Heiberg and Ellesmere Islands where few bathymetric soundings have been made (Sobczak and Sweeney, 1978). The shelf is relatively narrow and ranges between 20 and 160 km wide. The depth of the shelf break generally increases eastward from about 70 m north of Alaska to over 650 m northwest of Ellef Ringnes Island (Figure 1) and is said to reach 800 m to the west of the Mackenzie Delta (Grantz et al., 1979). North of Ellesmere Island the shelf break may not exist where the the Alpha and Lomonosov ridges approach the continent.

Overall the shelf bathymetric gradient is quite gentle and does not exceed 4 m/km. The gradient of the continental slope is more variable and ranges between 9 and 60 m/km (Sobczak and Weber, 1973).

Magnetic Anomalies

Aeromagnetic data over the Canadian Arctic margin have been collected mainly at moderate to wide line spacings of at least 20 km at high altitudes (3.5 to 5.5 km) (King et al., 1966; Haines, 1967; Riddihough et al., 1973; Haines and Hannaford, 1974; Coles et al., 1976) and at low altitudes (300 m) (Ostenso and Wold, 1971). The coverage extends from the Lincoln Shelf west to Banks Island (Figure 2). A poorly defined weak (<200 nT) magnetic low, locally prominent, lies over much of the polar margin and separates a landward relatively featureless anomaly field from a seaward zone of moderately intense short

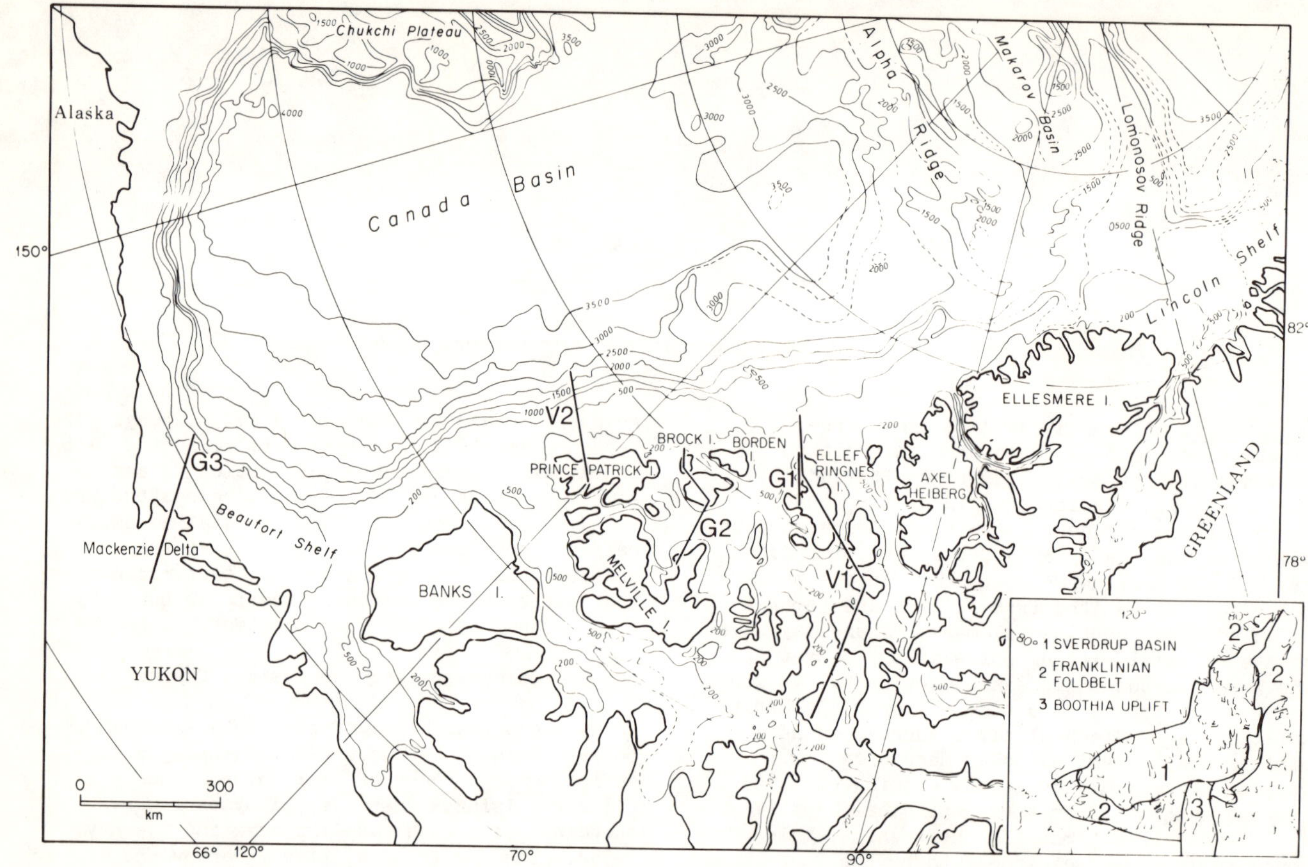

Fig. 1. North American polar margin. Bathymetric contour interval 500 m. Also shown is 200 m contour (from Sobczak and Sweeney, 1978). Profile locations of Figures 5, 6, 7, 8 and 9 are given.

wavelength broadly irregular anomalies (Coles et al., 1976). Detailed low level coverage north of Ellef Ringnes and Borden islands indicates that the shoreward flank of the margin anomaly low is locally bounded by a weak (< 200 nT) magnetic high along the outer shelf (Bhattacharyya, 1968).

North of Ellesmere Island a zone of very intense magnetic anomalies over the Alpha Ridge passes uninterrupted onto the continental shelf (Riddihough et al., 1973; Hood and Bower, 1976; Coles et al., 1976) and the polar margin low is displaced to the south over the exposed belt of Devonian and earlier Paleozoic rocks of the Franklinian foldbelt (inset, Figure 1). The continuation of this low over the continental margin to the southwest may indicate that Franklinian rocks are present in the basement below the polar shelf edge from Axel Heiberg to Banks Island. Bhattacharyya (1968), however, thought that the anomaly low was produced either by sediment filled grabens along the polar continental slope or by a fossil oceanic trench.

Gravity Anomalies

The free air gravity field over the polar margin forms a sinuous belt of elliptical highs that closely follows the trend of the continental shelf break from the Alpha Ridge to the Chukchi Plateau (Figure 3). The anomalies have pronounced gradients with amplitudes in excess of 80 mgal as is usual along passive margins (Worzel, 1968). The positive zone is flanked both seaward and shoreward by parallel but more diffuse belts of negative anomalies with -20 to -50 mgal amplitudes.

The positive gravity anomalies are produced by a combination of two factors. Always present is an edge effect associated with the zone of crustal thinning between the continent and the ocean basin. To this is added the gravity effect of either a basement ridge or a high density zone along the outer shelf or a seaward thickening sediment wedge across the margin (Sobczak, 1975a).

Wold et al. (1970) reproduced the gravity

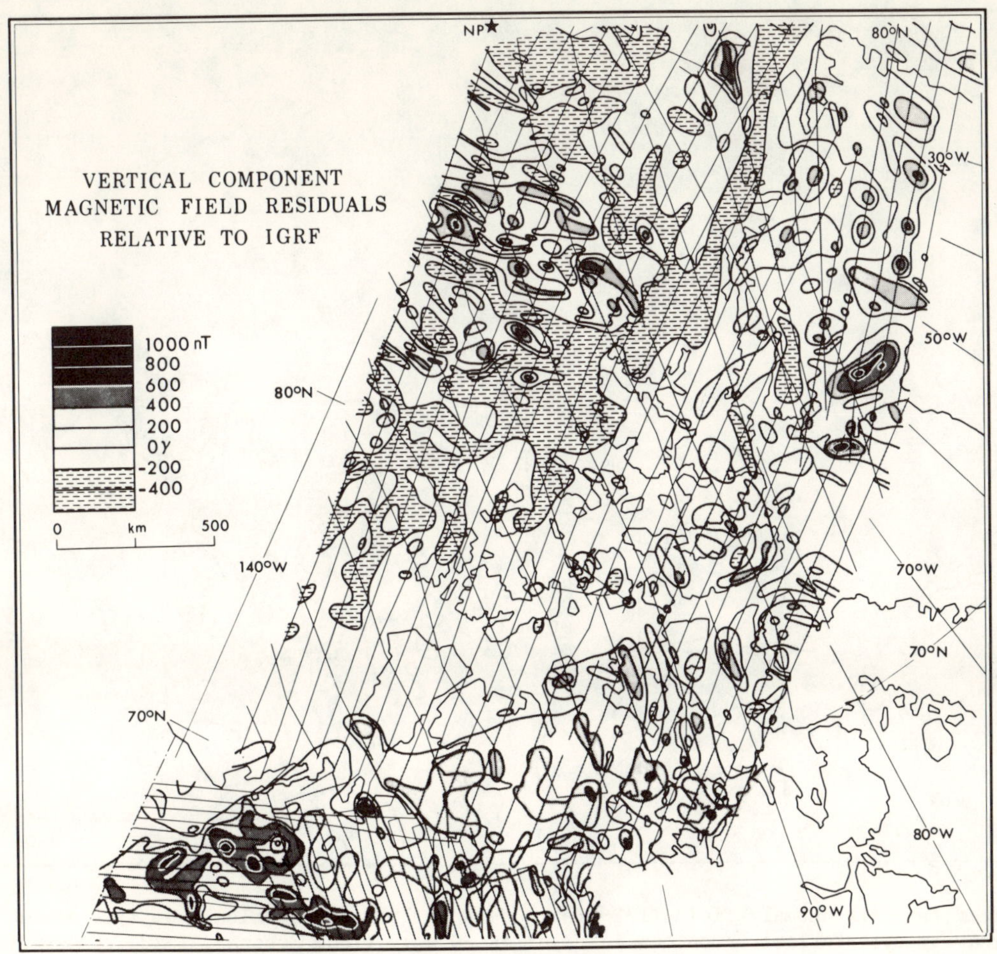

Fig. 2. Vertical component magnetic field residuals, relative to the IGRF, on Lambert Conformal Conic Projection, standard parallels 49°N and 77°N. Contour interval 200 nT (from Coles et al., 1976).

highs across the Beaufort Shelf with two-dimensional models of a basement ridge combined with an edge effect. This interpretation requires a very limited (less than 100 km wide) transition zone that is 50 to 100 km shoreward from its usual location, the 2000 m isobath (Worzel, 1968), along passive margins. The shoreward displacement of the gravity peak, Wold et al. suggested, may be caused by the negative gravity effects of large accumulations of low density sediments seaward of the "structural" margin. This explanation implies either that the rate of sediment outbuilding into Canada Basin has been faster than that along other passive margins or that this margin has been receiving sediments over a much longer period of time (i.e., Canada Basin is quite old). Sobczak (1975b, 1977), on the other hand, treated known thicknesses of prograded Cenozoic sediments as a source of positive gravity by considering them as an

isostatically uncompensated load on an elastic lithosphere and generated about 80 percent of the observed Beaufort Shelf free air high by means of sedimentary outbuilding along the margin.

Yorath and Norris (1975) cited offshore truncation of Upper Cretaceous sediments and fault offsets of progressively younger units toward the outer Beaufort margin as evidence of an uplifted fault bounded ridge along the continental edge. Evidence of the ridge in the magnetic anomaly field is not apparent for, although magnetic lineations along the outer shelf and slope have been noted (King et al., 1966; Bhattacharyya, 1968; Riddihough et al., 1973), they are discontinuous and of low amplitude. Preliminary analyses of unpublished magnetic data indicate that magnetic basement is deeply buried beneath the Beaufort Shelf (Coles et al., 1978). Using density contrasts established from offshore boreholes, Sobczak

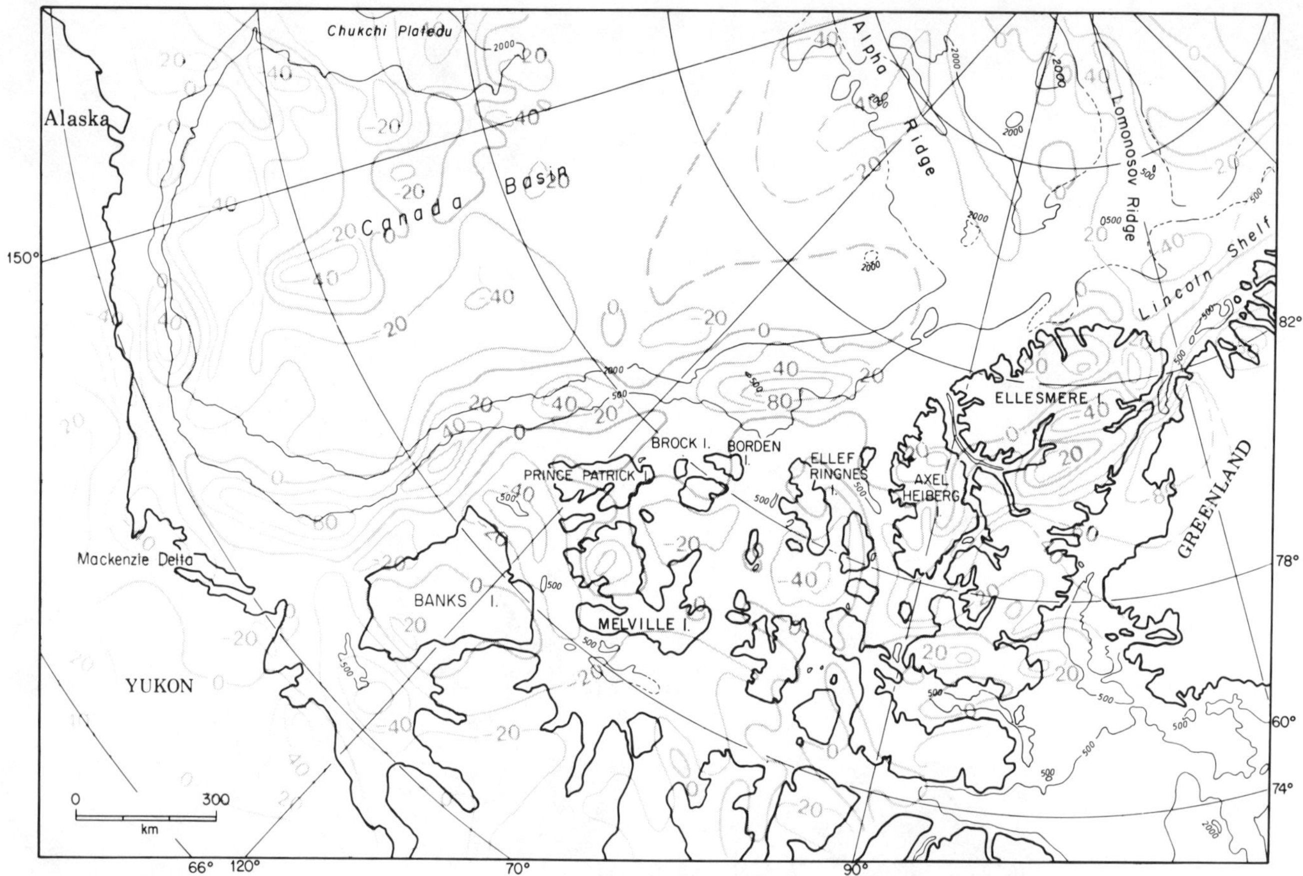

Fig. 3. Mean free air anomaly field (1/2° latitude x 2° longitude) calculated from about 15,000 offshore gravity observations. Contour interval is 20 mgal (from Sobczak, 1978). Also shown are 500 m and 2000 m isobaths.

(1977) showed that a shelf edge basement ridge can produce no more than a small fraction of the gravity high over the Beaufort margin, even for relatively extreme basement relief (4.2 km, Wold et al., 1970).

High density belts within the basement were proposed along the outer Norwegian shelf from the presence of large magnetic anomalies and the strong association of gravity and magnetic lineations at the margin (Talwani and Eldholm, 1972). The Canada Basin margin is characterized by low amplitude weakly lineated magnetic anomalies that correlate poorly with gravity anomalies along the shelf and slope. The available magnetic evidence therefore does not support the existence of dense basement zones along much of the North American polar continental boundary.

Basement ridges along the outer shelf to the east of the Mackenzie Delta are also unlikely in view of the magnetic low and the seismic evidence, described below, of thick sediment wedges along that margin. Very low heat flow values (mean = 19.2 mWm^{-2}) from the outer

shelf near Prince Patrick Island (Patterson and Law, 1966) may also reflect the presence of thick young sediments, but the significance of these results is not clear as heat flow values generally decrease away from the Sverdrup Basin in this region.

Seismicity

Clusters of low level seismic activity are located on the continental shelf of northeastern Alaska and on the continental slope seaward of the Beaufort Shelf and Ellef Ringnes Island (Basham et al., 1977; Wetmiller and Forsyth, 1978) (Figure 4). Except for the events at Martin Point, which may be shallow focus, seismic activity originates in the lower crust - upper mantle and all but a few events are less than magnitude 5. Most epicenters lie along the steep gradients of the major gravity highs along the margin. A prograded Cenozoic clastic wedge that best accounts for the positive gravity anomaly along the Beaufort Shelf edge may produce the deviatoric stress field responsible

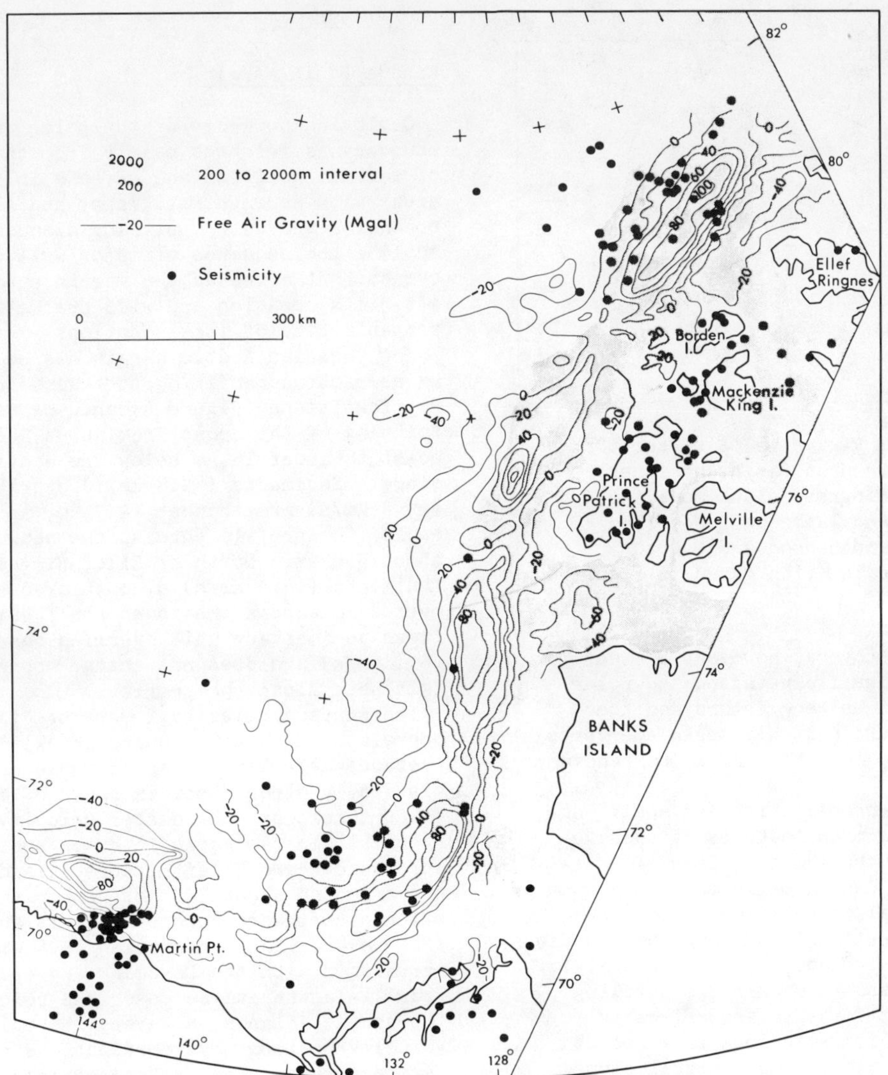

Fig. 4. Observed free air gravity in 20 mgal contours and seismicity for the period 1962-1974. Continental slope is shaded (from Wetmiller and Forsyth, 1978).

for the contemporary seismicity (Basham et al., 1977; Wetmiller and Forsyth, 1978).

A thick sedimentary load may also be present along the margin northwest of Ellef Ringnes Island. Hobson and Overton's (1967) refraction profile (northwestern third of Figure 5) indicates that a seaward thickening sediment wedge (< 4.4 km/s) of probable post-Jurassic age is at least 3 km thick across the inner shelf. Great thicknesses (19 km) of weakly magnetic material, much of which may be sediments, have been inferred along the outer parts of the margin from the presence of a major magnetic low in that region (Bhattacharyya, 1968), but the magnetic anomaly covers only the southwestern part of the gravity high.

Sediment loading is less obviously the cause of seismic activity in northeastern Alaska where seismicity correlates with the landward rather than the seaward flank of the gravity high along the shelf. Seismic events may be produced along reactivated basement structures below the sediment load or may result from offsets within the sedimentary sequence itself. It could be that suitable zones of weakness are absent beneath the main load of sediments north of Alaska but are present instead under the nearby inner continental shelf.

Flexural stresses associated with deglaciation are cited as a major cause of contemporary seismicity along the passive margin of eastern Canada (Stein et al., 1979) where over 150 m of post-Pleistocene uplift has already taken place. Stresses associated with

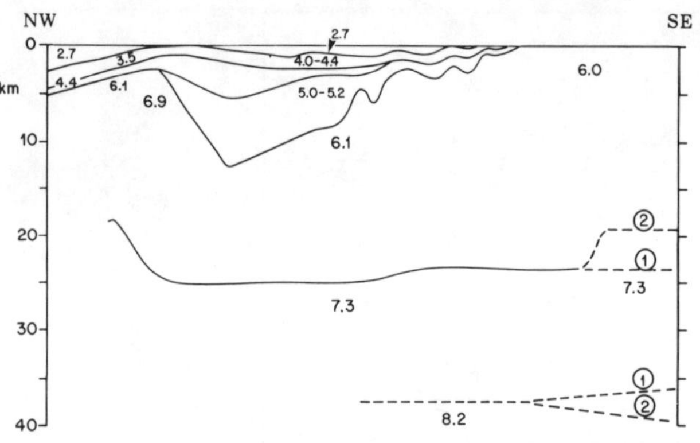

Fig. 5. Crustal P-wave velocity section V_1 across the Sverdrup Basin and inner shelf at Ellef Ringnes Island modified from Forsyth (1978) and based on Sander and Overton (1965) and Hobson and Overton (1967). Profile location given in Figure 1.

glacio-isostatic recovery, however, are not considered to be significant along the polar margin between Axel Heiberg Island and the Mackenzie Delta where post-Wisconsin emergence is documented to be less than 20 m (Farrand and Gajda, 1962).

Seaward sediment progradation and shelf edge gravity highs are common features of passive margins. If the outbuilding of Cenozoic sediments into the Canada Basin correctly explains the spatial correlation of low level seismicity with gravity highs along the Canadian polar margin, then the absence of seismicity along most other passive margins is puzzling. Two related reasons for this are offered. First, focal mechanism solutions from recent seismic events beneath the continental slope north of the Mackenzie Delta (Hasegawa et al., 1979) and in the western Canadian Arctic Islands (Hasegawa, 1977) indicate that the horizontal component of deviatoric tensional stress within the crust of this region is nearly normal to the polar margin on both its seaward and landward sides. A passive margin under such a tensional regime may have a lower threshold for seismic activity than margins (e.g., those of the North Atlantic, Sykes, 1978) not under tension. Second, the unusually thick (2 to 5 km, Berry and Barr, 1971; Hall, 1973; Grantz et al., 1979) sediment cover in Canada Basin may serve to downflex the underlying oceanic crust and produce a greater than normal state of stress across the adjacent margin which is presently coupled to the seafloor basement. The polar continental shelf may therefore require less additional stress input from other sources (e.g., prograding sediment wedges) than other passive margins to become a locus of seismic activity.

Canada Basin Margin

Geologic knowledge of the polar continental boundary is inferred mainly from the correlation of seismic horizons and offsets in offshore areas with known stratigraphy and structure from boreholes and outcrops. The discussion will outline the sequence of major well documented events that affected the margin both before and after its creation and will bracket the most probable time of its inception.

The Canadian Arctic margin has been traversed by refraction profiling northwest from Prince Patrick Island (Figure 6) and the results show a thinning of the crust from about 28 km at the coast to about 15 km below the continental slope. Sediments ($<$ 4.8 km/s) overlying basement ($\geq$ 5.4 km/s) are thinner, 2.7 km, at the landward end of the profile than at the seaward end, about 5.0 km. North of Ellef Ringnes Island sediments ($<$ 4.4 km/s) also thicken seaward to over 3 km across the inner shelf (Figure 5). A lower sedimentary unit overlies basement (6.1 km/s) and does not change appreciably in thickness along the profile.

The surface velocity layers of Figure 5 are correlated with a northward thickening latest Cretaceous-Tertiary clastic wedge which overlies distinctly older Mesozoic rocks of more or less fixed thickness, the deeper velocity layer of Figure 5, in two geologic cross-sections, one across northern Ellef Ringnes Island (Figure 7) and a second about 300 km further southwest that extends northward across Brock Island (Figure 8). Basement rocks ($\geq$ 5.4 km/s) are correlated with the Devonian and older Franklinian strata as described below. North of the Arctic Islands, however, the 6.1 km/s velocity horizon may represent the base of the Mesozoic succession as Upper Paleozoic rocks along that margin appear to be dominantly carbonates (Figures 7 and 8) and therefore close in velocity to the underlying lower Paleozoic and older rocks.

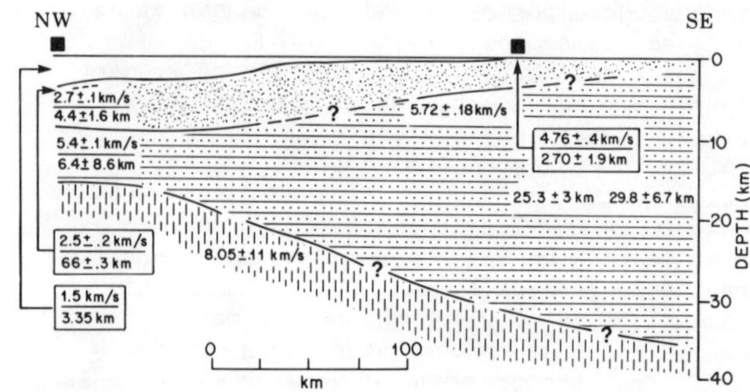

Fig. 6. Crustal P-wave velocity section V_2 across the polar shelf near Prince Patrick Island (from Berry and Barr, 1971). Profile location given in Figure 1.

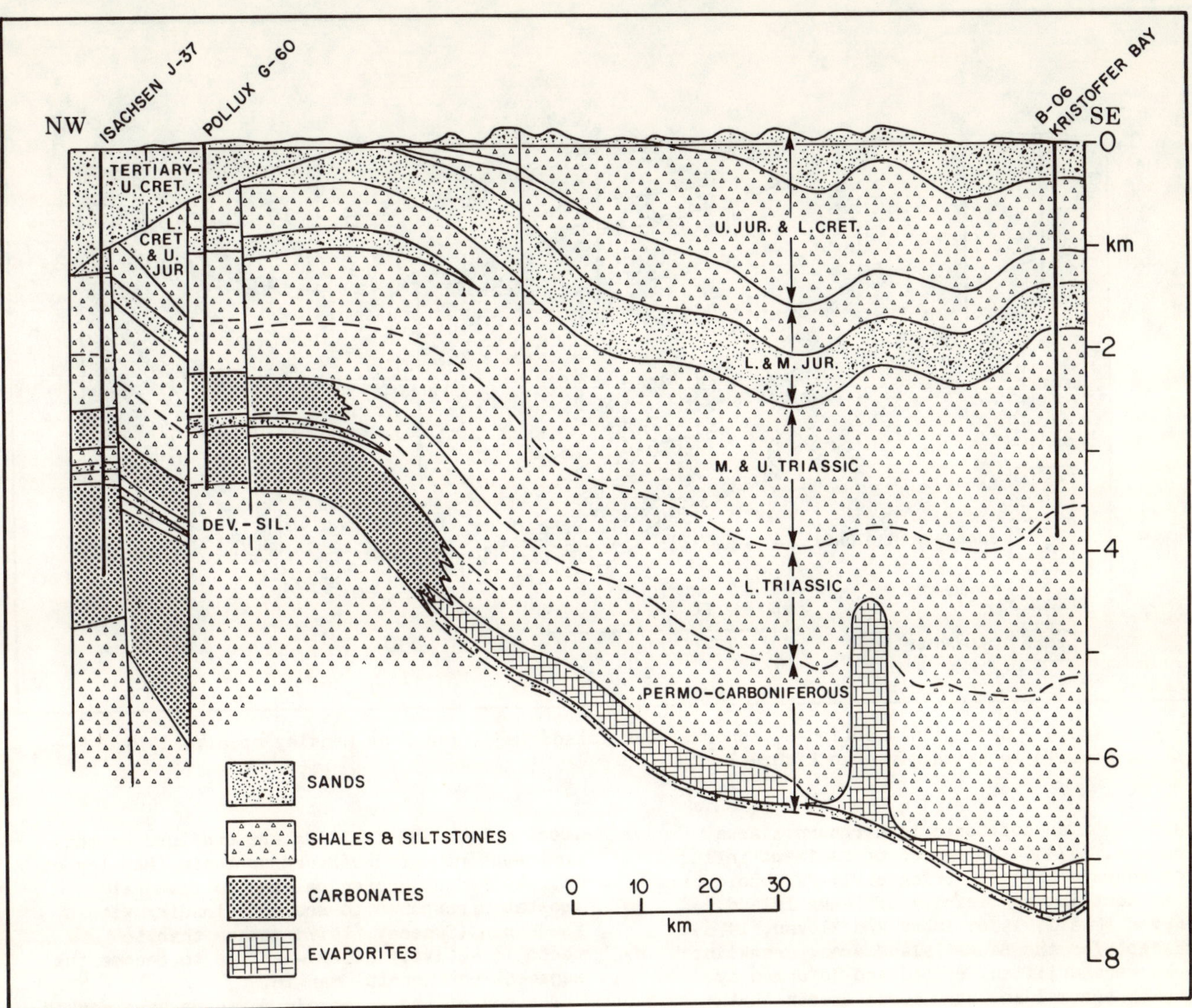

Fig. 7. Geological section G_1 across northern Ellef Ringnes Island (modified from Meneley et al., 1975). Profile location given in Figure 1.

These age versus velocity correlations are consistent with recent comparisons of borehole stratigraphy with a detailed seismic reflection profile just east of Ellef Ringnes Island (L.W. Sobczak, personal communication, 1979). An Early Cretaceous or younger age is inferred for the surface velocity layers as in Figure 5 and the underlying sediments are tentatively assigned a Triassic to Early Cretaceous age.

The Devonian and older Paleozoic Franklinian basement sequence may overlie rocks of Precambrian age. The presence of the latter cannot be confirmed from seismic velocity data but Precambrian rocks are locally present along the north coast of Ellesmere Island (Trettin and Balkwill, 1979) and along the southwestern shore of the Beaufort Sea (Yorath and Norris, 1975).

Franklinian rocks are exposed in northern Greenland (Dawes, 1976), northern Ellesmere and Axel Heiberg islands (Trettin et al., 1972) and are known in the subsurface from boreholes on northern Ellef Ringnes and Brock islands (Meneley et al., 1975), Banks Island (Miall, 1976), the Mackenzie Delta (e.g., Yorath and Norris, 1975) and north of Alaska (Churkin, 1975; Grantz et al., 1979).

Franklinian sediments along northern Greenland and the northeastern Canadian Arctic Islands appear to have been derived mainly from nearby terrestrial source areas seaward of the present polar coastlines (Dawes, 1976; Trettin et al., 1972). This seems to be true also for lower Paleozoic rocks along the southeastern part of the Beaufort Shelf (Lerand, 1973) and for at

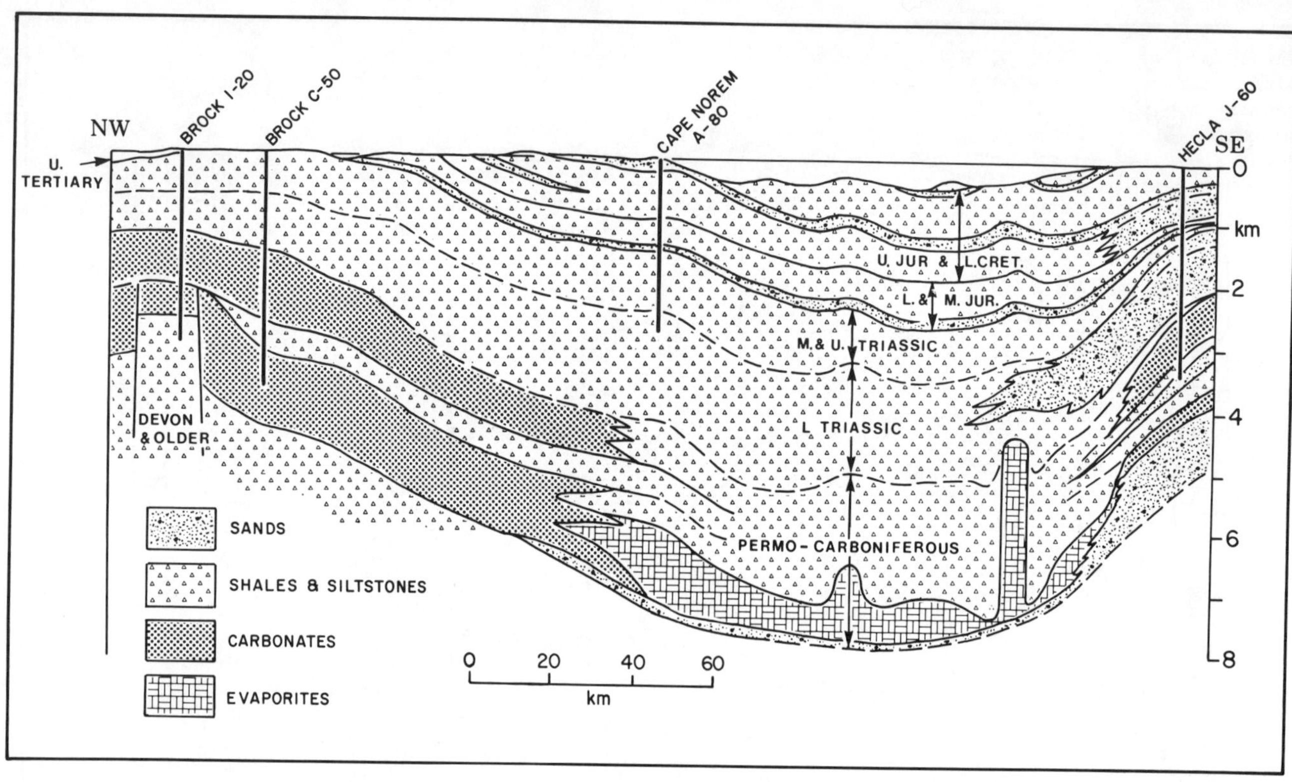

Fig. 8. Geological section G_2 across Brock Island (modified from Meneley et al., 1975). Profile location given in Figure 1.

least Devonian sediments in northern Alaska (Churkin, 1973). No record of sediment influx from seaward sources during early Paleozoic time is evident in the vicinity of Banks Island, however (Miall, 1976; Embry and Klovan, 1976).

Except for the Banks Island area, Franklinian rocks were uplifted, eroded and deformed by complex thermal and structural events that increased in intensity toward the north and culminated during Middle Devonian to Early Mississippian time (Ellesmerian orogeny in northeastern Canada and Greenland, Antler orogeny in northern Alaska). The major source of Franklinian sediments may have been offshore orogenic welts (Trettin et al., 1972; Tailleur, 1973) produced by subduction of oceanic crust beneath the adjacent continental mass (Trettin et al., 1972; Dawes, 1976) and followed in Devonian time by contact between the Laurentian and Siberian shields (Morel and Irving, 1978).

Stable conditions prevailed along this zone for the remainder of the Paleozoic. Sediments continued to be derived from sources further north (Lerand, 1973; Thorsteinsson, 1974; Grantz et al., 1979) and were deposited mainly in marine environments with carbonates a common rock type. A Pennsylvanian-Permian period of uplift and erosion is indicated, however, by an unconformity in shelf sediments north of Alaska (Churkin, 1973). Further east at this time,

local uplift and faulting are confined to the northwest rim of the Sverdrup Basin (Meneley et al., 1975) and appear related to flexural isostatic response to sediment loading within the basin (Sweeney, 1977) rather than to tectonic activity along what was to become the adjacent continental margin.

The end of the Paleozoic may have been marked by mild tectonism as indicated by a stratigraphic hiatus along much of the present polar continental boundary. Mesozoic rocks lie with angular unconformity on Paleozoic rocks across the outer shelf north of Alaska (Grantz et al., 1979) and along the Beaufort Shelf (Lerand, 1973; Miall, 1975; Young et al., 1976). This is not the case further east along the margin, but a fundamental change in tectonic conditions is indicated north of the Canadian Arctic Islands at the start of the Mesozoic era by events in the Sverdrup Basin. At the outset of Triassic time an abrupt and pronounced increase in rates of sediment accumulation within the basin was probably accompanied by the intrusion of mafic dikes and sills into the existing stratigraphic succession (Balkwill, 1978). These effects may be related to the development of a major graben beneath the basin axis parallel to the present polar margin (Sweeney, 1977; Balkwill, 1978). At this time significant sediment influx from the north into the Sverdrup Basin also ceased (Thorsteinsson, 1974). Sediments along parts of the present polar shelf of Alaska continued to be derived from sources further north until earliest Cretaceous time. The end of the Paleozoic is marked along the inner shelf north of Alaska by Middle Permian to Early Triassic deltaic progradation from northern sources (Churkin, 1973; Lerand, 1973; Grantz et al., 1979).

A rifting episode may have separated source areas to the north from much of Laurentia and produced crustal extension beneath the Sverdrup Basin (Yorath and Norris, 1975, p. 607). Contemporary weak structural deformation and possible uplift along the Beaufort Shelf plus normal block faulting of probable Triassic age parallel to the present margin along the coastal plain west of the Mackenzie Delta (Hawkings and Hatlelid, 1975, p. 637) suggests that crustal extension and perhaps arching and rifting extended as far as Alaska and possibly beyond as the presence of deltaic clastics is thought to be evidence of uplift to the north of Alaska (Churkin, 1973).

Rifting, if it occurred, was limited as the evidence suggests that a continental margin did not develop at the outset of Mesozoic time. First of all, widespread normal faulting of this age is not evident in late Paleozoic "basement" rocks along the existing continental boundary. Second, along the margin east of the Beaufort Shelf, a basal Mesozoic stratigraphic break is not apparent and Triassic rocks appear to be present and conformable in that region. Finally, no appreciable northward thickening of early Mesozoic sediments along the shelf is evident from the available seismic and stratigraphic data (Hobson and Overton, 1967; Lerand, 1973; Meneley et al., 1975; Grantz et al., 1979) (Figures 5, 7, 8 and 9).

The subsequent stratigraphic record northwest of the Sverdrup Basin appears to be unbroken, with local exceptions (e.g., Meneley et al., 1975; Miall, 1975), into Early Cretaceous time when the major period of rifting and margin development appears to have occurred. The event is signified and dated by three main geological artifacts: erosionally truncated basement rocks that are downfaulted toward the continental edge and unconformably overlain by a seaward thickening sedimentary prism. From west of Axel Heiberg Island to the Mackenzie Delta, pre-Upper Cretaceous rocks are downfaulted toward the Canada Basin and are truncated by an erosional unconformity that removed progressively older Mesozoic strata northward across the inner shelf (Meneley et al., 1975; Miall, 1975; Hawkings and Hatlelid, 1975; Young et al., 1976). Overlying this north of the Sverdrup Basin, as previously indicated, is a seaward thickening sediment wedge whose basal members are dated as Upper Campanian-Maastrichtian (about 70 Ma) (Balkwill, 1978). If one adds to this time the Sleep (1971) estimate of 50 Ma for the minimum

interval between arching-erosion associated with the initial stages of rifting to cooling-subsidence-initial deposition along the newly created rifted margin, then the continental boundary between Canada Basin and the Canadian Arctic Islands was initiated not later than about 120 Ma or late Early Cretaceous time. A similar minimum age has been estimated for the Canada Basin by Eittreim and Grantz (1979) who determined that, when corrected for the thickness and loading effects of overlying sediments, the seafloor basement depth is just over 6.0 km close to the margin north of Alaska. This corrected depth corresponds to a crustal age not less than 120 Ma using the Parsons and Sclater (1977) seafloor depth versus age relation.

Evidence of significant crustal extension in this region at this time is not confined to the polar shelf. Rejuvenation of late Early Cretaceous subsidence in the Sverdrup Basin is probably related to renewed downdropping along faults beneath the basin (Sweeney, 1977; Balkwill, 1978). This event is partly contemporary with emplacement of a major diabase dike swarm that extends across the basin from Melville Island to the shelf north of Ellef Ringnes Island (Balkwill and Haimila, 1978; Forsyth et al., 1979).

To the west, evidence for Early Cretaceous development of the margin is not so clear cut. Several unconformities and brief hiatuses mark this time interval along the Beaufort and Alaskan shelves and contemporary thrust faulting is documented in shelf basement rocks.

Seaward thickening prisms of Early Cretaceous (Aptian) and younger sediments may lie directly on Franklinian rocks along the inner shelf west of Banks Island (Miall, 1976) and along much of the outer shelf north of Alaska (Grantz et al., 1979). A relatively brief 10 to 20 Ma break is associated with a post-Albian (<100 Ma) unconformity seaward of Banks Island and this

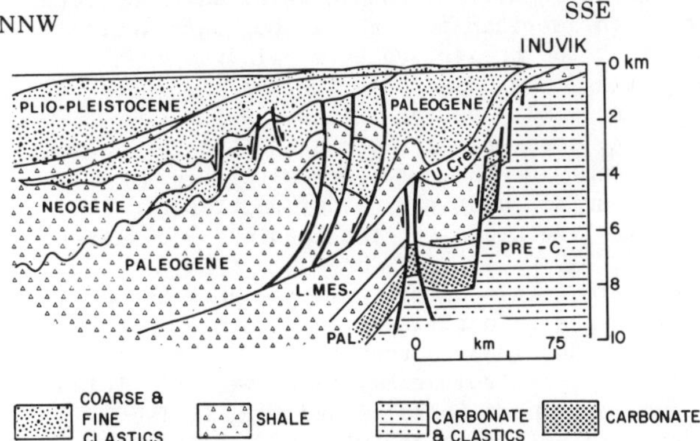

Fig. 9. Geological section G₃ across Mackenzie Delta and Beaufort Shelf (from Hawkings and Hatlelid, 1975). Profile location given in Figure 1.

may be partly contemporaneous with a major Late Cretaceous hiatus along the western polar margin of Alaska (Grantz et al., 1979) as well as a more abbreviated gap in northeastern Alaska (Detterman et al., 1975). A regional unconformity and hiatus of about 20 Ma is also present between Early and Upper Cretaceous rocks along the inner shelf near the Mackenzie Delta (Lerand, 1973; Hawkings and Hatlelid, 1975; Young et al., 1976).

To what extent the Aptian-Albian global high stand of the sea affected mid-Cretaceous stratigraphic breaks along the polar margin is not certain. If Aptian-Albian unconformity-bound sequences on west Banks Island and in northeastern Alaska are considered to be depositional interruptions along a mildly positive but temporarily submerged shelf, then the duration of marginal uplift at this time becomes much greater, at least 30 Ma, and the continental boundary may not have subsided until well into Late Cretaceous time.

Marginal stratigraphy can therefore be explained by a major Early Cretaceous rifting episode that produced uplift and erosion along a newly formed continental edge which later cooled, subsided and began to receive sediments during Late Cretaceous time. Several structural displacements within shelf sequences near the Mackenzie Delta, however, are compressional. Early Cretaceous south-directed thrust faults offset shelf sediments north of the Yukon (Yorath and Norris, 1975) while thrust faulting along the shelf of northeastern Alaska is north-directed and may be Late Cretaceous to Tertiary in age (Grantz et al., 1979).

Preliminary paleomagnetic evidence from the north slope of Alaska (Hillhouse, personal communication, 1981) and aeromagnetic data from southern Canada Basin (Taylor et al., 1979) indicate that this polar seafloor may have originated by rifting and rotation of a plate that included northern Alaska away from Arctic Canada about a pivot near the Mackenzie Delta. Structural and stratigraphic relationships along the North American continental boundary indicate only that the present polar margin was created by rifting, most likely during Early Cretaceous time. Proximity to the proposed pole of rotation, however, may be responsible for many of the Early Cretaceous complexities (e.g., thrust faults, multiple unconformities) in the geological record along the Beaufort and Alaskan shelves.

There is little direct evidence of Mesozoic or younger igneous activity along the margin but manganese nodules, thought to be related to nearby offshore volcanism, mark the post-Albian unconformity near Banks Island (Miall, 1975). Widespread extrusive activity of uncertain origin is indicated by bentonitic and tuff interbeds in Albian (106 Ma) through Campanian (70 Ma) shales south and east of the Beaufort Shelf (Miall, 1975; Young et al., 1976). If the source for these deposits was within the Canada Basin, then topographic remnants of once subaerial volcanoes should still be present in spite of the thick seafloor basement cover, but there is no bathymetric, gravity or magnetic evidence of seamounts. Geophysical data are sparse over much of the abyssal plain, however. Tuff from southern Banks Island is reported to be of intermediate composition, a type not usually produced by oceanic volcanism (Miall, 1979, p. 172).

A major Laramide episode of Maastrichtian to Eocene (70 to 37 Ma) uplift and rejuvenation of faults along the Beaufort Shelf (Miall, 1975; Young et al., 1976) coincides with the Eurekan Orogeny in the Canadian Arctic Islands (Trettin and Balkwill, 1979). These events produced great volumes of clastic detritus which were deposited along the polar shelf and formed thick sedimentary prisms that prograded the continental margin (Lerand, 1973). An episode of uplift and local normal faulting in mid-Tertiary time interrupted this process and produced a 10 to 15 Ma hiatus and unconformity along at least the inner shelf upon which a seaward thickening wedge of Miocene and younger clastics was later built (Young et al., 1976; Balkwill, 1978; Miall, 1979).

Margin East of the Alpha Ridge

Given the scarcity of data, estimates of the structure of the margin north of Axel Heiberg and Ellesmere islands are necessarily tentative. Balkwill (1978) showed that between the Beaufort and Lincoln shelves the polar margin can be divided into two parts, a southwestern segment that foundered in Late Cretaceous time and a northeastern segment that remained stable. The hinge zone between the two segments lies at the junction of the Alpha Ridge with the continental shelf, a point at which the magnetic character of the margin appears to change significantly (Figure 2).

To the northeast there is apparent bathymetric and magnetic continuity between the polar margin and the Alpha and Lomonosov ridges (Figures 1 and 2). The thickness of basement cover at each junction is unknown, however, and other geophysical parameters have yet to be measured. Trettin (1969) proposed structural continuity between the Lomonosov Ridge and the mainland on the basis of north-striking structural trends in rocks on line with the ridge on northernmost Ellesmere Island. It may be that a rifted continental margin did not develop along this part of the North American polar shelf. Furthermore, it is possible, although the evidence is very limited, that at least one of the two submarine ridges is structurally joined to the continent north of Ellesmere Island.

Final Words

This review has sketched the major Phanerozoic
tectonic events that shaped the polar margin of
North America and has attempted to identify the
factors responsible for its present geophysical
character. The data base for this purpose is
extremely limited and most of our geological
knowledge of this continental boundary comes
from a few widely separated seismic profiles
that traverse the shelf only at Prince Patrick
Island and to the west. Only a handful of
offshore boreholes exist and those are close to
the Mackenzie Delta. The thickness and
distribution of sediments along the margin is
poorly understood and the history of vertical
motions related to margin development is largely
unknown. Before these and other primary
questions related to shelf geodynamics may be
addressed, our basic understanding of the polar
margin needs to be greatly improved. A
significant first step in this direction would
be the undertaking of multiparameter
geotraverses completely across the continental
boundary in each of the two major shelf
segments; that is, north of the central Canadian
Arctic Islands into the Canada Basin and north
of Ellesmere Island onto both the Alpha Ridge
and the Lomonosov Ridge.

Acknowledgements

I thank R.L. Coles, H.S. Hasegawa, J.A. Mair
and L.W. Sobczak for assistance and helpful
suggestions. S. Cumyn drafted the figures.
Contribution 925 from the Earth Physics Branch.

References

Balkwill, H.R., Evolution of Sverdrup Basin,
    Arctic Canada, Am. Assoc. Petroleum Geol.
    Bull., 62, 1004-1028, 1978.
Balkwill, H.R., and N.E. Haimila, K/Ar ages and
    significance of mafic rocks, Sabine Peninsula,
    Melville Island, District of Franklin, Geol.
    Survey Can. Paper 78-1C, 35-38, 1978.
Basham, P.W., D.A. Forsyth and R.J. Wetmiller,
    Seismicity of Northern Canada, Can. J. Earth
    Sci., 14, 1646-1667, 1977.
Berry, M.J., and K.G. Barr, A seismic refraction
    profile across the polar continental shelf of
    the Queen Elizabeth Islands, Can. J. Earth
    Sci., 8, 347-360, 1971.
Bhattacharyya, B.K., Analysis of aeromagnetic
    data over the Arctic Islands and continental
    shelf of Canada, Geol. Survey Can. Paper
    68-44, 14p, 1968.
Churkin, M. Jr., Paleozoic and Precambrian rocks
    of Alaska and their role in its structural
    evolution, U.S. Geol. Survey Prof. Paper 740,
    64p, 1973.
Churkin, M. Jr., Basement rocks of Barrow Arch,
    Alaska, and circum-Arctic Paleozoic mobile
    belt, Am. Assoc. Petroleum Geol. Bull., 59,
    451-456, 1975.
Coles, R.L., G.V. Haines and W. Hannaford,
    Large scale magnetic anomalies over western
    Canada and the Arctic: A discussion, Can. J.
    Earth Sci., 13, 790-802, 1976.
Coles, R.L., W. Hannaford and G.V. Haines,
    Magnetic anomalies and the evolution of the
    Arctic, in Arctic Geophysical Review, edited
    by J.F. Sweeney, Pub. Earth Phys. Br., 45,
    51-66, 1978.
Dawes, P.R., Precambrian to Tertiary of northern
    Greenland, in Geology of Greenland, edited by
    A. Escher and W.S. Watt, Geol. Survey
    Greenland, 404-429, 1976.
Detterman, R.L., H.N. Reiser, W.P. Brosgé and
    J.T. Dutro Jr., Post-Carboniferous
    stratigraphy, northeastern Alaska, U.S. Geol.
    Survey Prof. Paper 886, 46p, 1975.
Embry, A.F., and J.E. Klovan, The Middle-Upper
    Devonian clastic wedge of the Franklinian
    geosyncline, Bull. Can. Petroleum Geol., 24,
    485-639, 1976.
Farrand, W.R., and R.T. Gajda, Isobases on the
    Wisconsin marine limit in Canada, Geograph.
    Bull. 17, Dept. Mines and Technical Surveys,
    Ottawa, Canada, 5-22, 1962.
Eittreim, S., and A. Grantz, CDP seismic sections
    of the western Beaufort continental margin, in
    Crustal Properties across Passive Margins,
    edited by C.E. Keen, Tectonophysics, 59,
    251-262, 1979.
Forsyth, D.A., Review of Arctic crustal studies,
    in Arctic Geophysical Review, edited by
    J.F. Sweeney, Pub. Earth Phys. Br., 45, 75-85,
    1978.
Forsyth, D.A., J.A. Mair and I. Fraser, Crustal
    structure of the central Sverdrup Basin, Can.
    J. Earth Sci., 16, 1581-1598, 1979.
Grantz, A., S. Eittreim and D.A. Dinter, Geology
    and tectonic development of the continental
    margin north of Alaska, in Crustal Properties
    across Passive Margins, edited by C.E. Keen,
    Tectonophysics, 59, 263-291, 1979.
Haines, G.V., A Taylor expansion of the
    geomagnetic field in the Canadian Arctic, Pub.
    Dominion Obs., 35, 115-140, 1967.
Haines, G.V., and W. Hannaford, A three-component
    aeromagnetic survey of the Canadian Arctic,
    Pub. Earth Phys. Br., 44, 209-228, 1974.
Hall, J.K., Geophysical evidence for ancient sea-
    floor spreading from Alpha Cordillera and
    Mendeleyev Ridge, in Arctic Geology, edited by
    M.G. Pitcher, Am. Assoc. Petroleum Geol.
    Memoir 19, 542-561, 1973.
Hasegawa, H.S., Focal parameters of four Sverdrup
    Basin, Arctic Canada, earthquakes in November
    and December of 1972, Can. J. Earth Sci., 14,
    2481-2494, 1977.
Hasegawa, H.S., C.W. Chou and P.W. Basham,
    Seismotectonics of the Beaufort Sea, Can. J.
    Earth Sci., 16, 816-830, 1979.
Hawkings, T.J., and W.G. Hatlelid, The regional

setting of the Taglu field, in Canada's Continental Margins, edited by C.J. Yorath, E.R. Parker and D.J. Glass, Can. Soc. Petroleum Geol. Memoir 4, 633-647, 1975.

Hobson, G.D., and A. Overton, A seismic section of the Sverdrup Basin, Canadian Arctic Islands, in Seismic Refraction Prospecting, edited by A.W. Musgrave, Soc. Explor. Geophys., Tulsa, Okla., 550-562, 1967.

Hood, P.J., and M.E. Bower, Arctic Ocean: Low-level aeromagnetic profiles obtained in 1975, Geol. Survey Can. Paper 76-1A, 421-424, 1976.

King, E.R., I. Zeitz and L.R. Alldredge, Magnetic data on the structure of the central Arctic region, Geol. Soc. Am. Bull., 77, 619-646, 1966.

Lerand, M., Beaufort Sea, in The Future Petroleum Provinces of Canada, edited by R.G. McCrossan, Can. Soc. Petroleum Geol. Memoir 1, 315-386, 1973.

Meneley, R.A., D. Henao and R.K. Merritt, The northwest margin of Sverdrup Basin, in Canada's Continental Margins, edited by C.J. Yorath, E.R. Parker and D.J. Glass, Can. Soc. Petroleum Geol. Memoir 4, 531-544, 1975.

Miall, A.D., Post-Paleozoic geology of Banks, Prince Patrick and Eglinton Islands, Arctic Canada, in Canada's Continental Margins, edited by C.J. Yorath, E.R. Parker and D.J. Glass, Can. Soc. Petroleum Geol. Memoir 4, 557-587, 1975.

Miall, A.D., Devonian geology of Banks Island, Arctic Canada, and its bearing on the tectonic development of the circum-Arctic region, Geol. Soc. Am. Bull., 87, 1599-1608, 1976.

Miall, A.D., Mesozoic and Tertiary geology of Banks Island, Arctic Canada, Geol. Survey Can. Memoir 387, 235p, 1979.

Morel, P., and E. Irving, Tentative paleocontinental maps for the early Phanerozoic and Proterozoic, J. Geol., 86, 535-561, 1978.

Ostenso, N.A., and R.J. Wold, Aeromagnetic survey of the Arctic Ocean: Techniques and interpretations, Marine Geophys. Res., 1, 178-219, 1971.

Parsons, B., and J.G. Sclater, An analysis of the variation of ocean floor bathymetry and heat flow with age, J. Geophys. Res., 82, 803-827, 1977.

Patterson, W.S.B., and L.K. Law, Additional heat flow determinations in the area of Mould Bay, Arctic Canada, Can. J. Earth Sci., 3, 237-246, 1966.

Riddihough, R.P., G.V. Haines and W. Hannaford, Regional magnetic anomalies of the Canadian Arctic, Can. J. Earth Sci., 10, 147-163, 1973.

Sander, G.W., and A. Overton, Deep seismic refraction investigations in the Canadian Arctic Archipelago, Geophysics, 30, 87-96, 1965.

Sleep, N.H., Thermal effects of the formation of Atlantic continental margins by continental break-up, Geophys. J.R. Astron. Soc., 24, 325-350, 1971.

Sobczak, L.W., Gravity and deep structure of the continental margin of Banks Island and Mackenzie Delta, Can. J. Earth Sci., 12, 378-394, 1975a.

Sobczak, L.W., Gravity anomalies and passive continental margins, Canada and Norway, in Canada's Continental Margins, edited by C.J. Yorath, E.R. Parker and D.J. Glass, Can. Soc. Petroleum Geol. Memoir 4, 743-761, 1975b.

Sobczak, L.W., Discussion on "The tectonic development of the southern Beaufort Sea and its relationship to the origin of the Arctic Ocean basin by Yorath and Norris, 1975", Bull. Can. Petroleum Geol., 25, 698-703, 1977.

Sobczak, L.W., Gravity from 60°N to the North Pole, in Arctic Geophysical Review, edited by J.F. Sweeney, Pub. Earth Phys. Br., 45, 67-74, 1978.

Sobczak, L.W., and J.F. Sweeney, Bathymetry of the Arctic Ocean, in Arctic Geophysical Review, edited by J.F. Sweeney, Pub. Earth Phys. Br., 45, 7-14, 1978.

Sobczak, L.W., and J.R. Weber, Crustal structure of Queen Elizabeth Islands and polar continental margin, Canada, in Arctic Geology, edited by M.G. Pitcher, Am. Assoc. Petroleum Geol. Memoir 19, 517-525, 1973.

Stein, S., N.H. Sleep, R.J. Geller, S-C. Wang and G.C. Kroeger, Earthquakes along the passive margin of eastern Canada. Geophys. Res. Letters, 6, 537-540, 1979.

Sweeney, J.F., Subsidence of the Sverdrup Basin, Canadian Arctic Islands, Geol. Soc. Am. Bull., 88, 41-48, 1977.

Sykes, L.R., Intraplate seismicity, reactivation of preexisting zones of weakness, alkaline magmatism, and other tectonism postdating continental fragmentation, Rev. Geophys. Space Phys., 16, 621-688, 1978.

Tailleur, I.L., Probable rift origin of Canada Basin, in Arctic Geology, edited by M.G. Pitcher, Am. Assoc. Petroleum Geol. Memoir 19, 526-535, 1973.

Talwani, M., and O. Eldholm, Continental margin off Norway: A geophysical study, Geol. Soc. Am. Bull., 83, 3575-3608, 1972.

Taylor, P.T., P.R. Vogt, L.C. Kovacs, J.A. Green and D.W. Handschumacher, West-Arctic Ocean basin: Aeromagnetic results (abstract), EOS, Trans. AGU, 60, 372-373, 1979.

Thorsteinsson, R., Carboniferous and Permian stratigraphy of Axel Heiberg Island and western Ellesmere Island, Canadian Arctic Archipelago, Geol. Survey Can. Bull. 224, 115p, 1974.

Trettin, H.P., A Paleozoic-Tertiary fold belt in northernmost Ellesmere Island aligned with the Lomonosov Ridge, Geol. Soc. Am. Bull., 80, 143-148, 1969.

Trettin, H.P., T.O. Frisch, L.W. Sobczak, J.R. Weber, E.R. Niblett, L.K. Law, J. DeLaurier and K. Whitham, The Innuitian Province, in Variations in Tectonic Styles in Canada, edited by R.A. Price and R.J.W. Douglas, Can. Geol. Assoc. Spec. Paper 11, 83-179, 1972.

Trettin, H.P., and H.R. Balkwill, Contributions to the tectonic history of the Innuitian Province, Arctic Canada, Can. J. Earth Sci., 16, 748-769, 1979.

Wetmiller, R.J., and D.A. Forsyth, Seismicity of the Arctic, 1908-1975, in Arctic Geophysical Review, edited by J.F. Sweeney, Pub. Earth Phys. Br., 45, 15-24, 1978.

Wold, R.J., T.L. Woodzick and N.A. Ostenso, Structure of the Beaufort Sea continental margin, Geophysics, 35, 849-861, 1970.

Worzel, J.L., Advances in marine geophysical research of continental margins, Can. J. Earth Sci., 5, 963-983, 1968.

Yorath, C.J., and D.K. Norris, The tectonic development of the southern Beaufort Sea and its relationship to the origin of the Arctic Ocean basin, in Canada's Continental Margins, edited by C.J. Yorath, E.R. Parker and D.J. Glass, Can. Soc. Petroleum Geol. Memoir 4, 589-611, 1975.

Young, F.G., D.W. Myhr and C.J. Yorath, Geology of the Beaufort-Mackenzie Basin, Geol. Survey Can. Paper 76-11, 63p, 1976.

THE PASSIVE MARGINS OF NORTHERN EUROPE AND EAST-GREENLAND

Olav Eldholm

Department of Geology, University of Oslo, Norway

Manik Talwani

Lamont-Doherty Geological Observatory and Department of
Geological Sciences of Columbia University, Palisades, New York 10964

Abstract. The state of knowledge about the
margins of the Norwegian-Greenland Sea varies
considerably. Off Norway comprehensive investigations have yielded detailed structural and
depositional data. A first order structural
framework has been developed for the margins off
the western Barents Sea, western and northern
(west of 20°E) Svalbard. The East-Greenland
margin is largely unexplored. The margins developed by rifting in a shallow sedimentary basin
in the late Paleocene. A late Jurassic extensional tectonic event which was terminated not
later than the mid-Cretaceous affected this
basin in the Mesozoic forming a system of basins
and ridges which later have been buried by late
Cretaceous and Cenozoic sediments. The margins
between the Greenland-Senja and Spitsbergen
fracture zones developed as a sheared-rifted
margin, the rifted margin being formed when the
relative plate motion changed in the early
Oligocene.

Introduction

In this paper we review the passive conti-
nental margins off Norway, the Barents Sea,
Svalbard and East-Greenland north of the Faeroe-
Iceland-Greenland Ridge.

The present state of knowledge varies considerably mainly due to weather conditions and
emphasis of the geophysical activity. A large
part of these margins lie at high latitudes and
are entirely or partly ice-covered throughout
the year. The exploration activity has also to
some extent been related to hydrocarbon discoveries in the North Sea. Thus, the eastern
margin, particularly south of 70°N, is by far
the most explored. A review of these margins has
been published by Talwani and Eldholm (1974),
who also have proposed a detailed plate tectonic
history of evolution of the Norwegian-Greenland
Sea (Talwani and Eldholm, 1977). Their work was
based on an integrated analysis of the various

kinds of geophysical data. However, the advent
of the multi-channel seismic reflection technique, deep sea drilling results and the increased exploration activity along the Norwegian
shelf have introduced new high quality data
which have enabled the study of important events
in the history of evolution of these margins.
Here, we shall first focus on the new data
inasmuch as they relate to fundamental questions
in the early history of the development of the
margin. Presently, an analysis of problems of
this kind is largely restricted to the most
thoroughly investigated area off Norway between
60° and 70°N. Then, we present the main structural and depositional framework of the other
margins as revealed by surveys of generally
exploratory nature.

Morphologically, the margins exhibit typical
shelves and slopes, whereas a continental rise
has not been developed anywhere in the Norwegian-
Greenland Sea. The slope gradient varies considerably, a phenomenon which can often be attributed to various events in the plate tectonic
evolution. A marginal plateau, the Vøring
Plateau, lies at a depth of about 1500 m off
Norway just north of Jan Mayen Fracture Zone.
A similar feature, which is less well known, the
Yermak Plateau, breaks the continental slope
northwest of Svalbard at a depth of about 1000 m
(Fig. 1).

Talwani and Eldholm (1973, 1974, 1977) have
shown that the Norwegian-Greenland Sea margins
may be structurally divided into two different
types, rifted and sheared margins. The division
reflects the existence of offsets, or transform
faults, at the original plate boundary. The
plate tectonic analysis based on a study of
magnetic anomalies shows that sea floor
spreading in the Norwegian-Greenland Sea was
initiated between anomaly 24 and 25 time (56-58
my), but that there are two distinct periods of
relative plate motion. Prior to anomaly 13 (36
my) there was no generation of oceanic crust

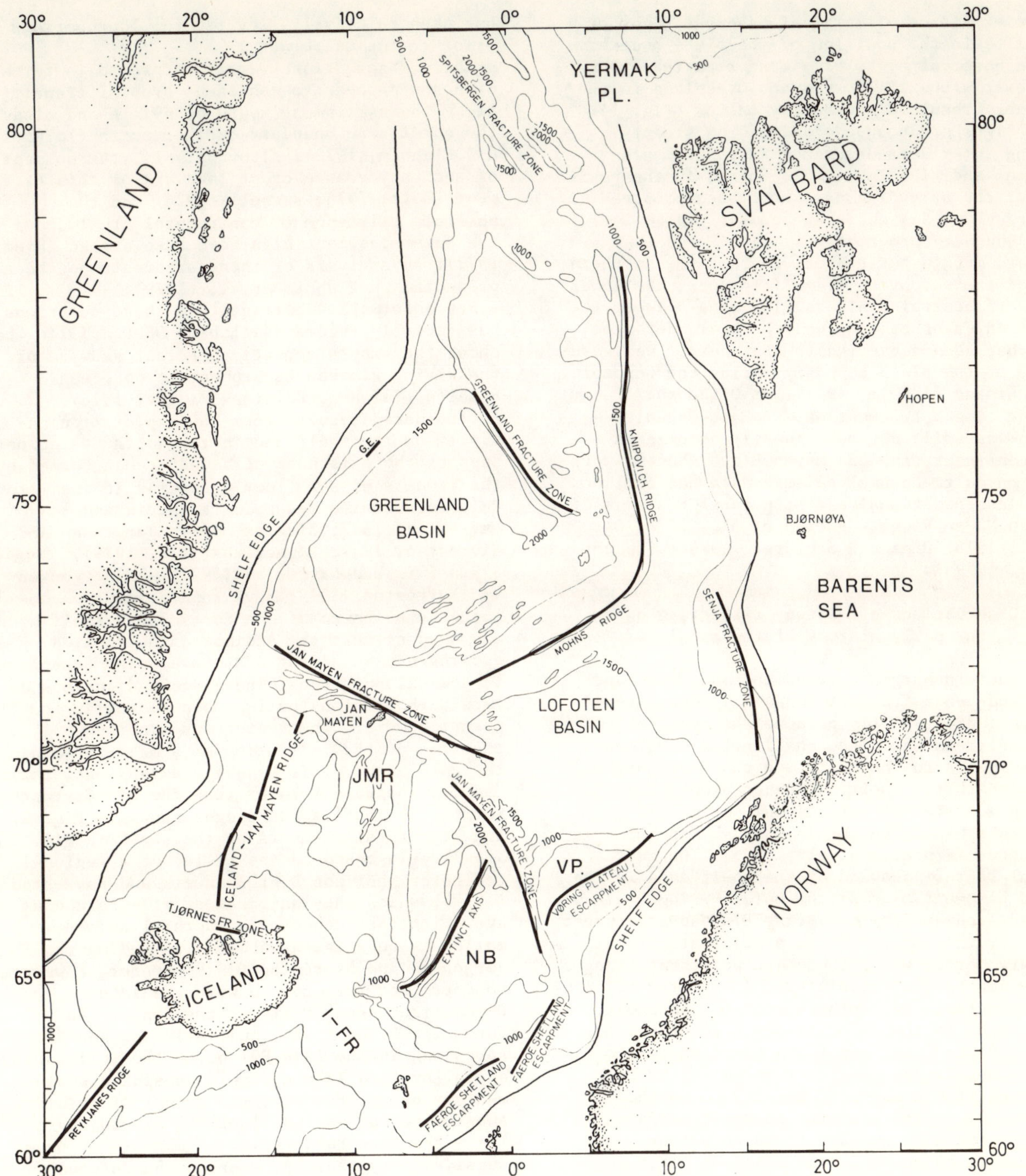

Fig. 1.   Regional bathymetry (contour interval 500 fms) and main structural features of the Norwegian-Greenland Sea. Based on Talwani and Eldholm (1977). GE, JMR, VP, I-FR and NB refer to Greenland Escarpment, Jan Mayen Ridge, Vøring Plateau, Iceland-Faeroe Ridge and Norway Basin, respectively.

between Greenland-Senja and Spitsbergen fracture zones. However, when the plate motion became west-northwest subsequent to anomaly 13 time the Greenland Sea started forming by the process of sea floor spreading. A consequence is that the margins off western Svalbard and northern East-Greenland should perhaps be classified as of a sheared-rifted origin.

In addition to this first order structural framework laid down at the time of initiation of sea floor spreading the later development of some of the margins has been influenced either

by the existence of marginal basement highs or
by shifts in the position of the plate boundary.
Buried basement highs have been reported along
the lower slope off Norway and Greenland in the
Lofoten, Greenland and Norway basins (Fig. 1).
These highs are bounded landward by steeply
dipping interfaces described as escarpments
(Talwani and Eldholm, 1972). Shifts in the position of the spreading axis have been proposed in
the Greenland Sea where at present the active
plate boundary progressively approaches the
Svalbard margin northwards, in fact the northern
part of the Knipovich Ridge intersects the lower
slope off central Svalbard. The area between the
Jan Mayen Fracture Zone and the Iceland-Faeroe
Ridge has been subjected to a number of westward
shifts of the plate boundary during the Cenozoic
(Talwani and Eldholm, 1977). At Oligocene/
Miocene time a fragment of the East-Greenland
margin was split off subsequently becoming a
microcontinent, the Jan Mayen Ridge. Recent work
based on multi-channel seismic data has outlined
the main depositional and structural features of
the Jan Mayen Ridge (Gairaud et al., 1978; Garde
et al., 1978; Hinz and Schlüter, 1978a; Sundvor
et al., 1979a).

Observations off Norway reflecting on
the early history of the margin

The eastern margin between the Jan Mayen and
Greenland-Senja fracture zones has been extensively studied, partly because the marginal
Vøring Plateau has posed challenging problems.
The sediments in the North Sea continue northward beneath the shelf and upper slope, re-
vealing a system of extensional basins and
ridges which had been established by the mid-
Cretaceous (Rønnevik et al., 1975). The most
typical feature seaward of the shelf edge is the
buried basement high at the outer Vøring Plateau
bounded landward by the Vøring Plateau Escarpment
(Fig. 2). The escarpment is a major structural
boundary across which there are prominent changes
in the geophysical parameters. These are: (1) a
sudden increase of depth to acoustic basement,
(2) changes in the velocity-depth function, (3)
a gravity gradient, and (4) a magnetic edge
anomaly separating sea floor spreading type
anomalies from a magnetic quiet zone on the
landward side. The acoustic basement surface
defining the high appears to be an extension of
the oceanic basement in the adjacent Lofoten
Basin, but its opaque nature and smooth surface
are clearly different from what is considered
typical for oceanic layer 2.

The area north of the Vøring Plateau, the
Lofoten-Vesterålen region, forms a unique part
of the margin. The continental slope becomes
progressively narrower northwards and the shelf
is underlain by high-velocity sediments at the
sea floor. It appears that the Tertiary and part
of the Mesozoic sedimentary sequence is missing
on the shelf, possibly reflecting an area that

has been relatively elevated for long periods
prior to the Quaternary.

These observations were interpreted in terms
of a sharp ocean to continent crustal transition
along the escarpment south of 69.1°N and along
the quiet zone boundary further north (Talwani
and Eldholm, 1972). Alternatively, the concept
of ancient oceanic crust landward of this escarpment (Bott, 1975; Russell, 1976) and the
possible existence of continental crust under
the outer basement high have been argued (Hinz,
1972). An analysis of these arguments has been
presented by Eldholm et al. (1979).

New information pertinent to these questions
has recently become available. DSDP drilling has
shown the smooth acoustic basement seaward of
the Vøring Plateau Escarpment to be oceanic
basalts (Talwani, Udintsev et al., 1976).
Furthermore, results from the exploration activity on the shelf and in the North Sea do not
support the existence of a deep ocean formed by
the process of sea floor spreading in the late
Paleozoic or the Mesozoic (Rønnevik and
Navrestad, 1977; Ziegler, 1977; Jørgensen and
Navrestad, 1979; Rønnevik et al., 1979).

Rønnevik and Navrestad (1977) and Jørgensen
and Navrestad (1979) have summarized the geologic development of the continental shelf areas
and suggest that the sedimentation started on an
epeirogenic subsiding post-Caledonian peneplane
in the Carboniferous. There are indications of
late Carboniferous half graben basins in the
north; however, the relief was smoothed out at
the time of the initiation of a major Mesozoic
tectonic event. This Mesozoic extensional regime
had terminated not later than the mid-Cretaceous
forming the basins and ridges shown in Fig. 2.
The structural elements on the shelf in Fig. 2
are largely based on the relief of a regional
reflector, horizon D of Rønnevik and Navrestad
(1977), which they dated as of mid-Cretaceous
age. Little evidence of structural deformation is seen in the abovelying sediments.
Jørgensen and Navrestad (1979) prefer, however,
to correlate reflector D with the late
Cimmmerian unconformity at the top of the
Jurassic. Prior to the initiation of sea floor
spreading in the late Paleocene the relief was
again levelled into a simple subsiding sedimentary basin. However, the amount of relative
subsidence has varied along the margin, particularly the Møre Basin (Fig. 2) appears to have
subsided very rapidly, whereas the Lofoten-
Vesterålen shelf has been an area which was
relatively elevated.

The Mesozoic structural regime off Norway
could well be related to similar episodes of
crustal extension in the North Sea, East-
Greenland (Surlyk, 1977) and on the northwestern
U.K. shelf, although the exact time and space
relationship has not yet been worked out. It is
particularly interesting to note that in the
West Shetland Basin thick post mid-Cretaceous
sediments form an outbuilt wedge over an Early

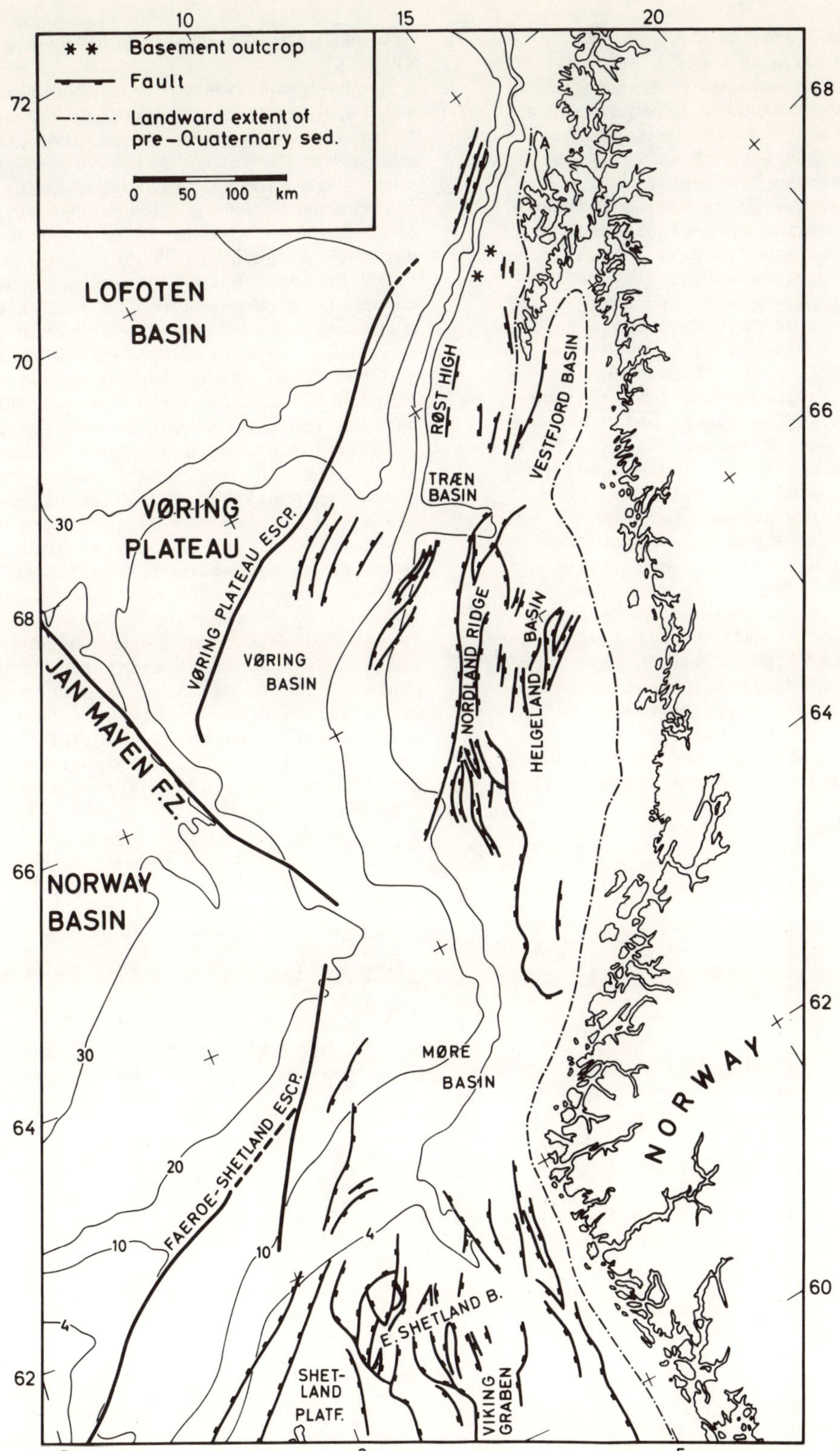

Fig. 2.   Main structural elements on the margin off Norway (60–70°N). "A" refers to local onshore exposure of late Jurassic/early Cretaceous sediments at Andøya (Dalland, 1979). Compiled from Talwani and Eldholm (1972), Rønnevik et al. (1979), Eldholm et al. (1979), Jørgensen and Navrestad (1979), Hamar (1979). Bathymetry in hundred meters (Perry et al., 1980).

Cretaceous block faulted relief (Cashion, 1975). Perhaps, the formation of the Faeroe-Shetland Channel also has to be related to these events.

In our view some of the most intriguing new observations have come from the multi-channel surveys which in places have shown series of generally seaward dipping reflectors below the acoustic basement at the outer Vøring Plateau and in the Lofoten Basin north of the plateau (Figs. 3,4) (Hinz and Weber, 1976; Talwani, 1978; Eldholm et al., 1979). Possibly, reflectors of this kind may originate from within oceanic layer 2, but the fact that these reflectors have only been obtained in a zone lying seaward of the Vøring Plateau Escarpment is puzzling. It has also been noted that the sub-basement reflectors always appear below areas of smooth opaque basement. Moreover, the smooth basement reflectors appear to lie landward of anomaly 23 (Eldholm et al., 1979).

The Vøring Plateau Escarpment has been mapped to about 69.1°N (Figs. 3,4), its seismic expression becoming progressively less distinct north of the plateau proper. The margin north of the escarpment is very narrow and is characterized by major block faults trending along the bathymetric contours (Figs. 2,4). In this area the acoustic basement appears to continue all the way from the Lofoten Basin to the shelf edge. South of Lofoten, the landward limit of the opaque basement reflector turns seaward and loses distinction approaching the northern flank of the Vøring Plateau (Eldholm et al., 1979) (Fig. 4).

Many of the observations appear to reflect events related to the rifting and early sea floor spreading phase of passive margin development. Many investigators have suggested that the normal sea floor spreading stage is preceded by a sequence of events associated with the break-up of continental crust. Thus, evidence of the early development of "non-failed" rifts is to be found at passive continental margins. Of special interest is the question of the extent of the continental crust and the structural and compositional changes associated with the change in crustal type. The geological model advanced to explain the data described above will have to address these fundamental questions.

The existence of a marginal basement high is not restricted only to the Vøring Plateau margin. A marginal basement high (Fig. 1) underlies the slope just to the south of the Greenland Fracture Zone (Eldholm and Windisch, 1974). Its geophysical signature is similar to the Vøring Plateau basement high and appears to be bounded landward by an escarpment (Talwani and Eldholm, 1974). The existence of this high is well established although it does not form a marginal plateau. The trend and configuration of the Greenland Escarpment (Fig. 1), however, is considerably less known. The Vøring Plateau and Greenland basement highs appear to have been continuous in a pre-anomaly 23 reconstruction

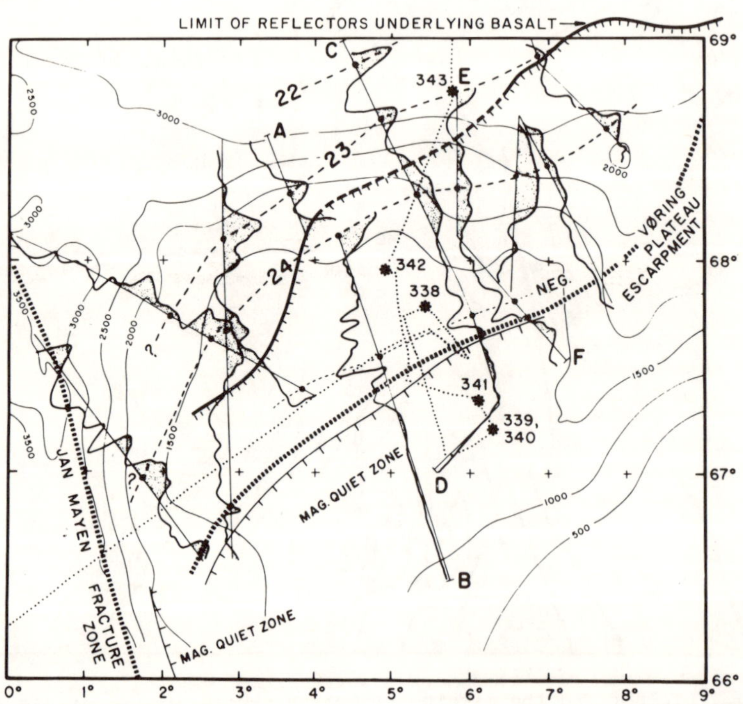

Fig. 3. Extent of sub-basement reflectors underlying the Vøring Plateau. Bathymetry in meters. DSDP drill sites 338-343 and magnetic anomalies plotted perpendicular to selected ship tracks are shown (Talwani, 1978).

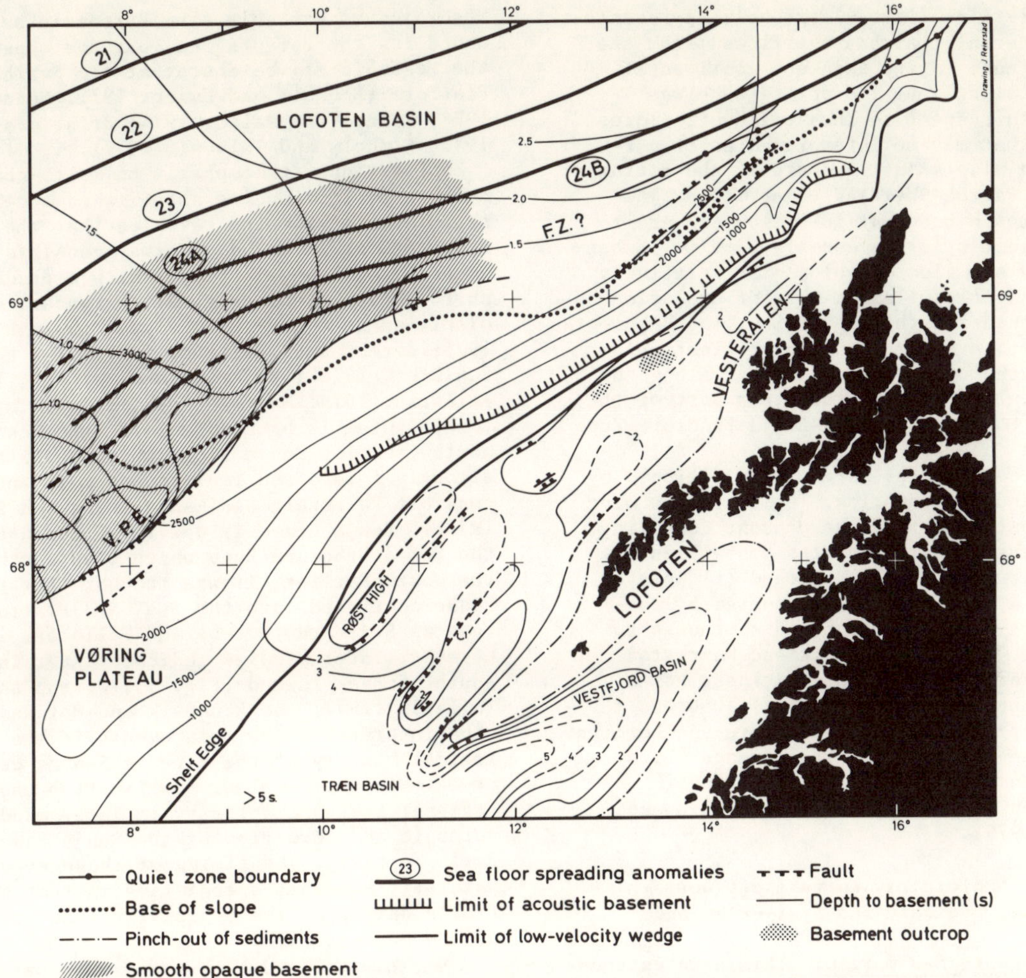

Quiet zone boundary   (23) Sea floor spreading anomalies   Fault
Base of slope   Limit of acoustic basement   Depth to basement (s)
Pinch-out of sediments   Limit of low-velocity wedge   Basement outcrop
Smooth opaque basement

Fig. 4.   Main geophysical and geological features on the Lofoten–Vesterålen margin (Eldholm et al., 1979).

(Talwani and Eldholm, 1977). South of the Jan Mayen Fracture Zone a prominent marginal high lies beneath the lower slope along the south-eastern flank of the Norway Basin. The high is bounded landward by the Faeroe-Shetland Escarpment. It is generally less clearly developed than the Vøring Plateau Escarpment and in many crossings appears as parallel scarps in a step-like pattern. It has been continued onto the northwestern flank of the Faeroe-Shetland Channel, but there is little agreement about the continent/ocean transition in this area (Bott et al., 1974; Talwani and Eldholm, 1972, 1977; Voppel et al., 1979).

In our opinion the existing data are most easily reconciled if one introduces a relatively narrow transition between continental and oceanic basement at the escarpment and the quiet zone boundary, noting that a sharp continent-ocean basement boundary does not rule out a much wider zone of modified deeper crust and upper mantle beneath the continental margin.

Although the area between Norway and Greenland was subjected to extension prior to the Cenozoic, accretion of oceanic crust by sea floor spreading did not start until the late Paleocene. It appears that the smooth opaque basement was formed prior to anomaly 23 time either by a sudden volcanic event or by a different mode of oceanic crustal generation of the type that has generated the Iceland-Faeroe Ridge (Eldholm et al., 1979). In both cases the smooth basement surface and the subbasement reflectors could represent flows and pyroclastic material possibly intermingled with terrigenous sediments, overlying slightly older oceanic crust seaward of the escarpment. The DSDP drilling data at the Vøring Plateau suggest extrusion at subaerial levels (Caston, 1976). Jacqué and Thouvenin (1975) have demonstrated that traces of basic volcanic activity are widespread in the lower Tertiary of North Sea wells. The major episode is dated as 53.5–55 my. The existence of the acoustic basement reflector on the slope off

Lofoten-Vesterålen (Fig. 4) probably represents
volcanic material that has overflowed from the
active rift zone to the adjacent continental
crust, indicating that the active zone was
elevated with respect to the neighbouring crust.
Reflectors that may be interpreted as of volcanic origin also exist landward of the Vøring
Plateau and Faeroe-Shetland escarpments. The
period of smooth basement formation appears to
have ended just prior to anomaly 23 time. Subsequently, new sea floor developed at relatively
normal depths. A similar history of evolution
may be proposed for the Faeroe-Shetland basement
high. In fact, a smooth basement reflector is
observed on the oldest oceanic crust along the
entire margin from the southwestern part of the
Hatton Bank to the Greenland-Senja Fracture Zone
(Eldholm and Sundvor, 1980).

Since Eocene time the margin has developed by
subsidence and progradation. The sediments have
been built out away from the present coastline,
but it appears that the Lofoten-Vesterålen area
has been emerged or partly emerged (Eldholm et
al., 1979), a fact which is supported by outbuilding both from the north and northeast to
the south of Lofoten (Rønnevik and Navrestad,
1977). The Tertiary sediment thickness has a
maximum in the Møre Basin (Fig. 2), whereas the
thickness varies little between Møre and Lofoten.
Moreover, the entire Tertiary sequence shows
little evidence of folding and faulting
(Rønnevik and Navrestad, 1977; Jørgensen and
Navrestad, 1979).

Results from exploratory surveys off western
Barents Sea, Svalbard and East-Greenland

Western Barents Sea margin. This area extends
from about 70°N to the latitude of Bear Island
(Bjørnøya) at 74.5°N. Eldholm and Talwani (1977)
showed that the western Barents Sea could be
divided into three geological units; the Svalbard
Platform, the western Cenozoic wedge and the
main basins. The Svalbard Platform is an area
where the seismic velocity at the sea floor is
generally above 3.8 km/s. It extends along
the central shelf from Svalbard to Bear Island
and is bounded to the south by the northern
flank of the Bear Island Trough just south of
Bear Island. The main part of the Barents Sea,
however, is underlain by rocks with seismic
velocities in the range of 2.7-3.1 km/s at the
seafloor except for the western shelf and the
slope where an outbuilt low-velocity sedimentary
wedge has been deposited over the older rocks.
The dominant structural trend is northeast with
some north trending lineaments along the western
margin (Fig. 5). These observations have been
interpreted in terms of an emergent Barents Sea
during the main part of the Tertiary with progressive deposition of terrigenous material
forming the Cenozoic wedge onto a gradually
subsiding margin along the growing Norwegian-
Greenland Sea. It has been assumed that a thick

Mesozoic and probably also Paleozoic sequence
underlies the Barents Sea, but the upper part of
the Mesozoic may be absent at the Svalbard
Platform (Eldholm and Ewing, 1971; Sundvor,
1974; Renard and Malod, 1974; Briseid and Mascle,
1975; Eldholm and Talwani, 1977).

Exploration with emphasis on multi-channel
seismic reflection has established a detailed
structural system and revealed that the region
between Bear Island and Norway consists of a
number of basins and ridges with a predominantly
northeastern trend (Fig. 5). Moreover, major
diapirism is observed locally (Rønnevik et al.,
1975; Øvrebø and Tallerås, 1977; Hinz and
Schlüter, 1978b; Rønnevik and Motland, 1979;
Tallerås, 1979; Marty et al., 1979). This structural system is bounded towards the west by the
north trending Senja Ridge which consists of
strongly folded and faulted Mesozoic and
possibly Paleozoic sediments (Hinz and Schlüter,
1978b). Senja Ridge is not to be confused with
the Senja Fracture Zone which lies farther to
the west. Earlier, it was thought that the Senja
Ridge continued onto the Svalbard Platform
towards Svalbard (Øvrebø and Tallerås, 1977).
However, later studies indicate a termination
south of Bear Island (Fig. 5). Øvrebø and
Tallerås (1977) and Rønnevik and Motland (1979)
have interpreted the main events in the geological history of the Barents Sea as deposition
of salt in the Permian, mobilization caused by
regional tectonic activity in late/ middle
Jurassic and formation of the Senja Ridge in the
Early Tertiary. The timing of these events is
naturally somewhat speculative in that no definite dating of the regional reflectors exists
at the present time.

Structurally, this sheared margin was formed
during the Eocene and early Oligocene when the
Lofoten Basin grew wider and deeper. It has been
suggested that the Senja Fracture Zone (Figs.
1,5) demarcates the boundary between oceanic and
continental crust (Talwani and Eldholm, 1973),
This interpretation is based on two lines of
evidence, geophysical and geometrical. Firstly,
the fracture zone is defined by a prominent
elongated gravity anomaly near the base of the
slope which implies a major structural lineament. This lineament also defines a distinct
quiet zone boundary in the magnetic data (Åm,
1975). Secondly, a plate reconstruction to
anomaly 13 time (36 my) shows that the gravity
anomaly trend aligns with the corresponding part
of the Greenland Fracture Zone. However, the
seismic reflection data have not revealed any
relief associated with the fracture zone, but
the Cenozoic sediment thickness is indeed large,
perhaps as much as 3-5 km. The refraction data
show continuation of layering on either side of
the fracture zone, but no crustal refractors
have been recorded underneath the central part
of the gravity anomaly (Houtz and Windisch,
1977; Myhre, 1978). We note that Hinz and
Schlüter (1978a,b) imply that the oceanic crust

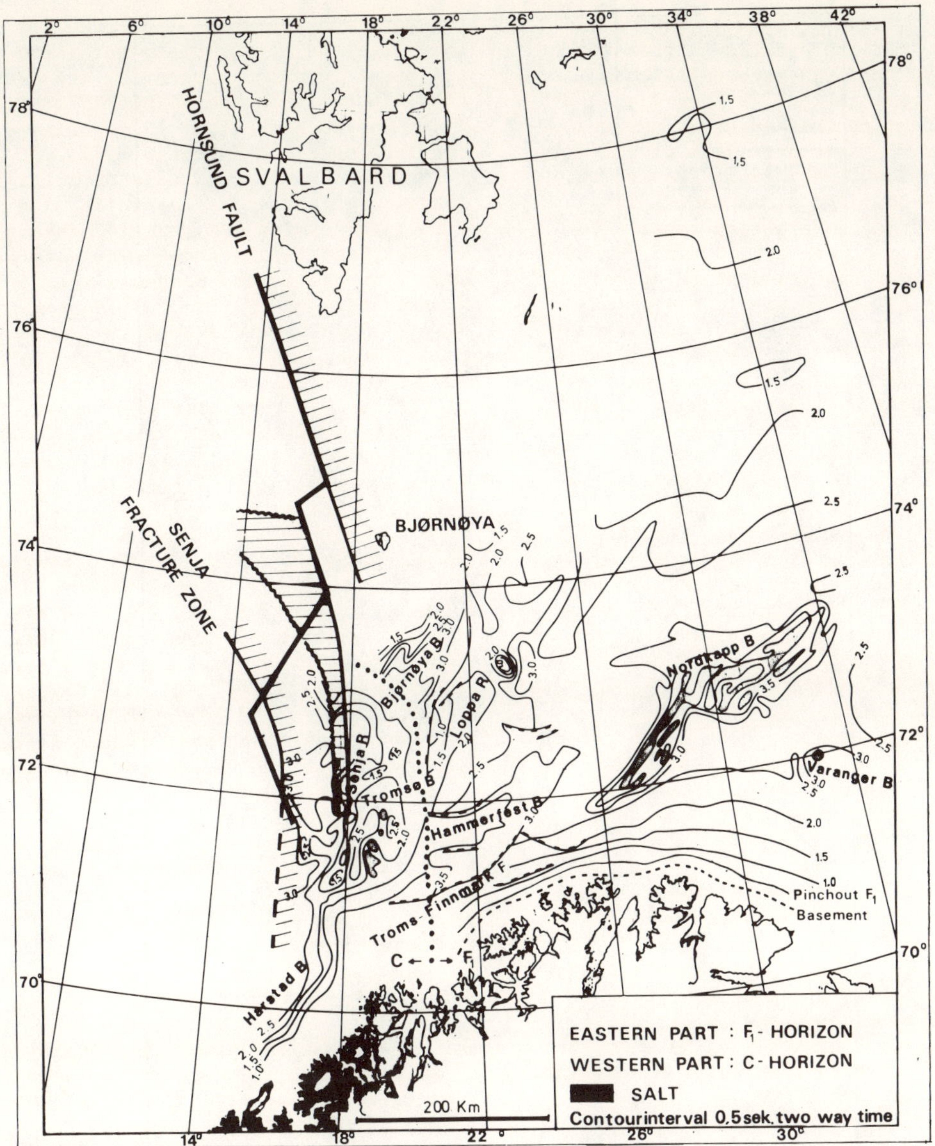

Fig. 5.   Main structural elements in the western Barents Sea (Rønnevik and Motland, 1979). Horizons C and F_1 represent regional reflectors and we refer to Rønnevik and Motland (1979) for a discussion of these features.

extends east all the way to the Senja Ridge. However, no data have been published documenting their suggestion.

Svalbard Margin. During the last few years several groups have conducted multi-channel surveys off western and northern Svalbard. Results on the western margin south of the Spitsbergen Fracture Zone have been published by Malod and Mascle (1975), Hinz and Schlüter (1978b), Schlüter and Hinz (1978), Sundvor and Eldholm (1976, 1979) and Sundvor et al. (1977, 1979b). The margin to the northwest and north of Svalbard has been investigated by Sundvor et al. (1977, 1978, 1979b). From these studies a first-order structural pattern has been developed.

Both the reflection data and the seismic velocity information obtained from sonobuoys on the western margin show that it may be divided into three geologic provinces by two regional north trending structures, the eastern flank of the elevated Knipovich Ridge rift block and the Hornsund Fault (Fig. 6). The active spreading axis, the Knipovich Ridge, has to be included in the analysis of the margin because it lies asymmetrically in the Greenland Sea progressively approaching the lower continental slope northwards. Between 78° and 79°N the ridge province transects the slope. It is believed that shifts in the plate boundary are responsible for this development. The timing for the latest axial

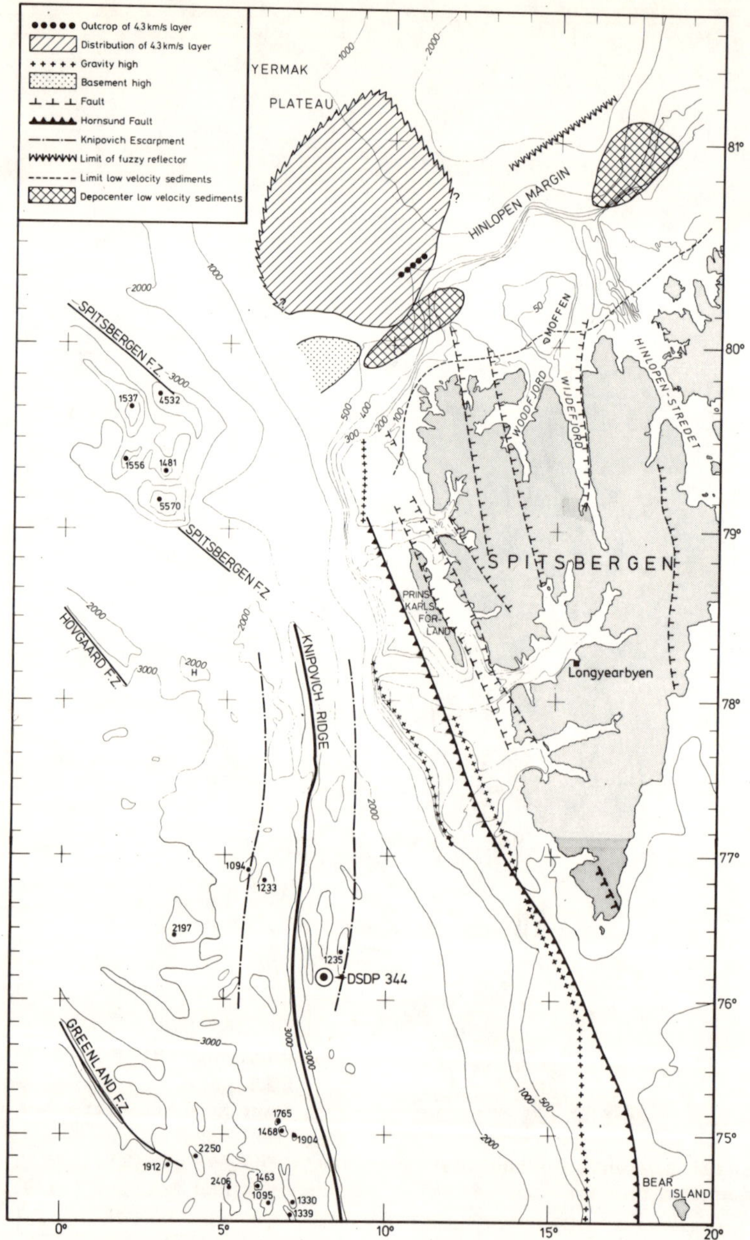

Fig. 6. Main geological and geophysical features on the Svalbard margin (Sundvor et al., 1979).

shift is somewhat uncertain, but is perhaps no older than 5-6 my (Eldholm and Sundvor, 1980) which is the time required to generate the elevated block.

The elevated basement block is bounded by steps (escarpments) in the oceanic basement surface on either side of the rift as indicated in Fig. 6. Oceanic basement can be followed for some distance landward of the eastern escarpment, but loses distinction and disappears under an increasingly thick sequence of sediments. Schlüter and Hinz (1978) indicate that the oceanic crust changes character from rough to smooth in the vicinity of this escarpment.

The Hornsund Fault, which is a major fault or sharp flexure at the central shelf, can be traced from just south of Bear Island to about 79°N (Figs. 6,7). It has been mapped as a single continuous feature although minor offsets cannot be excluded. Landward of the fault there are not any consistent seismic reflectors and and the velocity of the rocks at the sea floor is generally higher than 3.8 km/s.

The eastern Knipovich Ridge escarpment and the Hornsund Fault define a regionally elongated north-south trending sedimentary basin con-

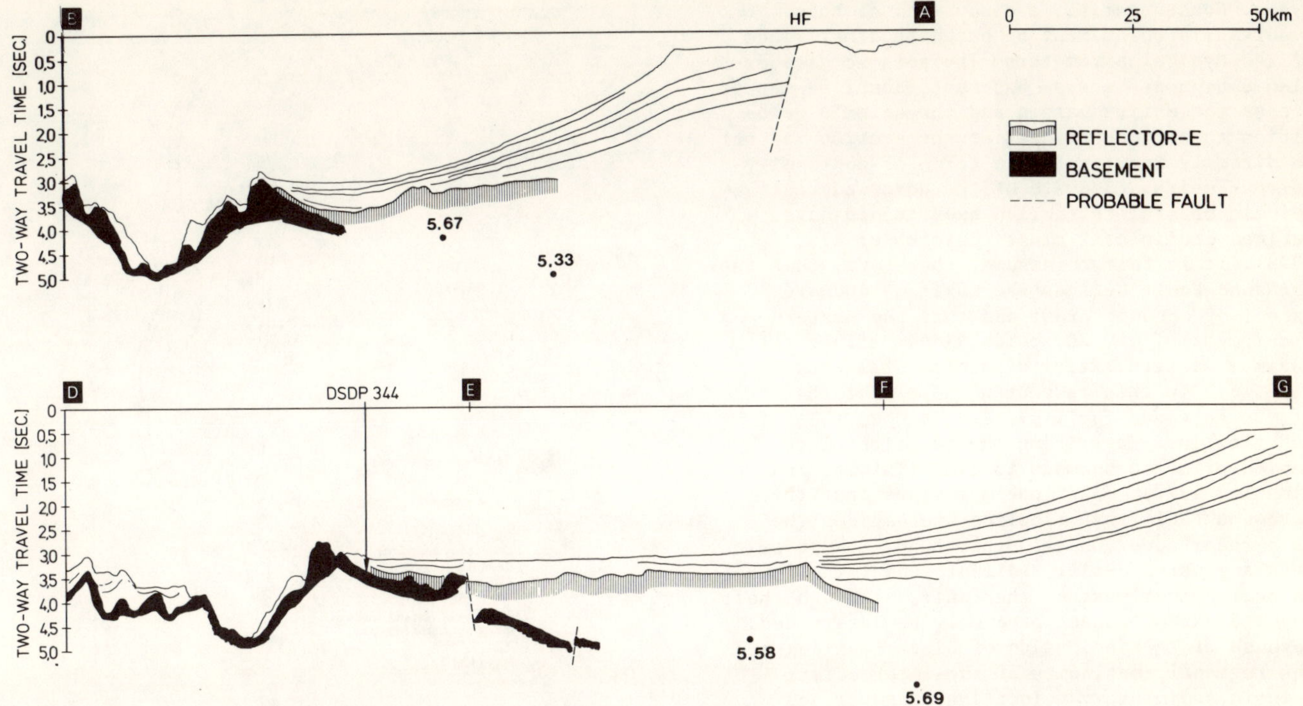

Fig. 7. Line drawings of profiler records from east-west tracks at 77 (AB) and 76°N (DG) west of Svalbard. Some seismic refraction velocities from basement have been plotted onto the sections (Sundvor and Eldholm, 1979).

sisting of an upper low-velocity unit, with layer velocities 1.9 and 2.2 km/s, and a lower sedimentary unit exhibiting average velocities of 2.7, 3.2, 3.7, and 4.4 km/s, respectively (Table 1) (Myhre, 1978). The low-velocity section (less than 2.7 km/s) is restricted to the region between the rift axis and the Hornsund Fault. A 5+ km/s velocity is defined as the seismic basement. This velocity distribution is representative of the entire margin between 75° and 79°N. Approaching the Knipovich Ridge, it is difficult to postulate the continuity of the velocity stratification because velocities in the range 4.5–5.5 km/s may be interpreted as oceanic layer 2, consolidated sediments and crystalline basement. We note, however, that the overall velocity data can be interpreted in terms of stepfaults in the deeper part of the section just seaward of the Hornsund Fault. Faults of this kind have also been observed by Schlüter and Hinz (1978) in the seismic reflection records. Their survey is concentrated to the area north of 76°N and the faulting is most easily recognized in the north where the basin becomes quite narrow.

There are two regional unconformities in the main basin. The upper unconformity (reflector E, Fig. 7) is quite prominent on the lower slope and onlaps the axial mountains (Eldholm and Windisch, 1974). An attempt to drill this unconformity was made during DSDP leg 38 (Talwani,

Udintsev et al., 1976). Glacial marine sediments of Pleisto-Pleiocene age with some turbidites just above basement were recovered. The homogeneous nature of the sampled sediments may indicate that the site has been located slightly off the unconformity.

Schlüter and Hinz (1978) have also mapped a deeper unconformity and speculate that it separates two sequences of Pliocene and pre-Middle Oligocene ages, respectively. The lower sequence lies partly on top of downfaulted rocks belonging to the sequence east of the Hornsund Fault, the Svalbard Platform. By extrapolation of sedimentation rates they infer downfaulting and consequently extension in the Greenland Sea region prior to anomaly 13 time. In view of Svalbard geology and because of regional plate tectonic considerations we find it difficult to accept the idea of absence of the entire Miocene and upper Oligocene sedimentary sequence on this margin, particularly because an uplifted Spitsbergen from Miocene (Kellogg, 1975) would be expected to be a prominent source area of terrigenous sediments.

The change from continental to oceanic crust along continental margins is often reflected in the signature of the magnetic field in many areas of the world ocean. Because of the low amplitude level in the Greenland Sea and adjacent margins the magnetic anomalies have not proved useful in defining the various crustal

types. Consequently,, a discussion of this
important problem must be based on other kinds
of geophysical parameters. The seismic data are
also ambiguous because basement cannot be mapped
across the entire margin and the seismic velocities in the deeper part of the section may not
be directly interpreted in terms of continuity
of rock units. Landward of the Hornsund Fault,
seismic crustal refraction experiments have
defined continental crust (Guterch et al.,
1978). It is fair to assume, therefore, that the
Hornsund Fault defines the maximum landward
extent of oceanic crust and that the eastern
escarpment of the Knipovich Ridge defines the
maximum western extent of continental crust
(Fig. 6). In this respect we note that the trend
of the Hornsund Fault closely follows that of
the flowlines describing the relative plate
motion prior to anomaly 13 time (Talwani and
Eldholm, 1977). Thus, one may argue that the
Hornsund Fault is a flowline and defines the
main shear zone. On the other hand, the seismic
data suggest pre-rift sediments under the low-
velocity cover west of the fault, although their
seaward extent cannot precisely be determined.
Because of the indication of block faulting and
the regional continuity of the intermediate
seismic sedimentary velocities, Sundvor and
Eldholm (1979) favour a continent-ocean transition somewhat west of the fault. A transitional area centered slightly seaward of the
shelf edge has been suggested by Schlüter and
Hinz (1978) north of 77°N. We stress, however,
that they base their interpretation on the
assumption that the smooth apparent basement
reflector landward of the Knipovich Escarpment
does indeed represent oceanic crust.

South of 78°N, the Hornsund Fault follows a
prominent elongate shelf gravity anomaly, but in
the north there is no such anomaly along the
seismic expression of the fault. There is also a
slight but well-defined angle between the
anomaly trend and the fault, but the relationship between these features has not yet been
explained.

In terms of plate reconstructions, the poles
of Talwani and Eldholm (1977) cause some overlap
west of Svalbard. This overlap becomes
negligible if the Hornsund Fault is taken as the
line of initial rifting. Nevertheless, we may
expect a crustal overlap here due to the fact
that the first post-anomaly 13 separation of
Greenland and Svalbard may have taken place as
thinning of the continental crust by normal
faulting prior to sea floor spreading, thus
permitting a crustal change to the west of the
Hornsund Fault.

The surveys on the <u>northern</u> Svalbard margin
have been restricted to the region between
Spitsbergen Fracture Zone and 20°E (Fig. 6).
The western area is characterized by the Yermak
Plateau, whereas a regular shelf and slope are
developed east of about 12°E. The seismic data
published by Sundvor et al. (1977, 1978) reveal

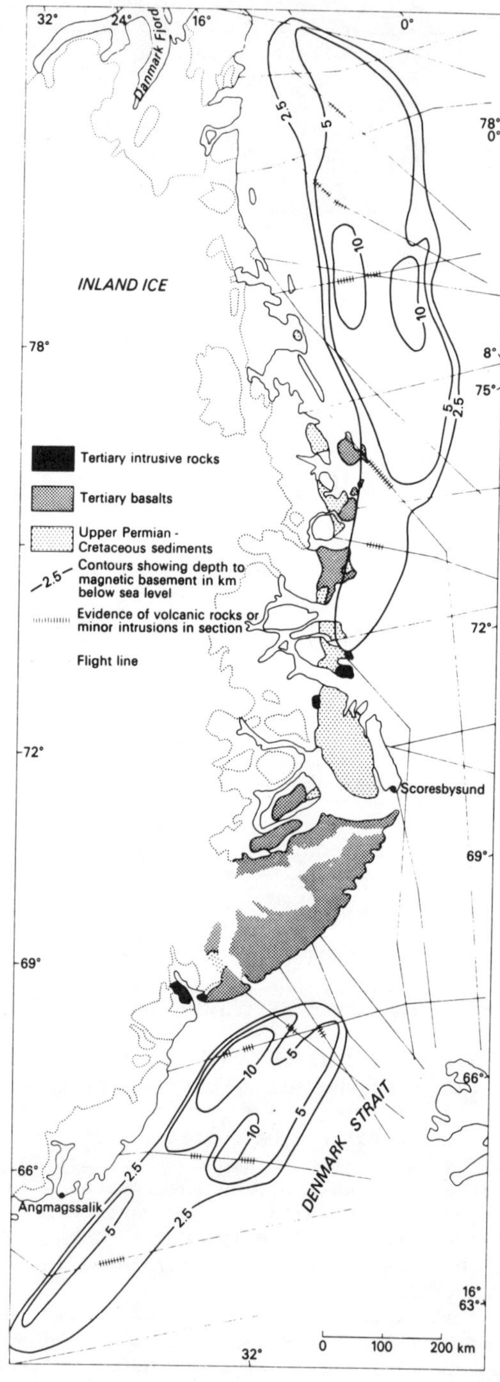

Fig. 8. Sedimentary basins on the East-Greenland
shelf. The contours are based on depth estimates
along aeromagnetic flight lines (Henderson, 1976).

that these parts of the margin also differ
structurally. Although only crystalline rocks
belonging to the Hecla-Hoek formation (Precam-
brian/ Ordovician) are exposed along the adjacent coastline a sedimentary wedge pinches out

on the innermost shelf overlying a seaward-dipping basement which is a continuation of the onshore rocks. The most conspicuous feature observed on the Yermak Plateau margin is an opaque shallow acoustic basement reflector. It is in places faulted, but the abovelying sediments appear undisturbed. The reflector terminates southward as shown in Fig. 6; however, its northern extent is unknown. Whereas the downdipping basement surface beneath the shelf exhibits an average velocity of 5.5 km/s the opaque reflector yields a velocity of 4.3 km/s. A sediment basin with a maximum thickness of approximately 5 km lies between the southern termination of the opaque reflector and the coastline. This basin continues beneath the shelf and slope also east of the Yermak Plateau margin. Here, there is evidence of progradation in the uppermost part of the sequence. The sediment basin consists of an upper low-velocity (less than 2.4 km/s) layer varying in thickness from .6 to 2.0 km overlying a series of layers with velocities similar to those in the main basin along the margin west of Svalbard.

The nature of the opaque reflector has not been determined. For a discussion we refer to Sundvor et al. (1979b) who indicate that it is most likely of volcanic origin. The idea of an Early Tertiary hot spot forming the Yermak Plateau and Morris-Jessup Rise has been advanced by Feden et al. (1979). Dating of lavas in North Greenland gives ages of 32-35 my (Dawes, 1976) and 63 my (Larsen et al., 1978). Moreover, Prestvik (1978) has dated plateau lavas on northern Spitsbergen as young as 10-12 my and he also hints at activity in the early part of the Tertiary. By anology, a history of development such as that of the Vøring Plateau may be tempting; however, Sundvor et al. (1979) stress that suggestions of this kind must be considered highly speculative at the present time.

East-Greenland margin. The geological data have been summarized by Talwani and Eldholm (1974), Johnson et al. (1975) and Larsen (1979).

Since then, few surveys have been carried out north of the Denmark Strait, and only little information has yet been published. During the next years, however, a large-scale surveying program will be instigated by the Danish authorities.

Several aeromagnetic surveys have been flown over parts of the margin. The shelf and slope are magnetically quiet with a typical change when the Cenozoic oceanic crust is approached. Larsen (1978) demonstrated that the early Tertiary dyke swarm of East-Greenland continues as a broad coast-parallel belt on the inner shelf between 63 and 69.8°N. The maximum width is about 80 km. A correlation between magnetic high amplitude on the inner shelf and lava flows is suggested by Johnson et al. (1975). They also pointed out that the anomaly patterns in the Denmark Strait resemble that on the Iceland-Faeroe Ridge which may imply a similar mode of evolution.

Depth estimates of the magnetic basement reveal two large elongate sedimentary basins on the shelf (Henderson, 1976). One lies south of 69°N and another between 72° and 81°N paralleling the coastline (Fig. 8). The maximum depth is 10 km with some indications of minor intrusions within the basin. Henderson (1976) proposed that the basin is an extension of the onshore sediments which contain numerous sills.

The marginal basement high south of the Greenland Fracture Zone (Fig. 1) has been described earlier. This high underlies the lower slope between 75.3° and 76.4°N at a depth of 500-700 m below the sea floor.

Analysis and correlation of refractor velocities yields average velocities of 2.0, 2.4, 3.0, 3.5, 4.3, and 5.4 km/s, a velocity stratification which is very similar to that on the Jan Mayen Ridge and off Norway between 62° and 68°N (Myhre, 1978) (Table 1).

Hinz and Schlüter (1978a) have described results from the margin south of 72°N, including multi-channel seismic data. A thick, up to 3 s,

TABLE 1. Average seismic refraction velocities (km/s). The velocities refer only to profiles overlying continental crust. Values in parenthesis represent profiles on the shelf only. From Myhre (1978).

Møre (62-65°N)	1.9	2.2	2.6	3.4		4.1	5.2
Nordland (64-68°)	1.9	2.2	2.7	3.4		4.2	5.2
Lofoten-Vesterålen (67-70°)			2.8	3.3	3.9		5.3
Western Barents Sea (70-75°N)	1.9	2.3	2.7	3.4		(4.2)	(5.2)
Svalbard (75-80°N)	1.9	2.2	2.7	3.2	3.7	4.4	5.5
East-Greenland	2.0	2.4	3.0	3.5		4.3	5.4
Jan Mayen Ridge	1.9	2.2	2.7	3.1		4.0	5.6

prograded sequence which thins beneath the slope
appears to have been deposited after the split-
off of the Jan Mayen Ridge, 18-22 my. A 50 km
wide rift zone on the central shelf extends
60-70 km north from Scoresby Sound. Here, they
have mapped a strongly reflecting horizon which
is believed to be of similar age as the early
Oligocene unconformity on the Jan Mayen Ridge.
Below this level there are reflectors that are
suggested to represent sediments older than the
late Paleocene opening of the Norwegian-Greenland
Sea.

Summary

At present the state of knowledge about the
margins of the Norwegian-Greenland Sea varies
considerably.
-Intensive investigations off Norway between
60° and 70°N have yielded results which make
it possible to develop a fairly detailed
depositional and structural history. The data
show features that appear to reflect significantly on processes and events related to the
rifting of the continental crust and the subsequent initial history of sea floor generation
and marginal subsidence.
-A first order geologic framework is in the
process of being developed for the margins
off Svalbard and the western Barents Sea.
-Off East-Greenland regional information is
based on little overall data supplemented
with some detailed information from local
surveys. The main regional geologic framework of this margin is not considered to
have been established.
The margins developed in conjunction with the
opening of the Norwegian Sea in the late Paleocene. The rifting took place within a large epicontinental sedimentary basin extending into the
Barents Sea and the North Sea. It appears that
the area of Cenozoic sea floor spreading had
been subjected to crustal extension during the
Mesozoic and possibly earlier, but that these
movements died out in the early Cretaceous. The
margins are classified into rifted and sheared
segments reflecting the geometry of the plate
boundary at various times of opening. Moreover,
the margins north of the Greenland-Senja Fracture
Zone may be of a combined sheared/rifted type.
To the south of this fracture zone there are
prominent marginal highs which probably reflect
important events during the first million years
after initiation of spreading. Possibly, an
Iceland-Faeroe Ridge type generation of oceanic
crust and/or regional volcanic events may relate
to many of the observations at or in the vicinity
of these highs. The Neogene history of the
margins is characterized by progradation over a
gently subsiding crust with structural modifica-
tions off Svalbard and East-Greenland associated
with migrations of the plate boundary.

Acknowledgments. We thank A.M. Myhre for per-
mission to reproduce Table 1. The work has been
supported by the Norwegian Petroleum Direc-
torate, The Norwegian Research Council for
Science and the Humanities (Grant D.40.31-46)
and National Science Foundation (Contract DPP
78-05733).

J. Thiede, W.C. Pitman III and J. Mutter reviewed the paper. Lamont-Doherty Geological
Observatory Contribution Number 2765.

References

Bott, M.P.H., Structure and evolution of the
Atlantic floor between Northern Scotland
and Iceland, Norges geol. Unders. , 316,
195-200, 1975.
Bott, M.P.H., J. Sunderland, J.P. Smith,
U. Casten, and S. Saxov, Evidence for
continental crust beneath the Faeroe Islands,
Nature, 248, 202-204, 1974.
Briseid, E. and J. Mascle, Structure de la
Marge continentale Norvegiènne au debouche
de la Mer de Barentz, Mar.Geoph.Res., 2,
231-241, 1975.
Cashion, W.W., The geology of the West Shetland
Basin, Offshore Europe, 216, 1-7, 1975.
Caston, V.N.D., Tertiary sediments of the
Vøring Plateau, Norwegian Sea, recovered
by leg 38 of the Deep Sea Drilling Project,
Init. Rept. Deep Sea Drilling Project, 38,
761-782. 1976.
Dalland, A., The sedimentary sequence of Andøy,
Northern Norway - Deposition and structural
history, Norw.Petrol.Soc., NSS/26, 31 pp. 1979.
Dawes, P.R., Precambrian to Tertiary of Northern
Greenland, in: Escher, A., and W.S. Watt,(eds.),
Geology of Greenland, Geol.Surv.Greenland,
248-303, 1976.
Eldholm, O. and J. Ewing, Marine geophysical
survey in the southwestern Barents Sea,
J.Geophys.Res., 76, 3832-3841, 1971.
Eldholm, O. and E. Sundvor, The continental
margins of the Norwegian-Greenland Sea:
recent results and outstanding problems,
Phil.Trans.Roy.Soc.Lond., A. 294, 77-86, 1980.
Eldholm, O. and M. Talwani, Sediment distri-
bution and structural framework of the Barents
Sea, Geol.Soc.Am.Bull., 88, 1015-1029, 1977.
Eldholm, O., and C.C. Windisch, Sediment dis-
tribution in the Norwegian-Greenland Sea,
Geol.Soc.Am.Bull., 85, 1661-1676, 1974.
Eldholm, O., E. Sundvor and A.M. Myhre,
Continental margin off Lofoten-Vesterålen,
Mar.Geophys.Res., 4, 3-35, 1979.
Feden, R.H., P. R. Vogt, and H.S. Fleming,
Magnetic and bathymetric evidence for the
"Yermak" hot spot northwest of Svalbard
in the Arctic Basin, Earth Planet.Sci.Lett.,
44, 18-38, 1979.
Gairaud, H., G. Jaquart, F. Aubertin, and P.
Beuzart, The Jan Mayen Ridge synthesis of
geological knowledge and new data, Oceano-
logica Acta, 1, 335-358, 1978.
Garde, S.S., Sur geologischen entwicklung des
Jan Mayen Ruckens nach geophysikalischen

daten, Dissertation, Univ.Clausthal, 74 pp. 1978.

Guterch, A., J. Pajchel, E. Perchuc, J. Kowalski S. Duda, J. Kamber, G. Boydys, and M. A. Sellevoll, Seismic reconnaissance measurement on the crustal structure in the Spitsbergen region 1976, Univ.Bergen, Seism.Obs., 61 pp. 1978.

Hamar, G., Tectonic map – North Sea, Map printed by Statoil A/S, Stavanger, Norway, 1979.

Henderson, G., Petroleum geology, in, Escher, A., and W.S. Watt, (eds.), Geology of Greenland, Geol.Surv.Greenland, 488–505, 1976.

Hinz, K., The seismic crustal structure of the Norwegian margin in the Vøring Plateau, in the Norwegian deep sea, and on the eastern flank of the Jan Mayen Ridge between 66° and 68°N, 24th. Int.Geol.Congr., Section 8, 28–36, 1972.

Hinz, K. and H.U. Schlüter, Der Nordatlantik – Ergebnisse geophysikalischer Untersuchungen der Bundesanstalt für Geowissenschaften und Rohstoffe an nordatlantischen Kontinenträndern, Erdoel-Erdgas Zeitschrift,94, 271–280, 1978a.

Hinz, K. and H.U. Schlüter, The geological structure of the Barents Sea, Mar.Geol., 26, 199–230, 1978b.

Hinz, K. and J. Weber, Zum geologischen Aufbau des Norwegischen Kontinentalrändes und der Barents-See nach reflexionseismischen Messungen, Erdöl und Kohle, Erdgas, Petrochemie, 3–29, 1976.

Houtz, R. and C.C. Windisch, Barents Sea continental margin sonobuoy data, Geol. Soc.Am.Bull., 88, 1030–1036, 1977.

Jacqué, M. and J. Thouvenin, Lower Tertiary tuffs and volcanic activity in the North Sea, in: Woodland, A.W. (ed.), Petroleum and the continental helf of north-west Europe, 1, Appl. Sci. Publ. London, 455–465, 1975.

Johnson, G.L., N. J. McMillian, and J. Egloff, East Greenland continental margin, in: Yorath, C.J., E.R. Parker and D.J. Glass, (eds.), Canada's Continental Margin and Offshore Petroleum Exploration, Can. Soc. Petrol. Geol. Mem., 4,205–224, 1975.

Jørgensen, F. and T. Navrestad, Main structural elements and sedimentary succession on the shelf outside Nordland (Norway), Norw.Petrol. Soc., NSS/11, 20 pp. 1979.

Kellogg, H.E., Tertiary stratigraphy and tectonsim in Svalbard and continental drift, Am.Assoc.Petrol.Geol.Bull., 59, 465–485, 1975.

Larsen, H.C., Geological perspectives of the east Greenland continental margin, (preprint).

Larsen, H.C., Offshore continuation of the East Greenland dyke swarm and its bearing on North Atlantic ocean formation, Nature, 274, 220–223, 1978.

Larsen, O., P. R. Dawes and N.J. Soper, Rb/Sr age of the Kap Washington Group, Peary Land, North Greenland, and its geotectonic implication, Grønland Geol.Unders.Repts., 90, 115–119, 1978.

Malod, J. and J. Mascle, Structures geologiques de la marge continentale a l'ouest du Spitsberg, Mar.Geophys.Res., 2, 215–229, 1975.

Marty, A.,J. Rabate, M. Cazes, G. Jaquart, and J.L. Auxietre, CEPM Barents Sea seismic survey. Norw.Petrol.Soc., NSS/16, 16 pp. 1979.

Myhre, A.M., Analyse av seismiske refraksjonsdata fra Norskehavet og omliggende kontinentalmarginer, Cand.real. thesis, Univ.Oslo, 135 pp. 1978.

Perry, R.K., H. S. Fleming, N.Z. Cherkis, R. H. Feden, and P.R. Vogt, Bathymetry of the Norwegian-Greenland and western Barents Sea. Naval Res.Lab., Map, 1980.

Prestvik, T., Cenozoic plateau lavas of Spitsbergen – a geochemical study, Årbok Norsk Polarinst.1977, 130–143, 1978.

Renard, V. and J. Malod, Structure of the Barents Sea from seismic refraction, Earth Planet.Sci.Lett., 24, 33–47, 1974.

Russell, M.J., A possible lower Permian age for the onset of ocean floor spreading in the northern North Atlantic, Scott.J.Geol. 12, 315–323, 1976.

Rønnevik, H.C. and K. Motland, Geology of the Barents Sea, Norw.Petrol.Soc., NSS/15, 34 pp. 1979.

Rønnevik, H. and T. Navrestad, Geology of the Norwegian shelf between 62° and 69°N, Geo-Journal, 1, 33–46, 1977.

Rønnevik, H.C., F. Jørgensen and K. Motland, The geology of the northern part of the Vøring Plateau, Norw.Petrol.Soc., NSS/12, 21 pp. 1979.

Rønnevik, H., E. I. Bergsaker, A. Moe, O. Øvrebø, T. Navrestad, and J. Stangenes, The geology of the Norwegian continental shelf, Petroleum and the continental shelf of Northwest Europe, 1, Appl.Sci.Pub., 117–129, 1975.

Schlüter, H.U., and K. Hinz, The continental margin of West Spitsbergen, Polarforschung, 48, 151–169, 1978.

Sundvor, E., Seismic refraction and reflection measurements in the southern Barents Sea, Mar.Geology, 16, 255–273, 1974.

Sundvor, E., and O. Eldholm, The western and northern nargin off Svalbard, Tectonophysics, 59, 239–250, 1979.

Sundvor, E. and O. Eldholm, Marine geophysical survey on the continental margin from Bear Island to Hornsund, Spitsbergen, Univ.Bergen, Seism.Obs.Sci.Rept., No.3, 28 pp., 1976.

Sundvor, E., A. Gidskehaug, A. M. Myhre, and O. Eldholm, Marine geophysical survey on the northern Jan Mayen Ridge, Univ.Bergen Seism. Obs. Sci.Rept.No. 6, 17 pp. 1979a.

Sundvor, E., A. M. Myhre, and O. Eldholm, The Svalbard continental margin, Norw.Petrol. Soc., NSS/6, 13 pp. 1979b.

Sundvor, E., A. Gidskehaug, A.M. Myhre and O. Eldholm, Marine geophysical survey on the northern Svalbard margin, Univ.Seism.Obs., Sci.Rept.,No. 5, 46 pp., 1978.

Sundvor, E., O. Eldholm, A. Gidskehaug, and A.M. Myhre, Marine geophysical survey on the western and northern continental margin off Svalbard, Univ.Bergen, Seism.Obs.Sci.Rept., No 4, 35 pp. 1977.

Surlyk, F., Mesozoic faulting in East Greenland, Geol.Minjnbouw, 56, 311-327, 1977.

Tallerås, E., The Hammerfest Basin - an aulacogen?, Norw.Petrol.Soc., NSS/18, 13 pp. 1979.

Talwani, M., Distribution of basement under the eastern North Atlantic Ocean and the Norwegian Sea, Geol.J., Spec. issue No. 10, 347-376, 1978.

Talwani, M. and O. Eldholm, Evolution of the Norwegian-Greenland Sea, Geol.Soc.Am.Bull., 88, 969-999, 1977.

Talwani, M. and O. Eldholm, The margins of the Norwegian-Greenland Sea, in: Burk, C.A. and C.L. Drake, (eds.), Geology of continental margins, 361-375, 1974.

Talwani, M. and O. Eldholm, The boundary between continental and oceanic crust at the margin of rifted continents, Nature, 241, 325-330, 1973.

Talwani, M. and O. Eldholm, The continental margin off Norway: a geophysical study, Geol.Soc.Am.Bull., 83, 3575-3608, 1972.

Talwani, M., G. Udintsev, et al., Init.rept. Deep Sea Drilling Proj., 38, 1256 pp., 1976.

Voppel, D., S.P. Srivastava, and U.Fleischer, Detailed magnetic measurements south of the Iceland-Faeroe Ridge, Deutsch.Hydrogr.Zeitschr., 32, 154-172, 1979.

Ziegler, P.A., Geology and hydrocarbon provinces of the North Sea, GeoJournal, 1, 7-32, 1977.

Øvrebø, O. and E. Talleraas, The structural geology of the Troms area, (Barents-Sea), GeoJournal, 1, 47-54, 1977.

Åm, K., Magnetic profiling over Svalbard and surrounding shelf areas, Norsk Polarinst Årb. 1973, 87-100, 1975.

THE CONTINENTAL MARGINS OF EASTERN CANADA: A REVIEW

C. E. Keen

Atlantic Geoscience Centre, Geological Survey of Canada
Bedford Institute of Oceanography, Dartmouth, Nova Scotia, Canada

Abstract. The Atlantic-type continental margins off eastern Canada exhibit both rifted and transform tectonic styles and range in age from Jurassic to Tertiary. The plate motions responsible for these margins involved the separation of four major plates from North America: the African plate in the Jurassic, the Iberian plate in the Cretaceous, and the European and Greenland plates in the late Cretaceous and early Teriary. The record of these complex plate motions is fairly well determined in some regions but disputed in others, such as the Labrador Sea. Cross-sections of the margins from seismic data, combined with the magnetic and gravity signatures across them, show evidence of substantial crustal thinning of the continental crust over a zone 100–350 km wide. In some areas the crust may have also been intruded by basaltic magma during the rift stage, but evidence of this process is less widespread than that for crustal thinning. Oceanic crust does not appear to thicken significantly near the ocean-continent transition and there is little evidence that oceanic or continental basement ridges are general characteristics of these margins. The geophysical data suggest that crustal thickness and magnetic anomaly trends are often good indicators of the nature of the crust (oceanic or continental) and that the magnetic slope anomaly marks the approximate position of the ocean-continent transition zone. Gravity anomalies situated near the edge of the shelf are not always associated with this transition zone in a simple way and have a complicated relationship to density changes in both the crust and upper mantle across the margins and to the way in which the margins are isostatically compensated. On many of the younger margins the high amplitude gravity anomalies are associated with a zone of very thin continental crust beneath the outer shelf. The thinned crust in these zones appears to have been produced by crustal stretching during rifting, since similar thinning is not seen across the transform margin south of the Grand Banks. Models of crustal thinning in the early history of margin formation have been used to explain the observed subsidence of the margins. This subsidence is an approximately linear, function of $\sqrt{age}$ and is thought to be due to the combined effects of cooling of the lithosphere and sediment loading. The cooling model can also be used to predict the paleotemperature history of the sediments and relate this history to hydrocarbon generation.

Introduction

The passive continental margins of eastern Canada display as great a diversity of geologic styles and ages as found anywhere around the North Atlantic. The margins span an age of about 100 Ma from Lower Mesozoic in the south to Tertiary in the north, and exhibit both rifted and transform tectonic styles. The large volume of geophysical data, in combination with the data from deep exploratory wells drilled on the shelves, enable a critical appraisal of the processes which have been proposed for the evolution of passive continental margins.

Several comprehensive reviews of these margins and adjacent oceanic areas have recently appeared [Hall, 1979; Haworth and Keen, 1979; Keen and Hyndman, 1979; Srivastava, 1978; Williams, 1979; Umbleby, 1979]. This paper, which deals with our state of knowledge to the end of 1979, will focus on specific topics of particular relevance to the geodynamic processes at the present margins. No attempt will be made to discuss the equally important ancient plate margins, lying immediately landward of the present margins, or to discuss problems concerned the origin of the oldest oceanic crust adjacent to the margins, except where necessary to clarify specific points. The three main topics discussed are:

(1) Plate Motions and Margin Formation. The separation and interaction of at least five plates were involved in the formation of the margins of eastern Canada. The timing and geometry of these motions is fairly well established [Pitman and Talwani, 1972; Laughton, 1972; Kristoffersen and Talwani, 1977; Haworth and Lefort, 1979; Jackson et al., 1979; LePichon et al., 1977; Keen et al., 1977; Srivastava, 1978] and will be briefly described.

(2) The Nature of the Ocean-Continent Transition. The position, width and composition

of the transition region, as well as the criteria by which it is defined, are discussed. The characteristics of magnetic and gravity anomalies across the margins and their importance in defining the transition are reviewed.

(3) The Early History of the Margins. The cycle of early continental rifting, followed by subsidence as the margins separated from the new mid-ocean ridge is recorded in the sedimentary history which can be studied in the numerous deep exploratory wells drilled on the continental shelves. The role of heat as the controlling factor in this cycle has been the subject of many recent investigations [de Charpal et al., 1978; Keen, 1979; McKenzie, 1978; Royden et al., 1980; Steckler and Watts, 1978; Hinz et al., 1979]. The evidence from wells on the Nova Scotian and Labrador Shelves for thermally driven subsidence is examined and related to crustal properties across these margins.

Plate Motions and Margin Formation

The earliest plate motion responsible for margin formation off eastern Canada was the separation of Africa from North America in the Early Jurassic. The rifted margin off Nova Scotia and the transform margin south of the Grand Banks were formed during this breakup (Fig. 1). The evidence for the timing of this separation from the oceanic magnetic record is poor. An extensive magnetic quiet zone occupies the oldest oceanic crust adjacent to the margins and consequently no magnetic anomaly chronology can be established. Extrapolation of spreading rates from younger magnetic anomalies east of the quiet zone, and tentative dating of two reversals within the quiet zone suggest an age of about 170 Ma for the oldest oceanic crust formed by sea floor spreading [Barrett and Keen, 1976]. The age of the sediments deposited upon the subsiding shelf after spreading began is consistent with this timing as discussed later.

The transform margin south of the Grand Banks [Auzende et al., 1970] is a striking example of a margin produced by shearing between two continental plates. The oceanic crust formed adjacent to this margin thus becomes progressively younger to the southeast. At its northwestern end it is inferred to be late Early Jurassic, while at its southeastern end it is mid-Cretaceous in age. The junction between the rifted margin off Nova Scotia and transform margin is not well mapped and requires further study. Furthermore, the transform margin does not appear to have formed along the seaward prolongation of a pre-existing zone of structural weakness in the continent, as was previously supposed [King et al., 1975]. Instead the transform margin lies at an oblique angle to a very prominent fault boundary (Fig. 2) in the continent which separates rocks of the Meguma and Avalon platforms [Haworth and Keen, 1979].

At the southeastern end of the transform margin lies a region of seamounts and volcanic basement ridges, indicating major volcanic activity in the past. This region includes the Fogo Seamounts, the Spur of J-Anomaly Ridge and the Newfoundland Ridge (Figs. 1 and 2). Various explanations for this complex region have been offered which include the possibility that it represents the trace of a fossil hot spot, a leaky transform fault [Burke et al., 1973; Sullivan, 1978], or a region mainly occupied by foundered continental crust [Gradstein et al., 1977; Grant, 1977]. The latter hypothesis would have serious consequences for the present pre-rifting reconstructions of the continents. Sullivan and Keen [1978] review some of the evidence on which these explanations are based.

The opening between Iberia and North America, probably in the mid-Cretaceous, formed the margin east of the Grand Banks. The oldest identified marine magnetic anomaly east of this margin is the J-anomaly, tentatively assigned an age of 115 Ma [Keen et al., 1977]. However, there is some evidence that a long period of rifting may have preceded sea floor spreading, and that a seaway extended northward between Iberia and the Grand Banks as early as Jurassic time [Jansa and Wade, 1975; Sullivan, 1978; Jansa et al., 1980]. The earliest direction of plate separation was roughly east-west, perpendicular to the margins, but may have changed about 15 Ma after spreading began, to accommodate the opening of the Bay of Biscay [Sullivan, 1978].

A major problem in reconstructing the original positions of the continents is encountered in this region. Most reconstructions require either an overlap of Flemish Cap and Galicia Bank (Fig. 1), or a gap in the southern part of the area of about 300 km. Studies by Sullivan [1978] of the margin east of the Grand Banks suggest that continental crust may extend somewhat further seaward in the southern part of the margin than previously supposed, which partly, but not completely, overcomes the problem. Further geophysical studies of the position of the ocean-continent boundary in this region are required.

During the late Cretaceous, spreading began between Europe and North America, forming the margins northeast of Newfoundland. The late Cretaceous was also the time of separation of Greenland from North America, creating the margins of the Labrador Sea. Rifting may have begun in the Baffin Bay region at that time, but sea floor spreading probably did not start until the Early Tertiary [Srivastava, 1978], when the direction of motion between Greenland and North America changed to allow significant plate separation in Baffin Bay. The marine magnetic anomaly record is well documented for the Labrador Sea region, as far south as Flemish Cap. Seaward of Flemish cap and Orphan Knoll (Fig. 1) the oldest marine magnetic anomaly is probably anomaly 34 (about 85 Ma; Cande and Kristoffersen, 1977) while further north the oldest identified anomaly is anomaly 32 (75 Ma; Srivastava, 1978). Reasonably good magnetic lineations have now been described in the Baffin Bay

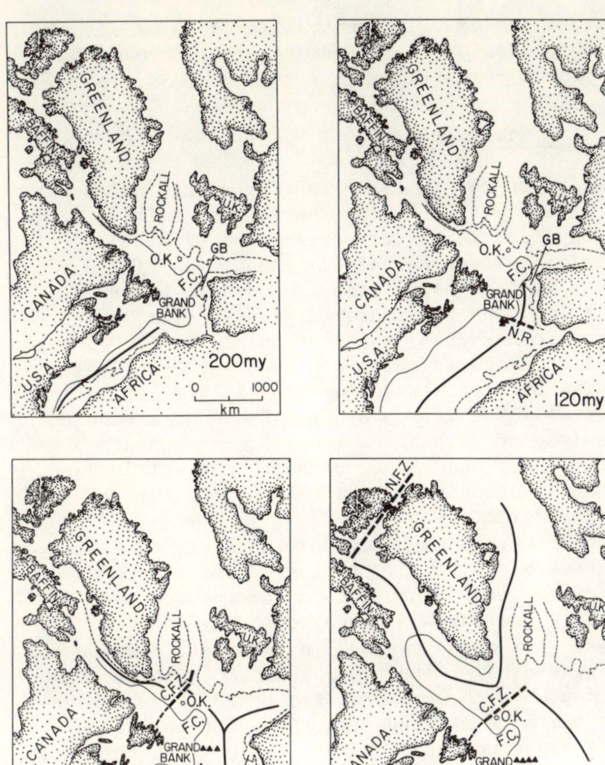

Fig. 1. A sketch of four stages of the plate tectonic opening of the North Atlantic, responsible
for the formation of the present margins. Solid
heavy lines markes spreading centres or incipient
spreading centres, heavy dashed lines denote
fracture zones, and triangles represent seamounts
(the Fogo and Newfoundland Seamounts south and
east of the Grand Banks respectively). The
lighter solid and dashed lines are bathymetric
contours which best delineate the edge of the
shelves. N.R.=Newfoundland Ridge; C.F.Z.=Charlie
Fracture Zone; N.F.Z.=Nares Strait Transform;
O.K.=Orphan Knoll; F.C.=Flemish Cap and
G.B.=Galicia Bank.

region and support a Tertiary age for the central
part of Baffin Bay [Jackson et al., 1979]. However, the magnetic anomalies decrease in coherence
and amplitude near the margins of the Bay, pre-
venting confident identification of the oldest
marine magnetic anomalies. A similar decrease has
been noted in the Labrador Sea for the older magnetic anomalies (anomalies 32-28, van der Linden
and Srivastava, 1975), leading some to question
the validity of their identification[van der
Linden, 1975a].

The late Cretaceous-Tertiary margins exhibit
several enigmatic features. South of the Charlie
Fracture Zone lie the continental fragments,

Orphan Knoll and Flemish Cap, with deeply subsided
continental crust occupying the area between them
and the adjacent shelf [Haworth and Keen, 1979].
North of the fracture zone no fragments or subsided crustal regions occur. Haworth [1977] suggested that the Charlie Fracture Zone is the seaward extension of the Dover Fault, a pre-rift
feature which separates rocks of the preCambrian
Avalon platform from the central Appalachian
mobile belt. It appears that the original pre-
rift geology may have controlled the position of
the Charlie Fracture Zone and the different geologic styles of the margin to the north and south
of the Fracture zone.

Separating the Labrador Sea from the Baffin Bay
basin is the Davis Strait, a shallow water sill
(water depth 700 m; Fig. 2). Its proximity to
Early Tertiary basalts on land which lie both east
and west of it [Clarke, 1977; Hall, 1979], the
resemblance of the underlying seismic crustal
structure to that of Iceland [Keen and Barrett,

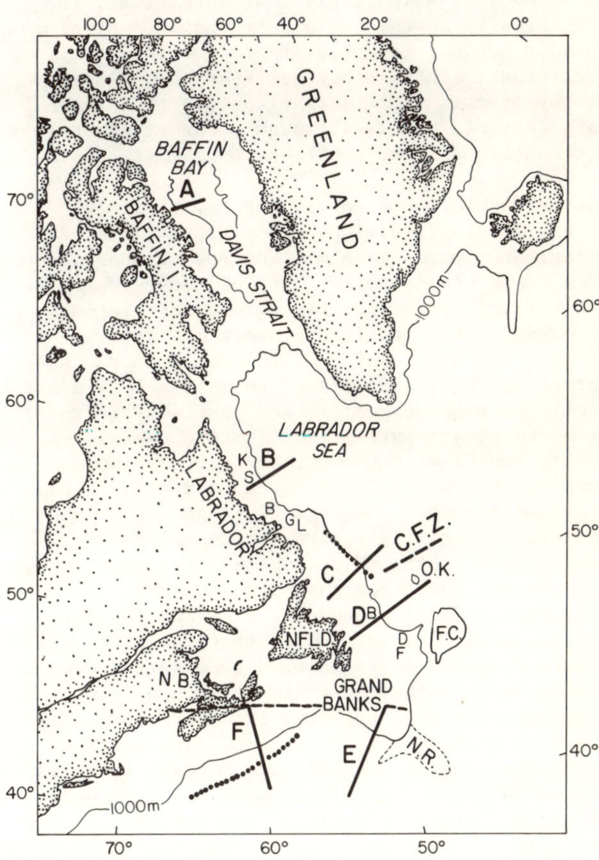

Fig. 2. Index map of the eastern Canadian offshore showing locations of the cross-sections
discussed in the text. The dotted lines delineate
the magnetic Slope Anomaly, where it is clearly
defined. The dashed line across the Grand Banks
is the fault boundary between the Meguma and
Avalon zones, discussed in the text. Letters are
deep exploratory well locations, show in Fig. 4.

1972], and the high amplitude magnetic anomalies
over part of the Strait [Hood and Bower, 1973;
Srivastava, 1978], have led to the hypothesis that
it may be a fossil hot spot, along which some
transcurrent motion may have also occurred.
Hyndman [1973] has shown plate motions consistent
with these hypothesis.

The timing and geometry of opening of the
Labrador Sea and Baffin Bay presented briefly
above have been criticized for a number of reasons. First, the plate tectonic motions require
some combination of transform motion along the
Nares Strait (up to 250 km) and compression
between Greenland and the Canadian Arctic Islands.
The geological evidence supporting these events is
either disputed or absent [e.g. Kerr, 1967; Dawes,
1971; Balkwell and Bustin, 1978; Trettin and
Balkwell, 1979; Newman and Falconer, 1978].
Second, alternative interpretations of the geophysical results from central Baffin Bay, the
Davis Strait, and the Labrador Sea [e.g. Umpleby,
1979; van der Linden, 1975a,b; Grant, 1979] have
been proposed, which imply that Baffin Bay, the
Davis Strait, as well as a substantial part of the
Labrador Sea may be underlain by subsided continental crust. This author believes this alternative hypothesis to be less plausible than the
plate tectonic model, although a concerted attempt
to reconcile all the available data must clearly
be made.

Of particular interest to the relationship
between margin formations and plate motions is a
comparison of conjugate margins now on opposite
sides of an ocean basin. Many of these conjugate
margin pairs exhibit different characteristics.
For example, the margin southwest of Great
Britain, in the vicinity of the Goban Spur
[Scrutton, 1979], appears to be marked by a sharp
transition between oceanic and continental crust,
while its counterpart on the eastern Canadian
margin, the Flemish Cap-Orphan Knoll region is
occupied by a wide region of thinned continental
crust and detached continental fragments. Other
apparently dissimilar pairs are Rockall—NE
Newfoundland, north of the Charlie Fracture Zone,
and possibly the eastern and western margins of
the Labrador Sea [Mayhew et al., 1970; van der
Linden, 1975a]. This implies that the final split
between the continents was not necessarily located
symmetrically with respect to structure formed
during rifting and that rift structures are often
preserved on only one of the conjugate margins.
Systematic studies of conjugate margins pairs is
an area in which very little work has been done
and where there is clearly a requirement for
future research.

The Nature of the Ocean-Continent Transition

The six cross-sections presented in Figure 3
[after Keen and Hyndman, 1979], the locations of
which are shown in Figure 2, are used to illustrate the variations in crustal structure and
geophysical characteristics across the margins of

eastern Canada. The central issue to be discussed
in this section is the nature of the ocean-
continent transition.

Crustal Structure

Evidence for two important processes which may
have been active during the initial rifting stage
is apparent in the cross-sections (Fig. 3).
First, the continental crust near the rifted margins has been significantly modified by crustal
thinning over horizontal distances of 100 km or
more (for example, see sections A, B, F). Second,
this decrease in thickness is sometimes accompanied by an increase in the velocity of the lower
crustal layers, from about 6.2-6.4 km/s beneath
the inner continent to 6.8-7.0 km/s beneath the
outer shelf and slope, perhaps indicating that
igneous intrusion of the thinned continental crust
has occurred.

There appear to be two distinct styles displayed
by crustal thinning. At the oldest rifted margin
off Nova Scotia thinning is accompanied by a
gradual decrease in depth to the M discontinuity
toward the ocean basin. In contrast, the younger
margins usually exhibit a rapid, abrupt decrease
in depth to the crust-mantle boundary, which, on
the basis of limited data, does not appear to
correspond to rapid changes in sediment thickness.
Thus the M discontinuity may be anomalously shallow beneath the outer shelves in these regions
(for example, sections A and B), and about 100 km
landward of the edge of the shelf rapidly increases to typical continental depths. Some
caution must be exercised, however, in relating
these two styles only to the age of the margins.
Northeast of Newfoundland, north of the Charlie
Fracture Zone (section C), crustal thinning more
closely resembles that off Nova Scotia, than that
observed beneath the other younger margins. The
continental crust beneath this region is part of
the central Appalachian mobile belt and perhaps
the fabric and constitution of the continental
crust is an equally important factor in
controlling the style which thinning displays.

High velocities in the lower crustal layers are
not found at all the rifted margins off eastern
Canada; they are mainly observed off Labrador and
Nova Scotia (sections B and F). Direct evidence
that these high velocities may be due to igneous
intrusions during early rifting comes from the
outer Labrador Shelf, where Early Cretaceous
(rift?) basalts have been sampled in some of the
deep exploratory wells beneath Late Cretaceous and
younger sediments [Umpleby, 1979]. Elsewhere, the
zone of thinned continental crust is not clearly
associated with high crustal velocities. In the
region west of Orphan Knoll the continental crust
has been attenuated from about 35-40 km to about
15 km (excluding the sediments of post-rifting
age), over a horizontal distance of 350 km, but
the crustal velocities appear to be similar to
those measured beneath the inner, unmodified
continent. Magnetic anomalies are also subdued in

this region, supporting (but not proving) the absence of magnetic, basaltic material [Haworth, 1977]. Similarly, normal continental velocities are measured beneath the outer shelf of western Baffin Bay, which otherwise closely resembles the structure beneath the Labrador Shelf. This suggests that igneous intrusion may not play as pervasive a role in modifying continental crust as does crustal thinning.

The transform margin south of the Grand Banks exhibits a very sharp transition from oceanic to continental crust (section E), with relatively little thinning of the continental crust. Although more seismic measurements of the continental crustal structure along this section are highly desirable, the structure shown in Figure 3 is consistent with the gravity anomaly measured across this margin [H. R. Jackson and C. E. Keen, unpublished data]. The very different nature of the crustal changes across the transform margin compared to those across the rifted margins suggest that crustal thinning may be related to differences in the tectonic forces acting across them in the early stages of margin formation.

The Ocean-Continent Transition

Definition of the ocean-continent boundary at rifted margins cannot be based on any single criterion, given the diversity of structural styles and the extensive modification of continental crust seen at the eastern Canadian margins. Crustal velocity changes from 6.1 (continental?) to 6.8 (oceanic?) km/s have been shown to be unreliable. Seismic reflection data may be useful in tracing the top of oceanic layer 2 from regions known to be underlain by oceanic crust landward toward the margin and have been used to place minimum limits on the landward extent of oceanic crust [Jansa and Wade, 1975; Barrett and Keen, 1976; Hinz et al., 1979; Keen et al., 1977; Jackson et al., 1975]. However, this reflector is often difficult to trace beneath the slope and upper rise, due to the great thickness of sediments and the pesence of water bottom multiples on the records. There are, however, two geophysical measurements which may provide fairly consistent discriminators between oceanic and continental crust.

The first of these is crustal thickness, omitting the sediments deposited after rifting occurred. Thinned continental crust appears to exhibit a minimum thickness of about 15 km, intermediate between the thickness of normal oceanic and continental crust. This observation is valid even where the continental crust has been attenuated over horizontal distance of several hundred kilometres, as in the Orphan Knoll region, and where basaltic intrusions appear to have accompanied crustal thinning, as on the Labrador Shelf. Perhaps these results typify the minimum thickness which continental crust can sustain before continental breakup and sea floor spreading begin. If this hypothesis is correct, crustal thicknesses

less than about 10 km would imply the presence of oceanic crust. Therefore, crustal thickness can be used as a guide to the nature of the crust near the margins except in a narrow transition region where oceanic and continental crust merge. This zone appears to be only tens of kilometres in width [Keen et al., 1975; Jackson et al., 1977; Sheridan et al., 1979; Fig. 3] but remains problematic with respect to the nature of the crust within it. The thickness criterion is useful in areas such as central Baffin Bay and the deep-water region adjacent to the Labrador margin where crustal thicknesses of less than 10 km are found, suggested that these areas are underlain by oceanic crust and not by attenuated continental crust as some investigators have suggested [van der Linden, 1975a; Kerr, 1967].

The second consistent geophysical indicator of the nature of the crust is the character of magnetic anomalies near the margins. Trends in continental or oceanic basement rocks, reflected as magnetic anomaly trends, may be very useful in mapping the extent of continental or oceanic crust. This approach has proved to be important in the Orphan Knoll region, where Haworth [1977] showed that magnetic trends associated with continental Precambrian and Paleozoic basement continue across the edge of the shelf into the deep water basin west of Orphan Knoll; thus providing the first indication that subsided continental crust occupies this region.

Although a similar detailed analysis has not been undertaken for the Labrador margin, the magnetic anomaly patterns appear to be different on the shelf and in the adjacent deep water region [Keen and Hyndman, 1979; Srivastava, 1978; van der Linden, 1975b]. In this region, the shelf anomalies exhibit high amplitudes and are segmented and less lineated, while the anomalies seaward of the edge of the shelf are subdued and lineated subparallel to the margin. These differences may support an oceanic origin for the latter region, in contrast to the hypothesis that both are underlain by thinned, intruded, and subsided continental crust [van der Linden, 1975a]. If the latter hypothesis were true, similar magnetic patterns in both regions would be expected.

The East Coast Magnetic Anomaly, or Slope Anomaly, is a clearly defined magnetic lineament which extends from Nova Scotia to Florida [Taylor et al., 1968; Emery et al., 1970; Keen et al., 1975; Grow et al., 1979]. Most investigations of this feature have concluded that it marks the ocean-continent transition but the models used to explain it differ. Off Nova Scotia it has been explained by an edge-effect model, consisting of a thick, relatively non-magnetic continental crustal layer adjacent to a thin, highly magnetic oceanic layer. Its position is consistent with the seismic evidence for the location of the transition zone [Keen et al., 1975].

There is some evidence from studies at other rifted margins that oceanic crust thickens towards the margins, mainly by an increase in thickness of

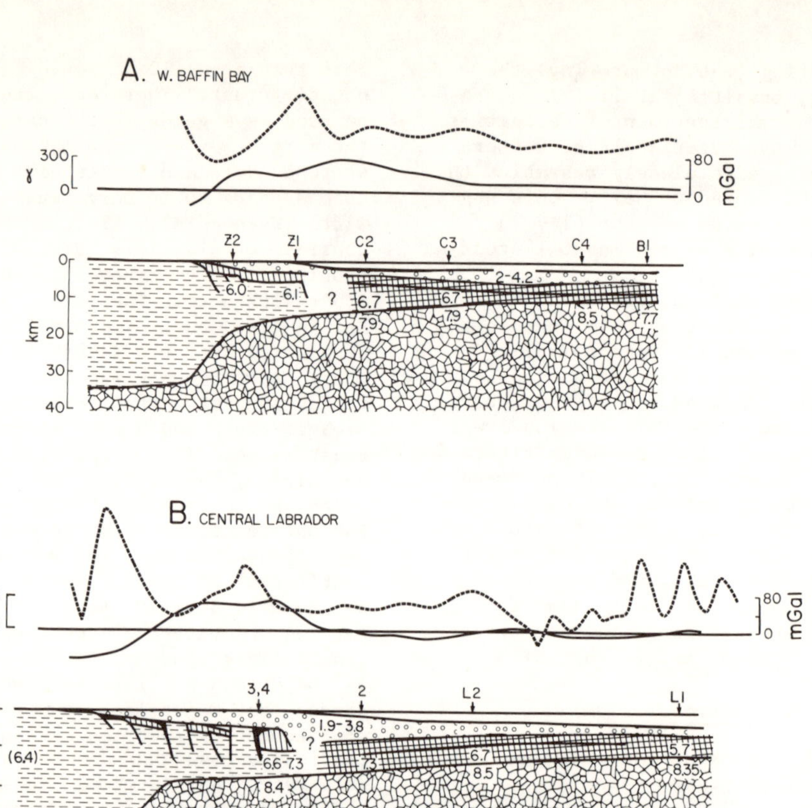

A. W. BAFFIN BAY
γ
300
0
80
mGal
0
Z2
Z1
C2
C3
C4
B1
km
0
10
20
30
40
6.0
6.1
?
6.7
6.7
7.9
7.9
8.5
7.7
2-4.2

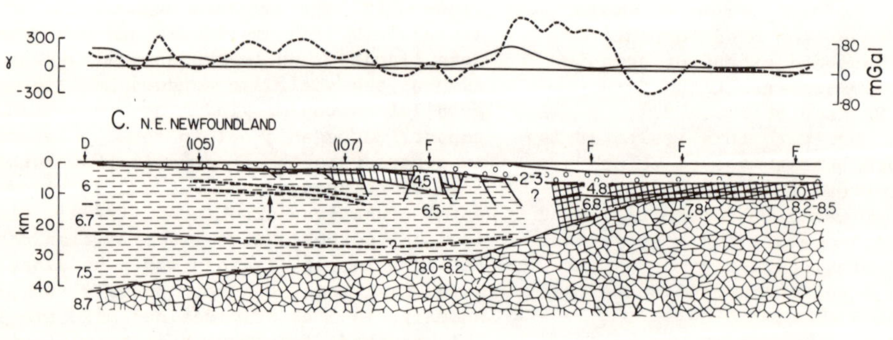

B. CENTRAL LABRADOR
γ
300
0
80
mGal
0
3,4
2
L2
L1
km
0
10
20
30
40
(6,4)
6.6-7.3
8.4
?
1.9-3.8
7.3
8.5
6.7
8.35
5.7
0
km
100

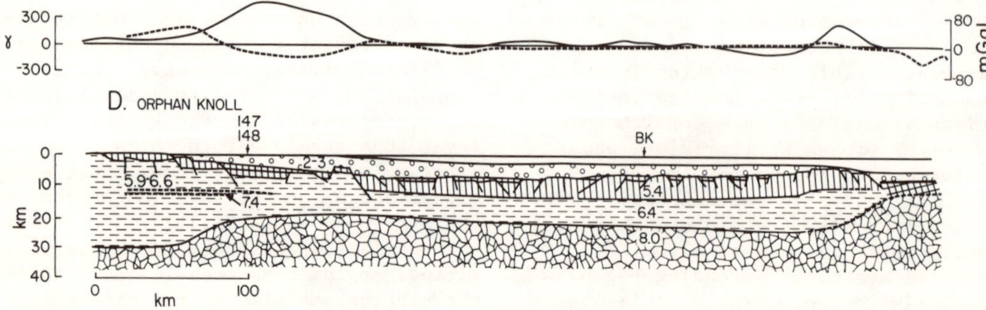

300
0
-300
γ
80
0
mGal
C. N.E. NEWFOUNDLAND
D
(105)
(107)
F
F
F
F
km
0
10
20
30
40
6
6.7
7.5
8.7
7
?
4.5
6.5
?
2-3
6.8
7.80-8.2
4.8
7.8
7.0
8.2-8.5
300
0
-300
γ
80
0
80
mGal
D. ORPHAN KNOLL
147
148
BK
km
0
10
20
30
40
5.9 6.6
7.4
2-3
5.4
6.4
8.0
0
km
100

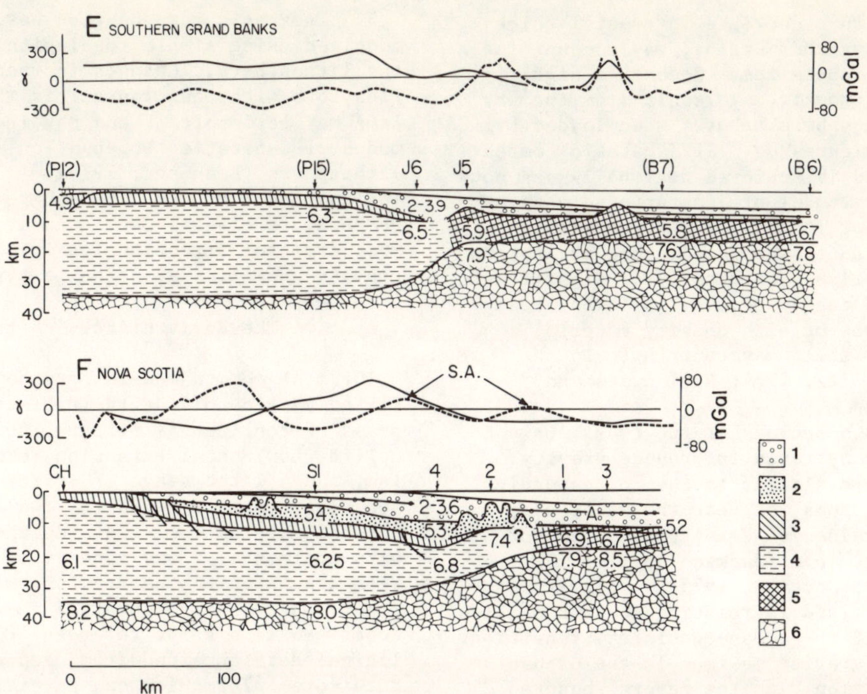

Fig. 3. Cross-sections along the tracks shown in Fig. 2. These are based mainly on seismic refraction data and the positions of the refraction lines are shown by arrows above the sections, with an identifying letter or number referring to the source of the data as follows: All lines Profile A, from Jackson et al. (1977). All lines Profile B, from van der Linden (1975a). Lines designated F, Profile C, from Fenwick et al. (1968). Lines 105, 107, 147, 148, Profiles C and D, from Sheridan and Drake (1968). Line BK, Profile D, from Barrett and Keen (1978). Lines J4, J6, Profile E, from Jackson et al. (1975). Lines P12, P15, Profile E, from Press and Beckman (1954). Lines B7, B6, Profile E, from Bentley and Worzel (1956). Lines D, CH and SI from Dainty et al. (1966, their lines 12, 1 and 3 respectively). Where the line designation is in brackets it has been projected onto the section. Velocities, in km s^{-1}, are shown where space permits. The sediment thicknesses were obtained from Geological Survey of Canada Map 1400A, from assorted seismic reflection profiles at Bedford Institute of Oceanography, and from deep exploratory well data. The gravity (solid) and magnetic (dashed) data shown above each section were taken from various sources including Jackson et al. (1977), Srivastava (1978), Haworth (1977), and Keen et al. (1975). S.A. indicates magnetic slope anomaly. The key for the various shadings, shown opposite profile F is as follows: 1. postrifting sediments; 2. sediments of Early Jurassic age and older on the Nova Scotian shelf, most of which may have been deposited during the rifting stage; 3. continental basement rocks – those which predate the formation of the margins; 4. continental crust; 5. oceanic crust including layer 2 and 3; 6. mantle.

oceanic layer 2. This layer is sometimes postulated to form a basement ridge adjacent to the continent and to contribute to the Slope Anomaly [Rabinowitz and LaBreque, 1977]. However, no oceanic basement ridge or thickening of the crust is consistently observed near the margins of Eastern Canada.

Similar features should be observed at all rifted margins, if the edge-effect model is the valid explanation for the Slope Anomaly. A very similar magnetic anomaly has been mapped northeast of Newfoundland, north of the Charlie Fracture Zone, and is also satisfied by an edge-effect model [Fig. 2; Fenwick et al., 1968]. Although the other rifted margins off eastern Canada do not exhibit such clear magnetic Slope Anomalies [see anomaly compilations of Srivastava, 1979; Keen and Hyndman, 1979; and Keen et al., 1974], edge effect models have been shown to satisfy some of the larger magnetic anomalies along the margins of the Labrador Sea and Baffin Bay [Keen et al., 1974]. The lack of clarity of these anomalies may be partly due to highly variable magnetizations within the crust, and to lesser magnetization contrasts between oceanic and continental crust.

Gravity Anomalies

The free air gravity anomaly maxima which usually occur immediately landward of the edge of the shelf and which are often accompanied by a minimum over the continental slope have been explained by three main contributions. First, the edge-effect produced by the increase in water depth combined with a decrease in crustal thickness seaward across the margins, may account for a large part of the anomalies. Second, a significant part of the anomaly may result from the way in which the lithosphere behaves when loaded by large sediment thicknesses. If isostatic compensation of the load is achieved regionally and not by the simple Airy model of isostatic adjustment [Walcott, 1972], or if the youngest sediments deposited on the outer shelf and slope are not in isostatic equilibrium [Sobczak, 1975] significant gravity anomalies can be produced. Third, basement highs or zones of high density material beneath the outer shelf may contribute to the anomalies [Rabinowitz, 1974; Rabinowitz and LaBreque, 1977].

Each of the cross-secions of the crust shown in Fig. 3 can be demonstrated to produce gravity anomalies which are similar to the corresponding observed gravity anomalies near the shelf edge when reasonable values of density are used for each of the layers [e.g. Jackson et al., 1977; Hinz et al., 1979; Sobczak, 1975]. However, caution must be exercised in relating these anomalies to the position of the ocean-continent transition. For example, the gravity maximum in the Orphan Knoll region (section D), lies several hundred kilometres inland of the transition zone and marks instead a change in water depth at the shelf edge which is not isostatically compensated in a simple manner by a change in crustal thickness [R. A. Folinsbee, personal communication, 1978].

An important problem in understanding the underlying cause of these anomalies is illustrated by sections A, B, and D, across the margins of western Baffin Bay, Labrador, and Orphan Basin respectively. In these regions, isostatic compensation is not maintained by changes in the depth of the crust-mantle boundary. This, and the Airy model of isostatic compensation have often been assumed in the analysis of marginal gravity anomalies elsewhere, [e.g. Rabinowitz and LaBreque, 1977]. The shallow depths to the M discontinuity beneath the outer shelf in these areas, which are primarily responsible for the large positive anomalies, are not compensated by, for example, a corresponding increase in water depth. These results suggest that in order to fully understand these gravity anomalies three factors affecting the entire lithosphere must be considered:

1) Lateral density differences in the upper mantle, with less dense mantle material beneath the outer shelf. Perhaps heating during the rifting stage has resulted in chemical change in the upper mantle. However, at the present time, there appears to be no evidence for this.

2) The thermal history of the lithosphere across the margsins has involved early heating followed by cooling. Residual high temperature may still exist in the lithoshere beneath the outer shelf region and further seaward. These would affect both the density and the rheology of the plate.

3) Isostatic compensation has usually been modelled using simple rheological properties for the lithosphere. Changes in rheological properties both with position across the margin and with time may be important and may indicate different modes of isostatic compensation at depth in the lithosphere [Beaumont, 1979]. A realistic rheological model of the lithosphere incorporating the temperature effects mentioned above seems to be most fruitful direction for future studies of the gravity anomalies at Atlantic type margins.

The Early History of the Margin

The most common models for the evolution of rifted margins divide their history into an early phase of continental rifting, perhaps involving uplift, horizontal extension, erosion, and volcanism, and a later stage of drifting, during which sea floor forms between the two continental blocks and the margins cool, subside and are blanketed with sediments. Some aspects of these models have been discussed in the previous sections of the paper, where the geophysical results were described with brief reference to some of the geological data obtained from deep exploratory wells. Much more information can be extracted from this geological data which provides some important constraints on the early development of the margin.

<u>Processes During the Rift Stage</u>

During the early rifting stage, some means of modifying the continental crust must have been active such that it later subsides below sea level, enabling the formation of the present marginal sedimentary basins. Crustal thinning, either by subaerial erosion during uplift or by horizontal extension, and densification of the crust by metamorphism or phase changes, have been suggested as ways in which the crust could be altered during rifting to produce the observed subsidence [Sleep, 1971; McKenzie, 1978; Falvey, 1974]. While these processes are not mutually exclusive, the data suggests that horizontal extension may play the most important role in crustal modification.

The data from the deep exploratory wells shows that there are regional unconformities on the Labrador and Nova Scotian rifted margins, which separate the younger post-rifting sediments from older rift sediments which were often deposited in a non-marine or marginal marine environment. This unconformity is Early Jurassic off Nova Scotia [Jansa and Wade, 1975] and mid-Cretaceous off Labrador [Umpleby, 1979]. The presence of non-marine rift sediments and of these unconformities suggest that uplift occurred during rifting. However, the amount of erosion which accompanied uplift is unlikely to have thinned the continental crust sufficiently to produce the observed crustal thinning of over 15 km. The presence of non-marine sediments of Early Cretaceous age beneath a mid-Cretaceous unconformity on the Labrador Shelf suggests little erosion occurred since no sediments of that age would be expected to remain if 15 km of erosion had occurred [Cutt and Laving, 1977; Umpleby, 1979].

The relative contribution of phase changes or crustal metamorphism to subsidence of the margins is difficult to estimate. There is no evidence from seismic measurements that such processes were once active. The high velocities beneath some outer shelf regions could be interpreted as support for such changes, but at least on the Labrador margin it seems more reasonable to relate the high velocities to the intrusion of basaltic material during rifting. The intrusion of basalts will, of course, also affect the subsequent subsidence and thermal history of the margins, but appears to play a less fundamental role in crustal modification than crustal thinning: high velocities are not prevalent in all regions, as discussed in the previous section of this paper.

The importance of crustal thinning is seen on all profiles across the rifted eastern Canadian margins, shown in Figure 3. It seems unlikely to be due to erosion. The occurrence of block faulting of the basement rocks, beneath the Labrador Shelf and the northeast Newfoundland Shelf and the presence of large grabens on the Grand Banks and the Nova Scotian Shelf, suggest that extensional forces were important during the early rift stage [Jansa and Wade, 1975; Umpleby, 1979]. The absence, or a very narrow zone, of crustal thinning along the transform margin south of the Grand Banks (Fig. 3, section E), compared to that along the rifted margins strongly reinforces the conclusion that crustal thinning may be directly related to horizontal extension during rifting. The consequences of this extension model will be discussed further below.

The duration of the rifting stage prior to the beginning of sea floor spreading appears to be quite variable. On the Labrador Shelf, the presence of Early Cretaceous basalts (about 125 Ma in age), overlying pre-Mesozoic continental basement rocks, suggests that the margin was subjected to tectonism at least 45 Ma before the drift phase. Further south, in northeastern Newfoundland, basic igneous intrusions of Jurassic to Early Cretaceous age have been mapped on land [Currie, 1976], probably indicating a somewhat longer rift phase than inferred for Labrador. The Triassic basalts in Nova Scotia, which have been dated at 190-200 Ma [Hayatsu, 1979], precede sea floor spreading by about 25 Ma, a significantly shorter time interval than that observed to the north. It is worth noting that some of these igneous intrusions occur several hundred kilometres landward of the zone of eventual continental separation, suggesting that the early rift phase affected a broad zone of continental crust in a very complex manner. Furthermore, it is puzzling that there appears to be a time gap between the ages of these basalts, presumed to have formed during the rift stage, and the onset of sea floor spreading. Either no basalts were intruded during this period, or

further sampling will reveal younger intrusions beneath the outer shelves.

The ages of the oldest marine sediments deposited on the subsiding shelves are generally consistent with the presumed times of onset of the drift stage, derived from geophysical data. Off Labrador and northeast of Newfoundland, these sediments are Late Cretaceous, while off Nova Scotia they are mid-Jurassic in age [Umpleby, 1979; Jansa and Wade, 1975; Barss et al., 1979]. The Grand Banks which have been affected by at least three different periods of plate separation exhibit a more complex history of sedimentation and further work is needed to relate this history to proposed plate motions in that area. The most striking and widespread feature is an early Cretaceous unconformity, often separating marine sediments of Jurassic age from those of Late Cretaceous and younger age [Jansa and Wade, 1975; Gradstein et al., 1975].

Subsidence and the Thermal History of the Margins

The subsidence history of the margins during the drift stage has been obtained from biostratigraphic data in deep exploratory wells on the Labrador and Nova Scotian shelves [Keen, 1979; Umpleby, 1979; Gradstein et al., 1975]. The subsidence can be attributed to underlying tectonic causes and to sediment loading. The tectonic subsidence can be obtained from the total subsidence by removing the sediment loading effect, assuming isostatic equilibrium is maintained [Watts and Ryan, 1976; Steckler and Watts, 1978]. Smaller corrections for other effects such as changes in eustatic sea level and paleo-water depth can also be made if these changes can be estimated [van Hinte, 1978].

The resulting tectonic subsidence curves on the eastern Canadian margin closely resemble the depth-age curves for oceanic crust [Keen, 1979; Fig. 4] as do the results from other margins [Sleep, 1971; Watts and Ryan, 1976; Mondadert et al., 1977; de Charpal et al., 1978]. This suggests that thermal cooling of the lithosphere is the primary cause of subsidence following the onset of sea floor spreading. If this hypothesis is correct, there will be a close relationship between the initial thermal conditions at which the lithosphere was maintained during early rifting, and the rates and amounts of subsidence measured, analogous to the relationship beteen the thermal conditions at mid-ocean ridges and the oceanic depth-age curve [Parsons and Sclater, 1977].

Two simple models which explain both the subsidence and thermal histories of the margins have been recently proposed based on two models of the behaviour of the lithosphere during early rifting [McKenzie, 1978; Royden et al., 1980]. The first involves horizontal extension and consequent thinning of the continental lithosphere during rifting, with passive upwelling of the asthenosphere from below [McKenzie, 1978]. This results

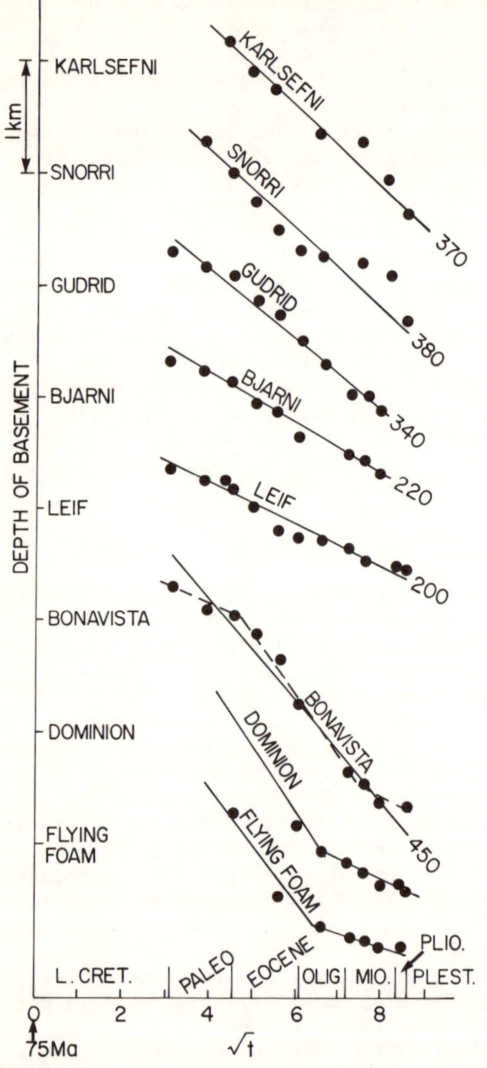

Fig. 4. Tectonic subsidence of the Labrador and Northeast Newfoundland shelves from deep exploratory well data (after Keen, 1979). The subsidence of basement, after removing loading effects, has been plotted as a function of $\sqrt{\text{age}}$. The names of the wells are shown and the slope of the straight lines drawn through the subsidence curves are given in (in m/$\sqrt{\text{ma}}$). The position of each well, designated by the first letter of its name, is showin in Fig. 2. Note the approximately linear relationship between subsidence and $\sqrt{\text{age}}$. Note also the anomalously high rates of subsidence for the region northeast of Newfoundland, which is adjacent to the continental fragments of Orphan Knoll and Flemish Cap. This has not been satisfactorily explained.

in higher initial temperatures and heat flow in the lithosphere during rifting, which subsequently cools and subsides. The rates of subsidence and heat flow are proportional to the amount of exten-

sion which has occurred. In the limit, if the plate were thinned to zero thickness, the subsidence and heat flow curves would be identical to those or oceanic crust. Royden et al. [1980] have used the same model to predict the paleotemperatures within the sediment deposited on the subsiding and cooling margins. These predicted paleotemperatures can be compared with paleotemperatures estimated from the thermal alteration of organic matter within the sediments, providing a means of relating model calculations to observational data. Royden et al. [1980] also describe a second model, in which higher initial temperatures and heat flow are produced by diapiric intrusion of mantle material into the lithosphere. This dyke intrusion model produces results which are very similar to the extenison model, except for the first few million years of the subsidence history.

Keen [1979] has used a similar model to those described above to explain the observed subsidence curves on the rifted margins off Nova Scotia and Labrador and to compute the paleotemperatures within the sediments. Results show that rates of tectonic subsidence, when plotted as a function of $\sqrt{\text{age}}$, are about 300 m/$\sqrt{\text{age}}$, not very different from the values for the subsidence of oceanic crust [350 m/$\sqrt{\text{age}}$; Parsons and Sclater, 1977]. The computed temperature-time history within the sediments is reasonable, when compared to measurements of the thermal alteration of organic material. This is an encouraging result and suggests that the models are consistent with the observational data. Such calculations may be important in estimating the petroleum potential within sedimentary strata of a certain age beneath rifted margins because the paleotemperatures may be directly related to the ability of the sediments for hydrocarbon generation.

The subsidence history and associated thermal history of the margins deserves further investigation. While it is unlikely that the initial conditions during rifting are as simple as those described here, the models provide a framework to which the observational data can be related and from which better models may evolve. Answers to the following questions should be sought in future investigations:

1) Given a simple model for the thermal history of the lithosphere at the margins, how does the rheology of the crust and upper mantle change with time? How do these changes affect subsidence due to sediment loading the margins?

2) The amount of continental crustal thinning measured at the rifted margins, up to about 50% of the original crustal thickness (Fig. 3), is in good agreement with the amount predicted by the subsidence curves [Royden and Keen, 1980], assuming that extension is responsible for thinning. However, these simple models do not explain the geological complexities of the rift phase, which includes volcanism and uplift. How can the model be modified to account for these complexities? Does the underlying mantle attenuate more than the

crust, perhaps by destruction of lithosphere from
below? Can a combination of the extension and
dyke intrusion models better account for the
observations? Are phase transitions required in
the deep crust or upper mantle to account for part
of the subsidence?

3) Can the subsidence curves be used to estimate the thickness of the lithosphere beneath the
older margins, in a manner similar to estimates of
the thickness of the oceanic plate [Parsons and
Sclater, 1977; Steckler and Watts, 1978]?

There is sufficient data now available in the
eastern Canadian margins to attempt to answer
these and other questions.

Concluding Remarks

While there has been a vast improvement in our
understanding of the continental margins of
eastern Canada during the past decade and sufficient data has become available to evaluate critically the geodynamic processes affecting them,
inevitably as many questions as answers have
arisen. Presented below is a short list of areas
which are considered most deserving for future
study.

1) In order to fully understand the rifting
stage, carefully planned studies of conjugate
margins should be undertaken. In order to reconstruct the tectonic setting of the margins during
the initial rift stage knowledge of both parts of
the rift zone is required.

2) Geophysical studies of variations of the
whole lithosphere across the margins should be
carried out, rather than studies of crustal
changes only. The possibility of deducing changes
in density and temperature distribution, or seismic velocity in the upper mantle across the margins and their relationship to the margin formation should be investigated. The interpretation
of gravity anomalies, incorporating realistic
temperature dependence effects in rheological
models of the lithosphere may provide an important
ingredient in future studies of the kind.

3) The relationship of the geophysical and
geological characteristics of the present margins
to the ancient plate margins now in the continental interior should be investigated more
thoroughly. In order to understand the present
margins we must know the properties of the
original continent which was split. Deep crustal
rocks are exposed at the ancient margins, which
are inaccessible at the present margins, and these
may provide ancient analogs which may be applied
to the present margins. Conversely, the results
on the present margins may provide guidelines for
understanding the ancient ones.

4) More accurate measurements of heat flow and
geothermal gradients in deep wells could be
extremely useful in testing some of the models
recently developed for explaining thermally driven
subsidence. Further refinements of these models,
as described in the last section of this paper,
may provide the best way of determining thermal

conditions during early rifting. They may also be
important in studies of petroleum potential at
passive margins.

Acknowledgements. I would like to thank all my
collegues at the Atlantic Geoscience Centre and
Dalhousie University who have contributed to the
results and ideas expressed in this paper. I am
grateful to R. K. H. Falconer, L. Jansa and H. R.
Jackson for critical review of the manuscript.

References

Auzende, J. M., J. L. Olivet, and J. Bonnin, La
marge du Grand Banc et la fracture de
Terre-Neuve. Comptes Rendues de l'Academic des
Sciences, Paris, Ser. D. 271, 1063-1066, 1970.

Balkwill, H. and R. M. Bustin, Late Cretaceous and
Tertiary structure, Queen Elizabeth Islands,
Arctic Canada, and Plate Motions of Arctic North
America, Geological Association of Canada, 1978
Joint Annual Meeting, Abstracts with Programs,
Volume 3, 362, 1978.

Barrett, D. L. and C. E. Keen, Mesozoic Magnetic
Lineations, The Magnetic Quiet Zone and Seafloor
Spreading in the northwest Atlantic, Journal of
Geophysical Research, 81, 4875-4884, 1976.

Barrett, D. L. and C. E. Keen, Ocean Bottom
Seismometer Studies of the Crust Near the Orphan
Knoll and Flemish Cap Continental Fragments, EOS
Transactions American Geophysical Union
(Abstract), 58, 322, 1978.

Barss, M. S., J. P. Bujak, and G. L. Williams,
Palynological Zonation and Correlation of
Sixty-Seven Wells, Eastern Canada, Geological
Survey Canada Paper 78-24, 118 pp., 1979.

Beaumont, C., On the rheological zonation of the
lithosphere during flexure. Tectonophysics, 59,
347-366, 1979.

Bentley, C. R. and J. L. Worzel, Geophysical
investigations of the emerged and submerged
Atlantic coastal plain. X. continental slope
and continental rise south of the Grand Banks,
Geological Society of America Bulletin, 67,
1-18, 1956.

Burke, K., W. S. F. Kidd, and J. T. Wilson,
Relative and Latitudinal motion of Atlantic hot
spots, Nature, 245, 133-137, 1973.

Cande, S. C., and Y. Kristoffersen, Late
Cretaceous magnetic anomalies in the North
Atlantic, Earth and Planetary Science Letters,
35, 215-224, 1977.

Clarke, D. B., the Tertiary volcanic province of
Baffin Bay, in Volcanic Regimes in Canada,
edited by W. R. A. Barager, L. C. Coleman, and
J. M. Hall, Geological Association of Canada,
Special Paper 16, 445-460, 1977.

Currie, K. L. The Alkaline Rocks of Canada,
Geol. Surv. Can. Bull., 239, 228 pp., 1976.

Cutt, B. J. and J. G. Laving, Tectonic elements
and gologic history of the South Labrador and
Newfoundland continental shelf, eastern Canada,
Bull. Can. Petrol. Geol., 25, 1037-1058, 1977.

Dainty, a. M., C. E. Keen, M. J. Keen, and J. E.

Blanchard, Review of Geophysical Evidence on Crust and Upper Mantle Structure on the Eastern Seaboard of Canada, in The Earth Beneath the Continents, J. S. Steinhart and T. J. Smith (Eds.), American Geophys. Union, 349-369, 1966.

Dawes, P. R., The North Greenland fold belt and environs, Geol. Soc. Denmark Bull., 20, 197-239, 1971.

deChapal, O., P. Guenoch, L. Montadert, and D. G. Roberts, Rifting, crustal attentuation and subsidence in the Bay of Biscay, Nature, 275, 706-711, 1978.

Emery, K. O., E. Uchupi, J. D. Phillips, C. O. Bowin, E. T. Bunce, and S. T. Knott, Continental rise off eastern North America, Amer. Ass. Petrol. Geol. Bull., 54, 1120-1139, 1970.

Falvey, D. A., The development of continental margins in plate tectonic theory, Aust. Journ. Petrol. Expl., 14, 95-106, 1974.

Fenwick, D. K. B., M. J. Keen, C. E. Keen, and A. Lambert, Geophysical studies of the continental margins northeast of Newfoundland, Can. J. Earth Sci., 5, 483-500, 1968.

Gradstein, F. M., G. L. Williams, W. A. M. Jenkins, and P. Ascoli, Mesozoic and Cenozoic Stratigraphy of Atlantic continental margin, eastern Canada, in Canada's Continental Margins and Offshore Petroleum Exploration, C. J. Yorath, E. R. Parker, and D. J. Glass (Eds.), Can. Soc. Petrol. Geol., Memoir 4, 103-131, 1975.

Gradstein, F. M., A. C. Grant, and L. F. Jansa, Grand Banks and J-Anomaly Ridge, A Geological Comparison, Science, 197, 1074-1076, 1977.

Grant, A. C., Continental Crust beneath the Newfoundland Ridge; Evidence from multi-channel seismic reflection data, Nature, 270, 22-25, 1977.

Grant, A. C., The continent-ocean crustal boundary in the western Labrador Sea. EOS, Trans. Am. Geophys. Union, 60, 375, 1979.

Grow, J. A., C. O. Bowin, and D. R. Hutchinson, The Gravity Field of the U.S. Atlantic Continental Margin, Tectonophysics, 59, 27-52, 1979.

Hall, J. M., The oceanic crust: Investigations by Canadian groups of the mid-Atlantic ridge crest and other areas, Can. J. Earth Sci., 16, 695-711, 1979.

Haworth, R. T., The continental crust northeast of Newfoundland and its ancestral relationship to the Charlie Fracture Zone, Nature, 266, 246-249, 1977.

Haworth, R. T. and C. E. Keen, The Canadian Atlantic Margin: A passive continental margn encompassing an active past, Tectonophysics, 59, 83-126, 1979.

Haworth, R. T. and J. P. Lefort, Geophysical evidence for the extent of the Avalon Zone in Atlantic Canada, Can. J. Earth Sci., 16, 552-567.

Hayatsu, A., K-Ar isochron age of the North Mountain Basalt, Nova Scotia, Can. J. Earth Sci., 16, 973-975, 1979.

Hinz, K., H.-V. Schlüter, A. C. Grant, S. P. Srivastava, D. Umpleby, and J. Woodside, Geophysical transects of the Labrador Sea Labrador to southwest Greenland, Tectonophysics, 59, 151-184, 1979.

Hood, P. And M. E. Bower, Low level aeromagnetic surveys of the continental shelves bordering Baffin Bay and the Labrador Sea, Geological Survey of Canada Paper 71-23, 573-598, 1973.

Hyndman, R. D., Evolution of the Labrador Sea, Canadian Journal of Earth Sciences, 10, 637-644, 1973.

Jackson, H. R., C. E. Keen, and M. J. Keen, Seismic structure of the continental margins and ocean basins of southeastern Canada, Geol. Surv. Can. Paper 74-51, 13 pp., 1975.

Jackson, H. R., C. E. Keen, and D. L. Barrett, Geophysical studies on the eastern continental margin of Baffin Bay and in Lancaster Sound, Canadian Journal of Earth Sciences, 14, 1991-2001, 1977.

Jackson, H. R., C. E. Keen, R. K. H. Falconer, and K. P. Appleton, New Geophysical Evidence for Sea Floor Spreading in Central Baffin Bay, Can. J. Earth Sci., 16, 2122-2135, 1979.

Jansa, L. F. and J. A. Wade, Geology of the continental margin off Nova Scotia and Newfoundland, in Offshore Geology of Eastern Canada, Volume 2, Regional Geology, W. J. M. van der Linden and J. A. Wade (Eds.), Geol. Survey Canada Paper 74-30, 51-105, 1975.

Jansa, L. F., J. Remaine, and P. Ascoli, Caplionellid and foraminiferal-ostracod biostratigraphy at the Jurassic-Cretaceous boundary, offshore Eastern Canada, Revisita Italiana di Paleontologia e Stratigrafia, 86, 67-126, 1980.

Keen, C. E., Thermal history and subsidence of rifted continental margins-evidence from wells on the Nova Scotian and Labrador shelves, Can. J. Earth Sci., 16, 505-522, 1979.

Keen, C. E. and D. L. Barrett, Seismic refraction studies in Baffin Bay: an example of a developing ocean basin, Geophys. J. Roy. Ast. Soc., 30, 253-271, 1972.

Keen, C. E., B. R. Hall, and K. D. Sullivan, Mesozoic Evolution of the Newfoundland Basin, Earth Planet. Sci. Letters, 37, 307-320, 1977.

Keen, C. E. and R. D. Hyndman, Geophysical review of the eastern and western continental Margins of Canada, Can. J. Earth Sci., 16, 712-747, 1979.

Keen, C. E., M. J. Keen, D. L. Barrett, and D. E. Heffler, Some aspects of the ocean-continent transition at the continental margin of eastern North America, in Offshore Geology of eastern Canada, W. J. M. van der Linden and J. A. Wade (Eds.), Geol. Surv. of Canada Paper 74-30, 189-197, 1975.

Keen, C. E., M. J. Keen, D. I. Ross, and M. Lack, Baffin Bay: Small Ocean Basin formed by sea floor spreading, American Association Petroleum Geologists Bulletin, 58, 1089-1108, 1974.

Kerr, J. W., A submerged continental remnant

beneath the Labrador Sea, *Earth and Planetary Science Letters*, 2, 283-289, 1967.

King, L. H., R. D. Hyndman, and C. E. Keen, Geological development of the continental margin of Atlantic Canada, *Geoscience Canada*, 2, 26-35, 1975.

Kristoffersen, Y., and M. Talwani, Extinct triple junction south of Greenland and the Tertiary motion of Greenland relative to North America. *Geol. Soc. Am. Bull.*, 88, 1037-1049, 1977.

Laughton, A. S., The southern Labrador Sea - a key to Mesozoic and early tertiary evolution of the North Atlantic, *in Initial Reports of the Deep Sea Drilling Project (U.S. Government Printing Office)*, 12, 1155-1179, 1972.

LePichon, X., J.-C. Sibuet and J. Francheteau, The Fit of the continents around the North Atlantic Ocean, *Tectonophysics*, 38, 169-209, 1977.

Mayhew, M. A., C. L. Drake, and J. E. Nafe, Marine Geophysical measurements on the continental margins of the Labrador Sea, *Canadian Journal of Earth Sciences*, 7, 199-214, 1970.

McKenzie, D. P., Some remarks on the development of sedimentary basins, *Earth and Planetary Science Letter*, 40, 25-32, 1970.

Montadert, L., D. G. Roberts, G. A. Auffret, W. Beck, P. A. DuPeuble, E. A. Hailwood, W. Harrison, H. Kagami, D. N. Lumsden, C. Muller, D. Schnitker, R. W. Thompson, T. L. Thompson, and P. P. Timofeev, Rifting and subsidence on passive continental margins in the Northeast Atlantic, *Nature*, 268, 305-309, 1977.

Newman, P. and R. K. H. Falconer, Evidence for movement between Greenland and Canada along Nares Strait, *Geological Association of Canada, 1978 Joint Annual Meeting, Abstracts with Programs*, Volume 3, 463, 1978.

Parsons, B. and J. G. Sclater, An Analysis of the Variation of Ocean Floor Bathymetry and Heat Flow with Age, *Journal of Geophysical Research*, 82, 803-827, 1977.

Pitman, W. C. III and M. Talwani, Sea Floor spreading in the North Atlantic. *Bull. Geol. Soc. Am*, 83, 619-646, 1972.

Press, F. and W. Beckmann, Geophysical investigations in the emerged and submerged Atlantic coastal plain, Part VIII; Grand Banks and adjacent shelves, *Geol. Soc. America Bull.*, 65, 299-314, 1954.

Rabinowitz, P. D., The boundary between oceanic and continental crust in the western North Atlantic, *in The Geology of Continental Margins*, C. A. Burk and C. L. Drake (Eds.), Springer, New York, 67-84, 1974.

Rabinowitz, P. D. and J. L. Labrecque, The isostatic gravity anomaly: key to the evolution of th ocean-continent boundary at passive continental margins, *Earth and Planetary Sciences Letters*, 35, 145-150, 1977.

Royden, L., and C. E. Keen, Rifting processes and thermal evolution of the continental margin of Eastern Canada determined from subsidence curves, *Earth Planet. Sci. Lett.*, 51, 343-361, 1980.

Royden, L., J. G. Sclater, and R. P. von Herzen, Continental Margin Subsidence and Heat Flow: Important Parameters in Formation of Petroleum Hydrocarbons, *Am. Assoc. Petrol. Geol. Bull.*, 64, 173-187, 1980.

Scrutton, R. A., Structure of the Crust and upper mantle at Goban Supr southwest of the British Isles - some implications for margin studies, *Tectonophysics*, 59, 201-216, 1979.

Sheridan, R. E. and C. L. Drake, Seaward extension of the Canadian Appalachians, *Canadian Journal of Earth Sciences*, 5, 337-373, 1968.

Sheridan, R. E., J. A. Grow, J. C. Behrendt, and K. C. Bayer, Seismic refraction study of the continental edge of the eastern United States, *Tectonophysics*, 59, 1-26, 1979.

Sleep, N. H., Thermal effects of the formation of Atlantic continental margins by continental breakup, *Geophys. J. Roy. Astro. Soc.*, 24, 325-350, 1971.

Srivastava, S. P., Marine Gravity and Magnetic anomaly maps of the Labrador Sea, *Geol. Sur. Can. Open File 627*, 1979.

Srivastava, S. P., Evolution of the Labrador Sea and its bearing on the early evolution of the North Atlantic, *Geophys. J. Roy. Astro. Soc.*, 52, 313-357., 1978.

Sobczak, L. W., Gravity anomalies and Passive Continental Margins, Canada and Norway, in *Canada's Continental Margins and Offshore Petroleum Exploration*, C. J. Yorath, E. R. Parker, and D. J. Glass (Eds.), *Can. Soc. Petrol. Geol.*, Memoir 4, 743-761, 1975.

Steckler, M. S., and A. B. Watts, Subsidence of the Atlantic-Type Continental Margin of New York, *Earth and Planetary Science Letters*, 40, 1-13, 1978.

Sullivan, K. D., Structure and Evolution of the Newfoundland Basin, *Ph.D. Thesis (unpublished)*, Dalhousie University, Halifax, Nova Scotia, 1978.

Sullivan, K. D., and C. E. Keen, On the nature of the crust in the vicinity of the Newfoundland Ridge, *Can. J. Earth Sci.*, 15, 1462-1471, 1978.

Taylor, P. T., I. Zeitz, and L. S. Dennis, Geologic implications of aero-magnetic data for the eastern continental margin of the United States, *Geophysics*, 33, 755-780, 1968.

Trettin, H. P. and H. R. Balkwill, Contributions to the tectonic history of the Innuitian Province, Arctic Canada, *Can. J. Earth Sci.*, 16, 748-769, 1979.

Umpleby, D. C., Geology of the Labrador Shelf, *Geological Survey of Canada Paper*, 79-13, 34 pp., 1980.

van der Linden, W. J. M, Crustal attenuation and sea floor spreading in the Labrador Sea, *Earth and Planetary Science Letters*, 27, 409-423, 1975a.

van der Linden, W. J. M., Mesozoic and Cainozoic opening of the Labrador Sea, the North Atlantic and the Bay of Biscay, *Nature*, 253, 320-324, 1975b.

van der Linden, W. J. M. and S. P. Svristava, The

crustal structure of the continental margin off central Labrador, in Offshore Geology of Eastern Canada, edited by W. J. M. van der Linden and J. A. Wade, Geol. Surv. of Canada Paper 74-30, 233-245, 1975.

van Hinte, J. E., Geohistory analysis-application of micropaleontology in exploration geology, Am. Assoc. Petrol. Geol. Bull, 62, 201-222, 1978.

Walcott, R. I., Gravity, flexure and the growth of sedimentary basins at a continental edge, Geological Society of America Bulletin, 83, 1845-1848, 1972.

Watts, A. b. and W. B. F. Ryan, Flexure of the Lithosphere and Continental Margin Basins, Tectonophysics, 36, 25-44, 1976.

Williams, H., The Appalachian Orogen in Canada, Can. J. Earth Sci., 16, 792-807, 1979.

CONTINENTAL MARGIN SUBSIDENCE: A COMPARISON BETWEEN THE EAST AND WEST COASTS OF AFRICA

R.V. Dingle

Marine Geoscience, Dept. of Geology, University of Cape Town, Rondebosch (South Africa)

Abstract. Sedimentation rates (corrected for compaction) from along the passive continental margin of Africa between the Equatorial Fracture Zone and Somalia are used to compare the rates of subsidence of the continental crust since early Mesozoic time. Three distinctive subsidence histories can be identified which correspond with basinal areas that have different structural styles: rifted (west coast), sheared (Equatorial and Agulhas fracture zones) and sunk (zones of vertical tectonics in eastern Africa). A comparison of subsidence rates with other tensional margins (NE USA and the North Sea) and a consideration of the plate tectonic history of the African margins leads to the proposal of a geo- and thermo-dynamic model that takes cognizance of the worldwide mid-Cretaceous rheological discontinuity between taphrogenic and epeirogenic basin formation recognized by Kent, and the more generally accepted, purely plate tectonic-driven model of margin subsidence. The new suggestion involves a lower Mesozoic worldwide rise in the geothermal gradient in the lithosphere which produces metamorphism of the base of the continental crust and initiates taphrogenesis along lineaments throughout Gondwanaland. A lowering of the geothermal gradient in the lower Cretaceous produces a switch to epeirogenic subsidence, driven solely by sediment loading and thermal contraction, by Aptian/Albian times.

The thermal event facilitated continental separation, and sea floor spreading commenced locally at various times along the active taphrogenic belts. Local thermal and tectonic aberrations associated with this phenomenon over print onto the worldwide pattern of marginal basin subsidence. A further rise in the geothermal gradient may have been responsible for renewed taphrogenesis in eastern Africa in Tertiary times.

Introduction

Here we will compare some aspects of the post-Permian subsidence history of the sedimentary basins around the African continental margin between the Equatorial Fracture Zones in the west and Somalia in the east. It is important to remember that although most of the basins in this area have oceanward extensions onto oceanic crust, it is the nature and history of their continental parts that concern us. Stratigraphical and structural data from the outer edges of the continental blocks are reviewed, and sedimentation/basement subsidence rates calculated. The results are discussed in the light of recent work done on rifted margins from NE America (e.g. Steckler & Watts 1978) and the general topic of tensional continental margin sediment basin formation (e.g. Falvey 1974, Kent 1977, Beaumont & Sweeney 1978).

Regional Distribution of Sediment Basins

In a previous publication (Dingle 1976) the west and east coast post-Permian basins on the edge of the African continental block were genetically classified on the grounds of tectonic style: all were tensionally formed, but the setting in which they are found was dominated by either normal tension, shear tension, or zones where vertical motion over long periods was particularly important. We will refer to these as rifted, sheared, and sunk margins respectively. The distribution of these various types is shown in Fig. 1 and they are clearly related to the breakup framework of Gondwanaland as originally proposed by Francheteau and Le Pichon (1972): sheared margins associated with the Equatorial and Agulhas fracture zones (Guinea and Outenique basins); rifted margins between these two original massive offsets (Niger to Orange basins), and following Kent's work in east Africa (e.g. 1974), sunk margins along the east coast between Durban and Somalia (Natal/Mozambique, Zambezi and Tanzania/Somalia basins).

Burke (1976) discussed the west coast basins and showed that, in common with other Atlantic Ocean marginal basins, they are typified by early graben-like structures with a trend approximately parallel to the present coast. Kent (1976 & 1977) has taken this discussion further and recognizes two phases of development: lower (taphrogenic) and upper (epeirogenic, miogeosynclinal, inter cratonic) which he claims is a world-wide phenomenon unrelated to the local onset of continental disruption.

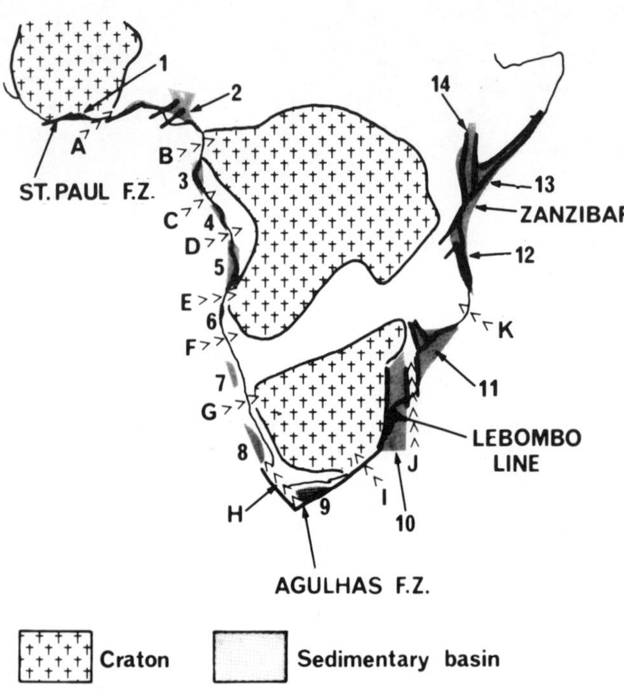

Fig. 1. Sedimentary basins around the edges of the African continent. Basins: 1-Guinea, 2-Niger, 3-Gabon, 4-Cabinda, 5-Cuanza, 6-Mocamedes, 7-Walvis, 8-Orange, 9-Outeniqua, 10-Natal/Mozambique, 11-Zambezi, 12 & 13-Tanzania/South Somalia. Basement Arches: A-Cape Three Points high, B-Cameroun line, C-South Gabon high, D-North Angola high, E-Lunda axis, F-Walvis Ridge, G-Lüderitz Arch, H-Columbine/Agulhas Arch, I-Transkei Swell, J-Mozambique Ridge, K-North Mozambique high. Rifted margins - 2, 3, 4, 5, 6, 7 & 8, sheared margins - 1 & 9, sunk margins 10, 11, 12 & 13.

These earlier works indicate that although many of the basins in question have present day onshore extensions they are essentially boundary effects related to rheological phenomena that controlled, or were controlled by, loci of continental separation or attempted separation. Their spatial distribution shows that around west and east Africa, post-Palaeozoic subsidence of the continental edges has been the norm and buoyancy the exception: subsided sectors account for about 68% of the total length of the margins along both east and west coasts. One important contrast between the two sides of Africa is that in the west the crests of buoyant sectors are typically 400-500 km apart, whereas in the east this distance is of the order of 1000 km. In other words, the west coast is characterized by relatively small, closely-spaced sediment basins, whilst in the east subsided areas are much wider

and farther apart. This is probably a reflection of the lower "subsidence potential" of the heterogeneous (i.e. alternating rigid-weak) structure produced by the numerous cratons along the west coast compared with the more homogeneous and coast-parallel grain of the pan-African metamorphic belts that dominate the basement along the east coast (Fig. 1 after Grant 1974 fig. 14).

Basin Stratigraphy and Geometry

Using published commercial borehole and borehole controlled seismic data, we have constructed longitudinal profiles along the west and east coasts of Africa (Fig. 2) and present stratigraphic sections for each basin (Fig. 3). Tables I & II show sediment thicknesses for various stratigraphic units.

West Coast

The stratigraphy of the west coast basins is remarkably similar over a distance of nearly 6000 km and can be expressed in terms of 5 lithostratigraphic units that are bounded by major hiatuses or important facies changes (Fig. 3). The latter are, within the limits of available palaeontological resolution, apparently synchronous.

The succession starts everywhere with thick continental (fluviatile and lacustrine) and volcanic sequences that rest on block faulted basement and can be typified by the Cocobeach group in Gabon (Unit 5). Locally, (for example in the Gabon, Cabinda and Cuanza basins) this sequence is subdivided by a major hiatus. Maximum thicknesses (corrected values in parenthesis) vary from about 1000 m (1600 m) in Guinea to 2500 m (4000 m) in Gabon, and estimates of age suggest that it ranges from Upper Jurassic to Neocomian (e.g. de Klasz 1978). These sediments fill in the earliest, graben-like (taphrogenic) parts of the west coast basins and are abruptly overlain by an Aptian/Albian (locally early Cenomanian) sequence which consists of combinations of black shales, evaporites and carbonates (Unit 4). This major facies discontinuity is so well-developed throughout the region that it has been called the "South Atlantic Unconformity" (Wenger 1972, quoted by de Klasz 1978). Evaporites (halite, anhydrite, and potash salts) are most extensively developed in the Gabon-Cuanza sector, and locally they probably exceed 400 m in an undeformed state. Although in some basins evaporite precipitation continued into lower Albian times, it is mostly thought to be of Aptian age. Maximum thicknesses of the Aptian/Albian Unit 4 vary from 500 m (725 m) in Cabinda, to 1800 m (2826 m) in Guinea. A regional Cenomanian regression is marked by disconformities and/or a return to continental sedimentation (although locally a restricted marine facies of black shale and Mg-rich car-

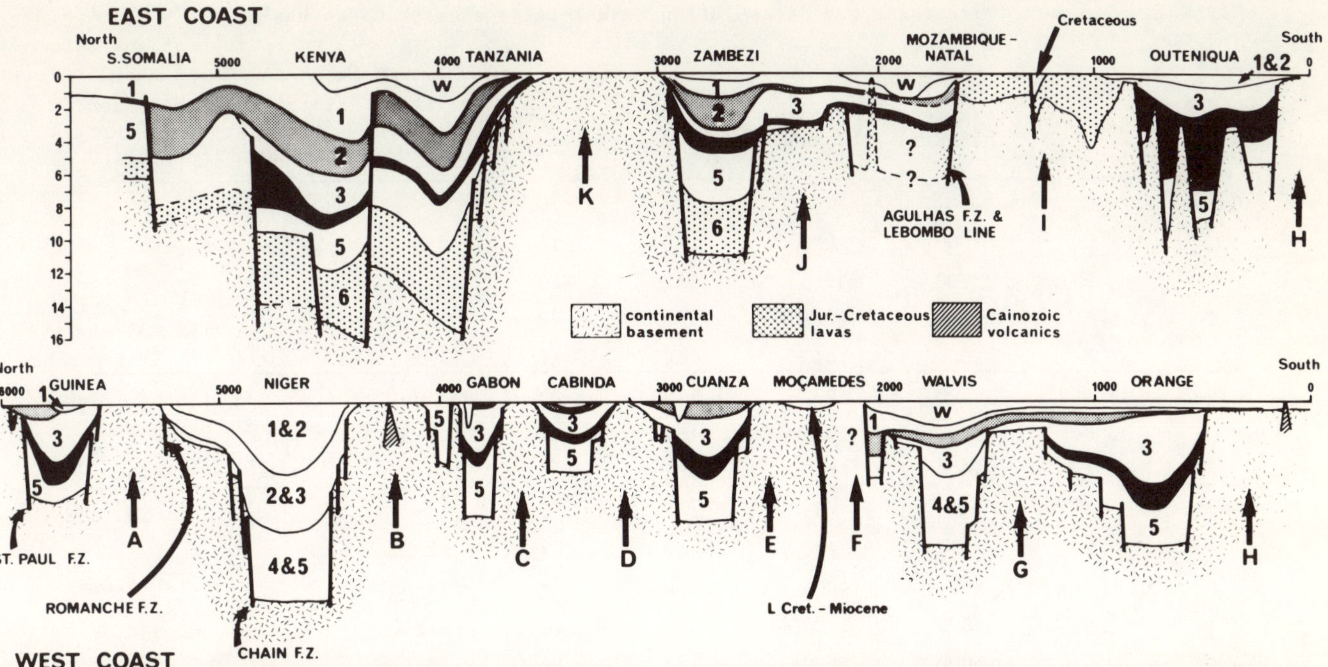

Fig. 2. Longitudinal sections across the African continental margins between the Guinea and Orange basins (west coast), and Outeniqua and Somalia basins (east coast). Scales in kilometers, vertical exaggeration approximately X75. Basement arches A to K are lettered as in Figure 1. W — water column in offshore areas. Sediment fill subdivided and notated according to lithostratigraphic schemes shown in Figure 3. Units 6, 4 & 2 are shaded for identification. Thick lines show main boundary fault lines and are generally schematic. Data sources are cited in the text, and these carry full bibliographies for detailed descriptions of individual basins.

bonates persisted into the lower Cenomanian in Guinea).

Upper Cretaceous sediments (Unit 3) are generally clays, sands and limestones, and mark the establishment of open ocean conditions. Locally, regressive sequences occur in the Turonian/Coniacian, and in the Niger Basin, Santonian. Sedimentation is typically a deltaic facies with the seaward advance of the main depocentres, and local maximum thicknesses range from 1050 m (1481 m) in Cabinda, to 4500 m (6795 m) in the Orange Basin. In common with all the world's ocean basins, there is a major late Maastrichtian/early Palaeocene hiatus along most of the west coast, although it has not been recognized in the Niger and Cuanza areas. Palaeogene sediments (Unit 2) comprise Palaeocene – late Eocene shallow water marine rocks which range in thickness from 300 m (420 m) in Gabon to 1000 m (1400 m) in Guinea and Cuanza.

A late Eocene – early Miocene hiatus separates the Palaeogene from the Neogene rocks (Unit 1), which are only locally developed, often in shallow water facies. Pliocene sediments are generally missing and most of the Neogene is represented by Miocene strata, often in the form of deep, narrow canyon infill (the canyons were presumably cut during Oligocene sea level

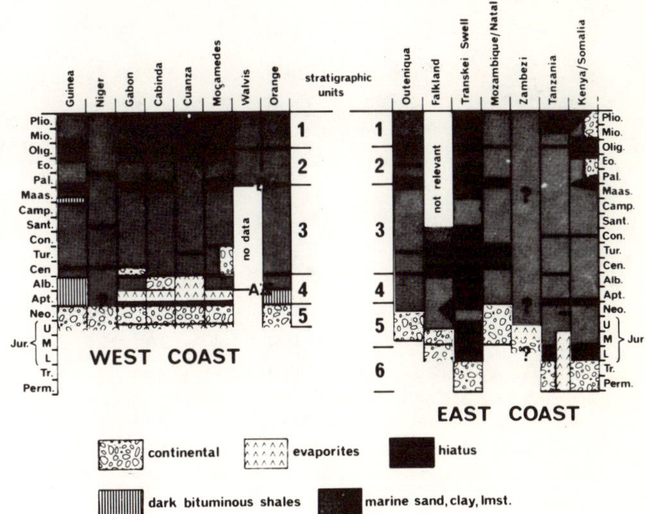

Fig. 3. Summary litho- and biostratigraphy of the African west and east coast sedimentary basins. Data sources cited in the text. Main hiatus boundaries or facies changes are shown by thick lines. Stratigraphic units 1-6 are those shown in Figures 2 and 4 and have been used to calculate sediment accumulation rates in Tables I, II & III.

TABLE 1. Sediment thicknesses and accumulation rates - African west coast basins

Basin	Sed.unit	measured thickness	corrected thickness	sed.accum. rate m/my	% total sed.	time interval my BP	total sed. thickness measured	total sed. thickness corrected
Guinea	1	0	0	0	0	–		
	2	1000	1400	33	14	65–22.5		
	3	2700	3915	112	40	100–65		
	4	1500	2826	188	29	115–100		
	5	1000	1610	64	17	140–115	6200	9751
Niger	1 & 2	4400	6380	119	35	53.5–0		
	3 & 2	3500	5636	161	31	100–53.5		
	4 & 5	4000	6440	161	35	140–100	11900	18455
Gabon	1	0	0	0	0	–		
	2	300	420	10	4	65–22.5		
	3	2800	4004	114	38	100–65		
	4	800	1208	81	12	115–100		
	5	3000	4830	193	46	140–115	6900	10462
Cabinda	1	400	560	25	10	22.5–0		
	2	400	560	13	10	65–22.5		
	3	1050	1481	42	26	100–65		
	4	500	725	48	13	115–100		
	5	1500	2265	91	41	140–115	3850	5591
Cuanza	1	0	0	0	0	–		
	2	1000	1400	33	13	65–22.5		
	3	2500	3625	104	33	100–65		
	4	1300	2041	136	18	115–100		
	5	2500	4025	161	36	140–115	7300	11091
Walvis	1	500	700	31	7	22.5–0		
	2	500	700	16	7	65–22.5		
	3	2000	2900	83	27	100–65		
	4 & 5	4000	6360	159	60	140–100	7000	10660
Orange	1	100	140	6	1	22.5–0		
	2	400	560	13	4	65–22.5		
	3	4500	6795	194	52	100–65		
	4	1500	2415	161	18	115–100		
	5	2000	3220	129	25	140–115	8500	13130

lows). Excluding the canyon fill, maximum Neogene thicknesses range from 0 to an estimated 500 m (700 m) in the Walvis Basin. Generally speaking the Cainozoic sediments are thickest beyond the continental edge, where depocentres migrated seaward by prograding.

As can be seen in Fig. 2, the Upper Jurassic-Neocomian continental Unit 5 fills a central graben-like depression in all the west coast basins where borehole data allow a stratigraphic subdivision to be made. Rapid thickness changes indicate that the boundary faults of all the sediment basins were active during this period and Upper Jurassic-Neocomian volcanic rocks have

TABLE 2. Sediment thicknesses and accumulation rates – African east coast basins

Basin	Sed.unit	measured thickness	corrected thickness	sed.accum. rate m/my	% total sed.	time interval my BP	total sed. thickness measured	total sed. thickness corrected
Outeniqua	1	100	140	6	1	22.5–0		
	2	500	700	26	5	65–38		
	3	2000	2820	81	20	100–65		
	4	4400	7040	261	50	127–100		
	5	2000	3220	81	23	167–127	9000	13920
Mozambique	1	716	1002	57	20	22.5–5		
Ridge	2	133	186	7	4	65–38		
	3	2035	2951	84	60	100–65		
	4	183	276	18	6	115–100		
	5	352	535	9	11	172–115	3419	4950
Zambezi	1	1000	1400	80	8	22.5–5		
	2	2000	2900	107	17	65–38		
	3	500	775	22	4	100–65		
	4	1250	1975	131	11	115–100		
	5	3150	5072	89	29	172–115		
	6	3300	5313	56	30	285–190	11200	17435
S.Tanzania	1	2000	2800	160	13	22.5–5		
	2	2500	3725	138	18	65–38		
	3	1500	2415	69	11	100–65		
	4	500	805	53	4	115–100		
	5	3100	4991	88	24	172–115		
	6	3900	6279	66	30	285–190	13500	21015
Zanzibar	1	3400	4930	282	20	22.5–5		
	2	2100	3234	120	13	65–38		
	3	2000	3220	92	13	100–65		
	4	500	805	53	3	115–100		
	5	2700	4347	76	18	172–115		
	6	4700	7567	80	31	285–190	15400	24103
Somalia	1	1400	1960	112	22	22.5–5		
	2	0	0	0	0	–		
	3	0	0	0	0	–		
	4	228	321	21	4	115–100		
	5	3155	4859	85	53	172–115		
	6	1219	1962	21	22	285–190	6002	9102

been reported as part of the graben-fill from the Cuanza, Mocamedes and Orange basins. The Aptian/Albian restricted marine sequence usually overlaps the main boundary faults, and represents regional expansions of the local depocentres.

This trend was well established by Upper Cretaceous times, and in all the basins these rocks are widely transgressive over the original graben boundaries.

Structurally, the west coast basins are (with

the exception of the Guinea and Niger) located
over coast-parallel graben with massive eastern
boundary faults, and end boundaries formed by
short marginal offset (small transform) fault
zones. The two exceptions are located along a
sheared margin and a large coast-normal graben,
respectively.

A striking feature of these basins is their
symmetry, regular lateral distribution, and (with
the exception of the Niger) their relatively
constant depth of maximum subsidence: 4-6 km
to the north, and 9 km to the south of the Walvis
Ridge. The latter, which are somewhat larger
structures, also differ from those farther north
in lacking present-day onshore extensions,
although there is evidence (at least adjacent
to the Walvis Basin) of mild onshore post-
Jurassic tectonism in the form of coast-
parallel faulting.

The largest of the west coast basins is the
Niger. It initially developed by taphrogenic
subsidence that was controlled by the landward
extension of several of the equatorial fracture
zones (e.g. Chain). It is unique along the west
coast in being at the distal end of a large
coast-normal graben (Benue Trough) that probably
owes its great width (and depth) to this structural control. No boreholes are reported to
have penetrated to basement along the outer edge
of the continental block and our profiles are
based on geophysical data given by Hospers
(1971), and Emery et al. (1974) and the geological summary of de Klasz (1978). The northern
flank of the Niger Basin is complicated by the
proximity of another of the equatorial fracture
zones (Romanche) which controls the Benin saddle-
basin, whilst the Guinea Basin lies between the
Romanche and St. Paul fracture zones.

Buoyant interbasinal areas all coincide with
major basement highs. The Walvis Ridge is the
most complex, being also at the crest of the
continental abutment of a large aseismic oceanic
ridge. At the abutment, the ridge is fault
bounded by large coast-normal fracture zones
(Dingle & Simpson 1976), but so far no evidence
has been found of these penetrating into the
continent. Multi-channel seismics show that
large-scale faulting has taken place on the
ridge as recently as Palaeogene times and has
produced large graben on its crest (Lehner &
Ruiter 1977). The basement seen today across the
Walvis barrier is, therefore, probably younger
than Unit 5 in the adjacent sediment basins, but
we do not know if sediments older than Cretaceous
lie beneath the abutment zone. To the north, the
continental margin is very narrow and the part of
the Mocamedes Basin on continental crust is probably less than 1 km deep. It contains only a
thin, proximal sequence of Lower Cretaceous and
younger sediments and was probably truncated by
a spreading ridge jump (du Plessis 1979), with
the result that the bulk of the original, taphrogenic basin now lies adjacent to the Brazillian
continental margin.

The other basement features (with the exception
of the Columbine-Agulhas Arch) are associated with
positive areas in the Pre-Cambrian metamorphic
and Archaen cratonic basement which caused small
offsets in the line of continental separation.
The Columbine-Agulhas Arch, which is the largest
buoyant block along the west coast, is unique in
being capped by folded Lower Palaeozoic sediments
that are draped over a late Pre-Cambrian batho-
lith.

Tertiary volcanism has been restricted to the
Cameroon basement high zone, and to minor intru-
sions south of the Walvis Ridge (Lüderitz and
Columbine-Agulhas arches, and in the Orange
Basin).

East Coast

Both the stratigraphy and tectonic styles of
the east coast basins are more complex than
found along the west coast. Not only are the
sediment basins larger, but they are also deeper,
and have a longer history of subsidence. Despite
this, the stratigraphy on a regional scale shows
certain similarities with the west coast succession (Fig. 3), and in most areas the same main
hiatuses are well-developed, although the associated lithofacies are different.

In three areas (Transkei Swell, Zambezi Basin,
and the Tanzania-Somalia sector) the earliest
sediments are a thick sequence of Permo-Triassic
rocks (Unit 6). In the north up to 4700 m
(7567 m) of continental and marine sediments
(with thick evaporites) occur in coast-parallel
faulted basins, whilst in central Mozambique
Kent (1974) indicates about 3300 m (5313 m) of
pre-Jurassic sediments under the Zambezi Cone in
a continuation of the Zambezi Graben. No pre-
Jurassic strata are known from the Natal/
Mocambique or Outeniqua basins, but about 4400 m
(6424 m) of continental red beds occur on the
Transkei Swell. Here, the original basins lay
NW-SE, approximately normal to the trend of the
Agulhas Fracture Zone which now truncates them.
Since the end of Triassic times this zone has
dramatically reversed its rheological character
and has remained strongly buoyant, except for its
crest which collapsed in Neocomian times. Here
about 2 km of ? Neocomian strata occur in two
half graben in the vicinity of Port St. Johns
(Fig. 2). Kent et al. (1971) show that locally
in South Tanzania the deposition of evaporites
and continental sediments persisted into Lower
Jurassic times.

Along the whole of the East coast there is evidence of widespread middle Jurassic tectonism and
local volcanic activity, as well as the initiation of widespread marine sedimentation (Unit 5).
This Unit probably has an older base than its
equivalent on the west coast. In the Tanzania-
Somalia sector, shallow marine middle Jurassic-
Neocomian sediments locally reach 3155 m (5000m).
They thin southwards against the north Mozambique
basement high but are about 3150 m (5000 m) thick

under the Zambezi Cone where Kent (1974) indicates thick interbedded evaporites. South of the Mozambique Ridge, Unit 5 consists of continental and lacustrine sediments (with possible minor marine interfingers) in SW Mozambique and in the Outeniqua Basin, with continental and marine facies on the Falkland plateau which lay adjacent to SE Africa at this time.

All the east coast areas have a mid- or upper Neocomian to early Aptian hiatus or major facies change which approximately coincides with the initiation of widespread marine conditions along the west coast (Unit 4). In the south (Outeniqua and Mozambique) this also marks the onset of extensive marine sedimentation, and locally (Outeniqua) very thick sequences were laid down: 4400 m (7040 m). Farther north shallow marine sediments were deposited in the Zambezi and Tanzania areas, but except locally, thicknesses were not great (Fig. 2). Upper Cretaceous (Unit 3), Palaeogene (Unit 2) and Neogene (Unit 1) sediments occur in all the east coast basins and are bounded by regional hiatuses that span late Albian to Cenomanian, late Maastrichtian to early Palaeocene, and late Eocene to early Miocene times. The earliest of these is the least well-defined and locally the non-sequence ranges into the Coniacian. All the post lower Cretaceous sediments are dominantly shallow marine, but the Upper Cretaceous (Unit 3) is generally thinner and volumetrically less important than the equivalent rocks on the west coast. In addition, there is a marked contrast in Cainozoic sedimentation between the two margins, where in the east, with one exception, sequences are very much thicker both in real terms, and compared to Cretaceous sequences (e.g. 5500 m (8000 m) in the Zanzibar compartment). The exception is in the Outeniqua Basin where only a thin Cainozoic succession is developed (about 600 m).

Turning to the tectonics and geometry of the east coast basins, three distinct styles are evident. The Outeniqua Basin is a sheared basin that is bounded to the SW by the Columbine-Agulhas Arch and to the SE by the Agulhas Fracture Zone. It is developed entirely on continental crust and consists of a lower taphrogenic part with massive half grabens up to 6000 m deep and an upper (post-Albian) epeirogenic basin that is partially dammed behind an outer marginal fracture ridge. The early basin complex was truncated by the Agulhas Fracture Zone, and its SE extension now lies between the Falkland Islands (equivalent to the Columbine-Agulhas Arch) and the Maurice Ewing Bank (equivalent to the Transkei Swell). Its tectonic grain is oblique to the margin outline and was controlled by the re-activation of the Cape Fold Belt tectonic trends that developed along the line of the Palaeozoic Cape geosyncline.

The Zambezi and Natal/Mozambique basins are large graben-like structures with steep boundary faults that are coast-normal continuations of

major lineaments in the SW Indian Ocean. The two are separated by a large basement high which is draped with Middle Jurassic-Lower Cretaceous lavas. This is the onshore extension of the Mozambique Ridge. In basin geometry they are most like the Niger Basin/Benue Trough, although they seemed to have started forming somewhat earlier. The western margin of the Natal/ Mozambique Basin lies along the northern continuation of the Agulhas Fracture Zone, which in western Mozambique is formed by the Lebombo Line, a major crustal fracture that cuts across the eastern end of the Kaapvaal craton (Fig. 1). Downward movement to the east probably started along this lineament in late Triassic time and was accompanied by massive outpourings of lavas (late Triassic to lower Cretaceous) which are draped over it in a structure that has been called the Natal monocline.

North of the north Mozambique basement high, the whole of the coastal fringe between Tanzania and Somalia (over 2000 km) is underlain by a deep narrow sediment basin that has been described by Kent et al. (1971), Kent and Perry (1973), Kent (1974), Walters and Linton (1973), Beltrandi and Pyre (1973) and Nairn (1978). It is bounded to the west by a series of large coast-parallel-faults (e.g. the Tanga fault in N. Tanzania) and is locally cut across by extensions of the East African Rift System (Fig. 1). Lengthwise, the basin can be divided into at least four compartments (Fig. 2) in which sediments locally reach a thickness of 13-15000 m. Major episodes of faulting occurred in Permo-Triassic, Middle Jurassic, late Neocomian/Aptian, and Miocene times. The coastal basin shallows into SW Somalia, but the main Mesozoic basinal structure strikes northwards from the coast and cuts across Ethiopia to the Red Sea area (Mandera Graben) (Fig. 1).

Temporal Continental Basement Subsidence

Variations in sediment accumulation rates with time are shown in Fig. 4, and the parameters used in its construction are given in Tables I, II & III. Corrections for sediment compaction have been made using an empirical curve constructed from porosity variation with depth for a deep continental margin borehole quoted by Steckler and Watts (1978). An assessment of crustal subsidence (Fig. 4) was obtained by reducing sedimentation rate by one third to allow for ongoing sediment compaction. Three distinct trends emerge, each typical of a particular margin type.

<u>Rifted Margins (West Coast)</u>

Basement subsidence rates for the rifted basins show a progressive decrease since Upper Jurassic times. During the earliest (Upper Jurassic-Neocomian Unit 5), taphrogenic phase of basin formation, the mean subsidence rate was 110m/my. This fell to between 80 and 90 m/my during the

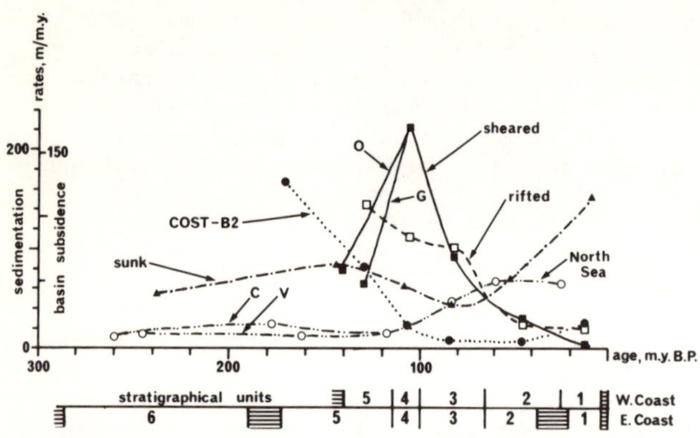

Fig. 4. Sediment accumulation/basement subsidence rates for the west and east African sedimentary basins. Values are means (Table III) and are plotted at the mid-point of each lithostratigraphic unit. Values for COST-B2 and the North Sea taken from Steckler & Watts (1978) and Kent (1977), respectively. Basement subsidence values are a nominal two thirds sedimentation rate to allow for ongoing sediment compaction. The early part of the North Sea curve is plotted separately for the Central (C) and Viking (V) grabens. The Natal/Mozambique Basin has not been included because maximum thicknesses below the top of the lower Cretaceous are not known.

rest of Cretaceous time (Units 4 & 3), while even lower rates (20 m/my) prevailed throughout the Tertiary (Units 2 & 1).

The tectonic setting (tensional rifted) of the west coast of Africa is similar to that off eastern USA (Wilson & Williams, 1979), and a comparison between the two areas is possible through the work of Steckler and Watts (1978) (Table III). The sediment accumulation/basement subsidence curve for the COST-B2 borehole (SE of New York) has been added to Fig. 4 and compares favourably with our curve for the rifted west African basins. Both curves show a steady decay during their early history, followed by low gradients in their old age. Steckler and Watts (1978), testing the ocean thermal cooling model of Parsons and Sclater (1977) concluded that during the first 90 my or so of the USA margin's history, the basement subsidence rate shows a linear relationship proportional to the square root of its age, with the younger history approximately exponential to a steady state. Whilst the early (Upper Jurassic-Lower Cretaceous) history of the west African basins may show a linear subsidence rate, and their Cainozoic subsidence could be fitted to an exponential curve, the mean sedimentation rate for Upper Cretaceous time deviates significantly from the theoretical mode (it is too high). Notwithstanding this aberration, the main inflection in the west coast curve falls at about 80 my after the oldest recorded

sediments from the area, and in this respect fits the model well.

There is no likelihood that Upper Cretaceous sediments could have accumulated above sea level, so that we must conclude that during this period basement subsidence was faster than predicted by the Parsons and Sclater model for cooling oceanic crust. This could have resulted from the high sea level stands (> 300 m) quoted by Steckler and Watts (1978), coupled with particularly rapid local sediment dumping that would have caused an isostatically-related spurt in basement subsidence. Alone, this could account for about 750 m of sediment (equivalent to about 1000 m corrected for compaction) and would have been responsible for 29 of the 107 m/my measured for this period, reducing the basement subsidence curve to the same shape as that found off the Eastern USA by Steckler and Watts (1978) (Fig. 4).

Low basement subsidence rates during Tertiary would have impaired the margin's ability to accommodate significant quantities of terrigenous material and biogenic production. In some areas (e.g. off the arid coast of southwestern Africa) sediment input was drastically curtailed during this period, but elsewhere the higher energy of the shelf environment together with a moderate sediment input promoted oceanward prograding of the margin and a westward shift in depocentres.

Lack of published deep borehole information from the Niger Basin does not allow a satisfactory comparison to be made with other west coast areas. There has certainly been a slowdown in the Tertiary subsidence rates (the massive Tertiary accumulations in the Niger Cone are on oceanic crust), but its early history (on the meagre data available) does not seem to fit the general west coast pattern (the Lower Cretaceous sedimentation rates are too low).

Sheared Margins (Guinea and Outeniqua Basins)

Two of the largest sheared continental margins in the world lie within the area of our survey: the composite Equatorial Fracture Zone, and the Agulhas Fracture Zone, and the Guinea and Outeniqua basins provide us with good examples of sheared margin sediment basins. Both have remarkably similar sedimentary and tectonic histories, although their internal structures are somewhat different. A mean of their sedimentation/basement subsidence rates is plotted on Fig. 4, and the curve shows a very different shape compared with those for the rifted and sunk margins.

Both basins have an Upper Jurassic-Neocomian history of graben fill with continental sediments (the Outeniqua basin contains sediments as old as Middle Jurassic) which reflect moderate basement subsidence rates (48 m/my). This was followed by a short (Aptian to Albian in Guinea, and Valanginian to Albian in Outeniqua) period of very rapid subsidence (148 m/my). Subsequent basement

TABLE 3. Mean sediment accumulation rates (using corrected sediment
thicknesses)

Rifted margins (Gabon, Cabinda, Cuanza, Walvis and Orange basins)

stratigraphic unit	accumulation rate (m/my)	bulk accumulation (% of total sediments)
1	12.4	4
2	17	8
3	107	35
4	117	24
5	147	42

Sheared margins (Guinea and Outeniqua basins)

stratigraphic unit	accumulation rate (m/my)	bulk accumulation (% of total sediments)
1	3	1
2	30	9
3	97	30
4	224	40
5	73	20

Sunk margins (Zambezi, Tanzania, Kenya and Somalia basins)

stratigraphic unit	accumulation rate (m/my)	bulk accumulation (% of total sediments)
1	158	16
2	91	12
3	46	7
4	64	6
5	84	31
6	56	28

E. USA COST-B2

	accumulation rate (m/my)
Mio-Pliocene	25
Palaeogene	7
U. Cretaceous	9
Apt.-Albian	22
U. Jurassic-Neocomian	82
195-140 my	163

subsidence shows a steady decline which has reduced to low values ($<$ 5 m/my) during Neogene
times.

No study comparable to that of Steckler and
Watts (1978) has been published on sheared margin
basins, but the data presented in Fig. 4 show
that the history of subsidence in such basins
is not controlled by the oceanic cooling curve
predicted by Parsons and Sclater (1977). Peak
subsidence occurred 60 my (Outeniqua) and 35 my
(Guinea) after the oldest sediments reported
from each basin, and its synchronism suggests
a common cause. The most obvious explanation
is to correlate this episode with tectonism
associated with shearing along the two fracture
zones as South America and Africa separated, but,
as we will discuss below, the timing of this
spurt in basement subsidence may have been con-
ditioned by worldwide rheological events which
occurred in Aptian/Albian times (Kent 1977,
Whitten 1976). Finally, both areas have very
low Neogene sediment accumulation rates, suggesting that sheared margins soon reach a level
beyond which they will sink no more. This
would be understandable, at least for any
thermal contribution to their sinking history
which, because of their narrowness, ought to
be relatively short-lived.

Sunk Margins (Natal to Somalia)

Kent (1974, 1976, 1977 etc) has pointed out the
anomalous nature of the large sediment basins off
eastern Africa. It appears that the greater part
of this margin including the present day buoyant
Transkei Swell, commenced crustal subsidence
(about 50 m/my) in Permian or Triassic times and
accumulated sediment thicknesses locally exceeding 7000 m. An important phase of earth movements occurred in Middle Jurassic times and this
heralded the commencement of a further long phase
of crustal subsidence at a somewhat higher rate
(84 m/my).

Throughout middle and Upper Cretaceous times
there was a general decline in the rate of downward basement movement (35 m/my in U. Cretaceous
times), but this was followed by a spectacular
progressive increase through Palaeogene (91 m/my)
to Neogene (158 m/my) times. The latest phase
commenced with widespread Miocene tectonism.

Quite obviously the bimodal crustal subsidence
curve for this area (Fig. 4) sets the region
apart from the rifted and sheared margins, and
as Kent (1977) has pointed out places it in the
same category as the North Sea area and the
western margin of Australia. It is important
to reiterate that in the Somalia-Tanzania sector
at least, the crust has been subsiding on a large
scale in a miogeosynclinal setting for nearly 300
million years.

Although there is no consensus as to the timing
of the separation of India/Madagascar from Africa,
it must have been prior to anomaly M22 (mid-
Jurassic) and has been tentatively dated as

late Triassic/early Jurassic (Dingle & Scrutton
1974). If these dates are of the right order,
then large-scale subsidence commenced in Tanzania
and over the Transkei Swell up to 100 my prior to
continental separation, and reached its maximum
rates between 40 and 50 my after separation.
Whilst it is difficult enough to reconcile this
Mesozoic history to thermal events associated
with continental separation in a way suggested
by Steckler and Watts (1978) for eastern USA, it
is impossible to link the Cainozoic spurt with
such phenomena. The latter does, however, coincide with widespread rifting over large tracts
of continental eastern Africa. There is no doubt
that driving mechanisms other than the simple
thermal and loading models proposed by Steckler
and Watts (1978) are, and have been, at work in
this complex area.

Some General Observations and Conclusions

At present there are two schools of thought on
the temporal structural development of passive
continental margin sediment basins. Most workers
accept the direct relationship between the time-
table of basin formation and local continental
separation (e.g. Burke 1976, Falvey 1974, Steckler & Watts 1978, amongst many others), and whilst
there is no consensus as to the cause of continental basement subsidence (e.g. crustal extension
by hot creep or tensional necking, crustal erosion,
thermal metamorphism, thermal contraction, subaerial erosion) there is general agreement that
locally, the onset of sea floor spreading and
early basin formation are a process/response
relationship. An alternative philosophy has been
proposed by Kent, who in a series of publications
(1974, 1976, 1977) has pointed out the great similarity in tectonic style and subsidence histories
of the world's tensional (Atlantic, aseismic,
passive) continental margin basins.

Briefly, Kent shows that they all have a history
of taphrogenic (graben) formation that stretches
from at least Upper Jurassic (and in most areas
Permo-Triassic) to mid-Cretaceous times. Whilst
the commencement of this tectonic episode is not
everywhere synchronous (with the proviso that
dating is often poor in the continental deposits),
termination is remarkably so (varies between
Aptian to Albian). Thus, all tensional continental margins lie along lines of early Mesozoic
taphrogenesis, although the converse is not so,
as some areas, such as the North Sea, did not
develop as a spreading centre, and in others
spreading was aborted after an initial separation. After mid-Cretaceous times downwarping in
all the basins was epeirogenic (inter cratonic,
miogeosynclinal). It follows, therefore, that if
Kent's ideas are correct, then attempts to define
the tectonic history of passive margins on time-
tables strictly controlled by continental separation are likely to be unsound: the local
structure and stratigraphy will depend on the
superimposition of the local sea floor spreading

events onto the worldwide taphrogenic/epeiro-
genic cycle, and the phase shift in these two
will vary from place to place.

The review made here corroborates Kent's ideas,
and it has to be accepted that the "consensus"
explanation does not provide for the remarkable
mid-Cretaceous "fundamental rheological modification of the crustal rocks ...... which must .....
reflect a thermal event" (Kent 1977, p.15-16).
It is quite possible, as Kent (1977, p.16)
suggested, that the two phenomena outlined above
(lower Mesozoic taphrogenesis and sea floor
spreading) are related and that the driving
mechanism of the former provided favourable
conditions in which the latter could take place.
How, if at all, does knowledge of the African
basins help us to resolve the situation?

The west coast basins lie on continental margins that are thought to have come into being in
lower Cretaceous times (Valanginian, Larson &
Ladd, 1973), although other estimates put the age
at anywhere between about 165 and 140 my (e.g.
Emery et al. 1975). The major Aptian/Albian
facies change (i.e. the boundary between Units 5
& 4 in Fig. 3) has generally been accepted as
merely marking the earliest marine incursion
into the new ocean (e.g. Burke 1976, Lehner &
de Ruiter 1977), but to the writer this seems
unlikely to satisfactorily explain the widespread
changes in tectonic style that also occurred. It
is more reasonable to argue that the synchronous
entry of the sea into hitherto relatively isolated
non-marine basins spread over a distance of about
6000 km, was the result of epeirogenic downwarping along the length of the proto-South Atlantic.
In terms of Kent's proposed taphrogenic/epeiro-
genic cycle, the African west coast basement
subsidence curve shows no strong inflexion at
the mid-Cretaceous transition, whereas the curve
for the NE-USA continental margin does (Fig. 4).
In the latter, this period in time also coincides
with the cessation of taphrogenic block faulting
on the margin (e.g. Sheridan 1974). This suggests
that for rifted passive margins local plate tectonic events have an over-riding influence on
basement subsidence rates, but not on tectonic
style. Only areas such as the North Sea will
show an unbiased sinking history controlled
solely by the cycle proposed by Kent. Here,
(Fig. 4) the early, taphrogenic basement subsidence rates are low (< 20 m/my), whereas there is
a marked inflexion in mid-Cretaceous times and a
steep rise with the onset of the epeirogenic
phase, reaching values of about 50 m/my throughout Tertiary time.

The North Sea curve is most similar in shape to
that for the sunk margins of eastern Africa. In
the latter area, continental separation probably
took place in late Triassic/early Jurassic times
and there is no strong evidence to link the
crustal subsidence rates at any stage of its
history to a thermal cooling response as predicted by Steckler and Watts' work (1978). In
fact quite the opposite because subsidence

increased during Jurassic times and only declined
significantly during the Cretaceous, at least
100 my after continental separation. The very
strong subsidence recorded in the Tertiary rocks
is presumably a response to renewal of the taphrogenesis in eastern Africa and the Red Sea/Gulf
of Arabia areas along older lines of weakness
(e.g. McConnel 1977).

No theoretical model has been proposed for even
a simple plate tectonic/basement subsidence
relationship on sheared margins, although
Scrutton (1979) has recently discussed various
aspects of the problem. Steckler and Watts'
(1978) model cannot be expected to apply, because
of the unlikelihood of large-scale crustal heating during the early pull apart processes.
Frictional heat may be generated but until the
spreading ridge moves past any section of the
margin (and this may be several millions of years
after separation) hot ocean crust and an active
intrusive centre do not play a role in the margins' formation. The sudden collapse of continental basement in lower Cretaceous times in the
Guinea and Outeniqua basins may, therefore, be
purely the result of the lack of physical support,
or some dramatic response to the termination of
the episode of worldwide taphrogenic downwarping
proposed by Kent, presumably occasioned by the
special tectonic circumstances of the margins.

From our review of the African margins, a
compromise of the two conflicting hypotheses on
basement subsidence as exemplified by the work
of Kent (1977) and Steckler and Watts (1978) is
suggested along the following lines. A worldwide
rise in the geothermal gradient in the lithosphere commenced in Permo-Triassic times causing
an increase in metamorphic grade of parts of the
continental crust in a fashion similar to that
proposed by Falvey (1974) (i.e. the upward migration of the green schist/amphibolite facies
boundary). Old lines of crustal weakness (particularly those in inter-cratonic zones) were
preferentially affected on a worldwide basis, and
the isostatic readjustments caused by density
increases at the base of the continental lithosphere along these lineaments resulted in the
initiation of taphrogenesis as suggested by Kent
(1977). Crustal subsidence may not have been
synchronous everywhere, but the thermal event
continued until lower Cretaceous times, when a
lowering of the gradient "switched off" the
original metamorphic driving influence and
taphrogenesis on a large scale ceased. Subsequent subsidence was in an epeirogenic mode,
driven by crustal cooling and sediment loading.

Once formed, grabens may have behaved in a way
predicted by the mathematical models of Beaumont
and Sweeney (1978) who showed that a 100 km wide
graben progressively loaded with sediment on a
viscoelastic lithosphere with an exponential
decay rate of subsidence of 50 my, could produce
a basin 2500 m deep and 400 km wide. The metamorphic mechanism proposed here could provide the
extra driving force to produce the wider and

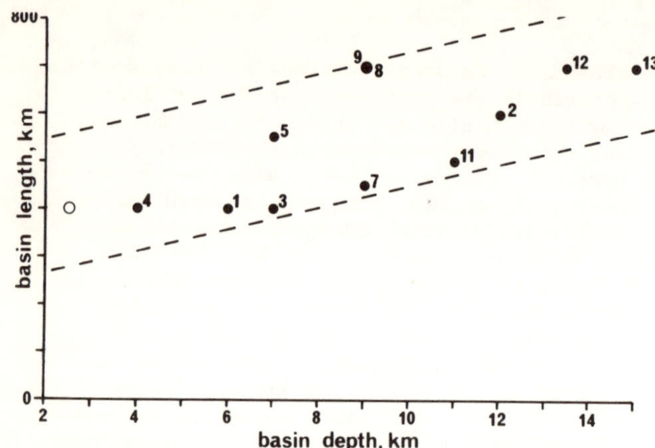

Fig. 5. Length vs maximum depth of west and east African sediment basins. Numbers as in Figure 1. Circle is the modelled example of Beaumont & Sweeney (1978, fig. 1b).

deeper basins found on passive continental margins (Beaumont & Sweeney 1978, p.T22). A plot of length vs depth of the African east and west coast basins shows a linear relationship, with the theoretical example quoted by Beaumont and Sweeney (1978, fig. 1) at the small-scale end of the field (Fig. 5). The Aptian/Albian rheological discontinuity described by Kent represents the cessation of metamorphic-induced subsidence and the start of a purely crustal cooling/sediment loading response.

The Triassic-lower Cretaceous thermal event also provided the mechanism for "unsticking" of the continental plates (Kent 1977, p.16) which set continental fragmentation in train. Some of the taphrogenic lineaments broke early (e.g. central Atlantic and parts of the Indian Ocean in late Triassic/early Jurassic time), others later.

The post-lower Cretaceous epeirogenic subsidence in the North Sea area has continued at a high rate presumably as a continued isostatic response to: a) cooling of the crust, b) readjustments due to sediment loading and c) higher densities caused by the metamorphism of the base of the crust. The fact that rifted margins do not show this continued response is presumably because continental separation has destroyed the symmetry and disrupted the heat budget of the original taphrogenic belt. The local effects of continental separation on the subsidence history of particular margins will vary depending on the local geology and pull apart style. It appears to be most pronounced on the rifted margins, where a fresh local thermal event is likely to be initiated by the establishment of a spreading centre adjacent to the margin. This results in a cooling model curve (similar to that proposed by Steckler & Watts 1978) being superimposed onto the margin's subsidence history. Sheared margins' responses to continental separation are dominantly of a tectonic nature with possible stress release

generated spurts, whilst the sunk margins of east Africa show no identifiable response at all. This is presumably due to fundamental basement characters that favour continued vertical movement which is not easily interrupted by local thermal events.

Renewed taphrogenesis in eastern Africa during Neogene times has been explained by McConnel (1977) in terms of a rise in the geothermal gradient causing thermal expansion and localized tension along the line of former taphrogenic trends. Whether this is merely a localized phenomenon or is a worldwide event on the scale of the Triassic-lower Cretaceous episode is difficult to decide at present, but increases in subsidence in the North Sea area and in the COST-B2 borehole during Neogene times suggest that it may have been of more than local significance.

Acknowledgements

This work has been funded by grants from the University of Cape Town, South African National Committee for Oceanographic Research, and the Geological Survey of South Africa.

References

Beaumont, C., and J.F. Sweeney, Graben generation of major sedimentary basins, Tectonophysics, 50, T19-23, 1978.

Beltrandi, M.D. and A. Pyre, Geological evolution of southwest Somalia, In: Blant, G. (ed.) Sedimentary Basins of the African coasts II, South and East Coast, Assoc. Afr. Geol. Surv. Paris, 159-178, 1973.

Burke, K., Development of graben associated with the initial ruptures of the Atlantic Ocean, Tectonophysics, 36, 93-112, 1976.

de Klasz, I., The west African sedimentary basins, In: Moullade, M. and Nairn, A.E.M. (eds.), The Phanerozoic geology of the World. The Mesozoic, A, Elsevier, Amsterdam, 371-399, 1978.

Dingle, R.V., A review of the sedimentary history of some post-Permian continental margins of Atlantic type, Ann. Brazil Acad. Sci., 48 (supplement), 67-80, 1976.

Dingle, R.V. and R.A. Scrutton, Continental break-up and the development of post-Paleozoic sedimentary basins around southern Africa, Geol. Soc. Am. Bull., 85, 1467-1474, 1974.

Dingle, R.V., and E.S.W. Simpson, The Walvis Ridge, a review, In: Drake, C.L. (ed.) Geodynamics: progress and prospects, Am. Geophys. Union, Washington, 160-176, 1976.

du Plessis, A., The evolution of the South-Eastern Atlantic Ocean: a review, Bull. Geol. Surv. S. Afr., 63, 1-27, 1979.

Emery, K.O., E. Uchupi, C.O. Bowin, J. Phillips, and E.S.W. Simpson, The continental margin off western Africa: Cape St. Francis (South Africa) to Walvis Ridge (South West Africa), Am. Assoc. Petrol. Geol. Bull., 59, 3-59, 1975.

Emery, K.O., E. Uchupi, J. Phillips, C. Bowin, and J. Mascle, The continental margin off

western Africa: Angola to Sierre Leone, Unpublished manuscript, Woods Hole Technical
Report, WHOI-74-99, 1974.

Falvey, D.A., The development of continental margins in plate tectonic theory, Austral. Petrol.
Expl. Assoc. Jl., 14, 95-106, 1974.

Francheteau, J., and X. Le Pichon, Marginal fracture zones as structural framework of continental margins in South Atlantic Ocean, Bull.
Am. Assoc. Petrol. Geol., 56, 991-1007, 1972.

Grant, N.K., Orogeny and reactivation of the west
and southeast of the West African craton. In:
Nairn, A.E.M. and Stehli, F.G. (eds.) The Ocean
Basins and Margins Vol. 1. The South Atlantic,
Plenum, New York, 447-492, 1973.

Hospers, J., The geology of the Niger delta area.
Inst. Geol. Sci. Rept., 70/16, London, 121-142,
1971.

Kent, P.E., Continental margin of East Africa -
a region of vertical tectonics, In: Burke, C.A.
and Drake, C.L. (eds.) Geology of Continental
Margins, Springer, Berlin, 313-320, 1974.

Kent, P.E., Major synchronous events in continental shelves, Tectonophysics, 36, 87-91, 1976.

Kent, P.E., The Mesozoic development of aseismic
continental margins, Jl. Geol. Soc. Lond., 134,
1-18, 1977.

Kent, P.E., J.A. Hunt, and D.W. Johnstone, The
geology and geophysics of coastal Tanzania,
Inst. Geol. Sci. Geophys. Paper, 6, 1-101,
1971.

Kent, P.E., and J.T.O'B. Perry, The development
of the Indian Ocean margin in Tanzania, In:
Blant, G. (ed.) Sedimentary Basins of the
African Coast II, South and East Coast, Assoc.
Afr. Geol. Surv. Paris, 113-131, 1973.

Larson, R.L., and J.W. Ladd, Evidence for the
opening of the South Atlantic in the early
Cretaceous, Nature, 246, 209-212, 1973.

Lehner, P., and P.A.C. de Ruiter, Structural
history of Atlantic margin of Africa, Am.
Assoc. Petrol. Geol. Bull., 61, 961-981, 1977.

McConnel, R.B., East African Rift System dynamics
in view of Mesozoic apparent polar wandering,
Jl. Geol. Soc. Lond., 134, 33-39, 1977.

Nairn, A.E.M., Northern and Eastern Africa, In:
Moullade, M. and Nairn, A.E.M. (eds.) The
Phanerozoic Geology of the World. The Mesozoic
A, Elsevier, Amsterdam, 329-370, 1978.

Parsons, B., and J.G. Sclater, An analysis of the
variation of ocean floor bathymetry and heat
flow with age, Jl. Geophys. Res., 82, 803-827,
1977.

Scrutton, R.A., On sheared passive continental
margins, Tectonophysics, 59, 293-305, 1979.

Sheridan, R.E., Atlantic continental margin of
North America, In: Burke, C.A. and Drake, C.L.
(eds.) Geology of Continental Margins, Springer,
Berlin, 391-407, 1974.

Steckler, M.S., and A.B. Watts, Subsidence of the
Atlantic-type continental margin off New York,
Earth Planet. Sci. Lett., 41, 1-13, 1978.

Walters, R., and R.E. Linton, The sedimentary
basin of coastal Kenya, In: Blant, G. (ed.)
Sedimentary Basins of the African Coast II,
South and East Coast, Assoc. Afr. Geol. Surv.
Paris, 133-158, 1973.

Wenger, R., Geologie du bassin sedimentaire
cotier gabonais, Internal Report ELF R.E.
(unpublished), 1972.

Wilson, R.C.L. and C.A. Williams, Oceanic transform structures and the development of Atlantic
continental margin sedimentary basins - a review, Jl. Geol. Soc. Lond., 136, 311-320, 1979.

AUSTRALIAN RIFTED MARGINS

J.J. Veevers

School of Earth Sciences, Macquarie University, North Ryde, NSW 2113

Abstract. The modern western and southern
rifted margins of Australia have evolved by plate
divergence from rift valley systems. From inception to continental breakup, the rift systems
lasted from the Late Carboniferous to the Late
Jurassic in the northwest, to the Early Cretaceous
in the southwest, and from the Late Jurassic to
the early Eocene in the south. The elements of
the rift system were (1) broad arches split by
single or multiple rift valleys that shed sediment into the rift valleys within the arches,
between them, or outside them, and (2) terrains
of diffuse rifting, called rift divergence zones.
In the south, such a zone was succeeded by an
arch split by a single median rift valley that
became the line of continental breakup. The
marginal half-arches or continental rims formed
in this way took up to 40 m.y. to subside below
sealevel by rotation about a hinge some 500 km
from the continent-ocean boundary.

The eastern margin is not readily interpretable
in these terms possibly because breakup was non-
axial, with none of the rift system, except
failed arms, being left on Australia, and during
the early phases of opening, the main motion, at
least along southeast Australia, was probably
shearing on a transform fault.

Introduction

The continental lithosphere of Australia-New
Guinea, part of the Indian Plate, is girt by the
lithosphere of rift or divergent oceans on its
western, southern, and eastern margins, and of
convergent oceans on its northern margin from
Timor to the Papuan Peninsula (Fig. 1). The
convergent boundary (Branson, 1978) is marked
by intense earth movement, seismicity, and
scattered vulcanicity; south of this boundary,
as befits its intra-plate position, Australia is
quiescent except for diffuse seismicity in the
southwest, in the central-west at 22°S, 127°E,
in the south about longitude 138°E, and in the
southeast, and two areas of vulcanicity, one in
northeast Queensland, the other on the southern
margin about longitude 141°E. Outside these
areas, earth movement is presumably very slow -

for much of the interior of Australia, one is
tempted to believe that the present shape is
essentially unchanged from that of the Aptian-
Albian, about 100 M.Y. (million years ago), as
suggested by the extent of marine sediments of
this age (Fig. 1). The perpetuation of this
gross shape to the present day poses the question:
what forces have maintained the uplift of the
eastern margin in spite of the subsidence expected to follow continental break-up in the Late
Cretaceous? This question will recur in the
review of the eastern margin.

The submarine features of the Australian rifted
margins have a certain bilateral symmetry about
the meridian half-way across Australia: complexes
of marginal plateaus and abyssal plains backed by
a broad shelf in the north, and narrow margins
expanding into long appendages (Naturaliste
Plateau and Tasmania/South Tasman Rise) in the
south. This symmetrical pattern extends into the
arrangement of seafloor spreading magnetic anomalies (Fig. 2), in that the azimuth of the anomaly
set off the northwest is reflected by that off
the northeast, as is that off the southwest by
that of the southeast, while the set south of
Australia is crossed at right-angles by the line
of symmetry. The symmetry does not extend beyond
the geometry, and the ages of these reflected
anomaly sets, as shown in Figure 2 by the ages of
the continent-ocean boundary (COB), are totally
different.

Authors have recognised rift valley complexes
along various parts of the rifted margin so that
altogether an almost continuous set of rift
valley complexes rings all but the north of
Australia, as seen clearest in the palinspastic
map of Figure 3A. The ubiquity of postulated
rift valley complexes along Australian rifted
margins, all allegedly inspired by actualistic
comparisons with modern rift valley complexes,
provides a focus for this review; that is, to
test these postulated structures and their geodynamic behaviour against the modern rift system
of East Africa, which, from its continuity with
the divergent oceans of the Red Sea and Gulf of
Aden, serves as a unique actualistic example of
this kind of structure.

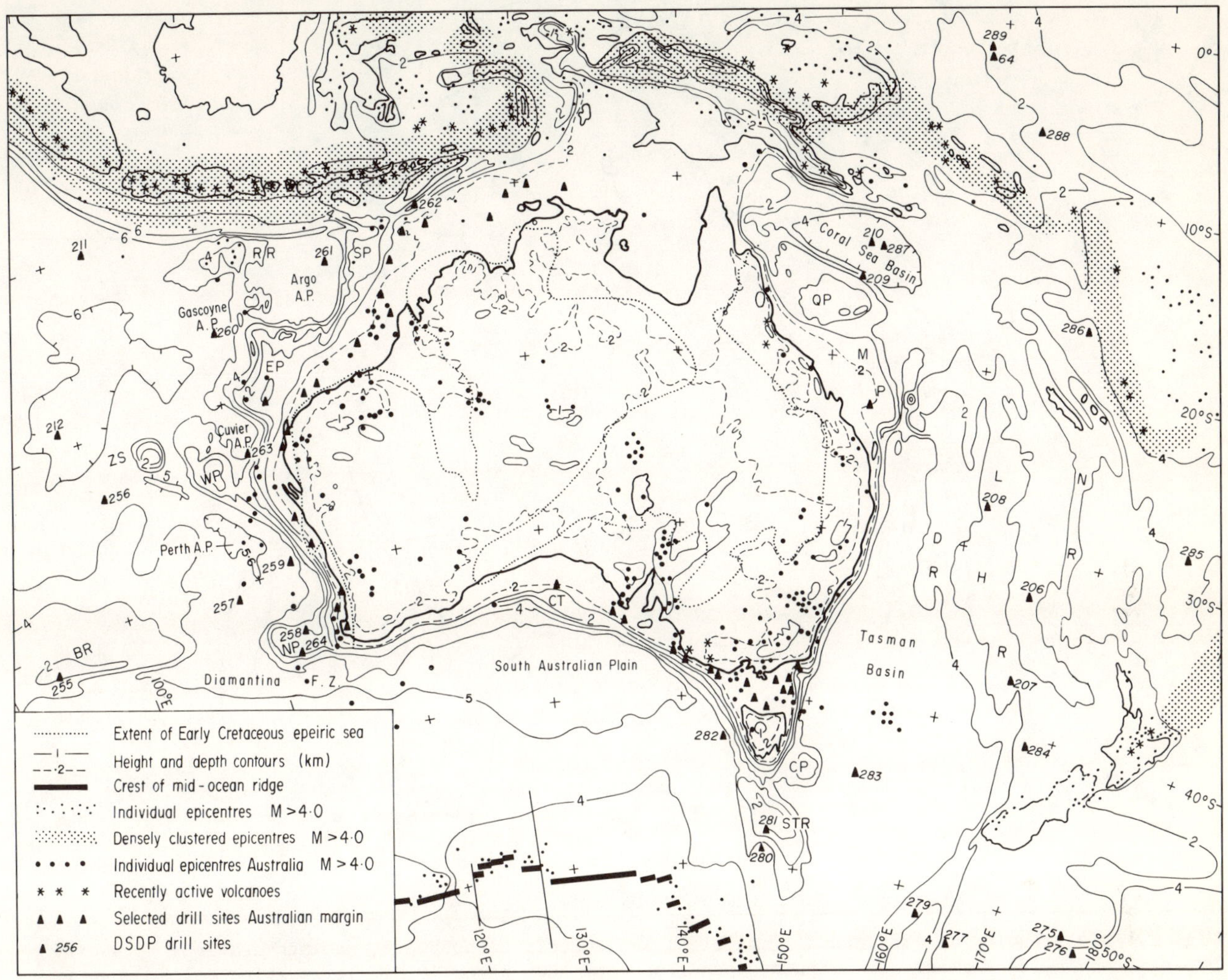

Fig. 1. Australia in its setting within the eastern part of the Indian Plate, showing topography
(Australia and New Guinea only) and bathymetry (that of the Australian margin from Symonds and
Willcox, 1976), extent of Early Cretaceous (Aptian-Albian) epeiric sea over Australia [Veevers and
Evans, 1975; Exon and Senior, 1976], seismicity in Australia [Denham et al., 1975] and elsewhere
[Barazangi and Dorman, 1969], modern volcanicity in Australia [Douglas and Ferguson, 1976; Stephenson
et al., 1978] and elsewhere [Holmes, 1965, fig. 740; Kuenen, 1950, plate B], selected exploration
wells on the Australian margin, and DSDP sites (Initial Reports of the DSDP, volumes 7, 21, 22, 26-
30). The minimum extent of the Early Cretaceous sea (dotted line) corresponds with the present 0.2 km
height contour (broken line) that delineates a major central-eastern depression and a minor one at
latitude 20°S in Western Australia. In the Aptian-Albian, shorelines on the east [Exon and Senior,
1976], on the western slopes of the Great Dividing Range, indicate correspondingly higher ground to
the east and south-east. Since the main drainage of eastern Australia in the Early Jurassic ran from
west to east [Veevers and Evans, 1975, fig. 40.8], the eastern margin must have been uplifted later
in the Jurassic. Submarine features at or oceanward of the Australian continental margin are (anticlockwise from northwest Australia)
RR Roo Rise; SP Scott Plateau; EP Exmouth Plateau; WP Wallaby Plateau; ZS Zenith Seamouth;
NP Naturaliste Plateau; BR Broken Ridge; CT Ceduna Terrace; STR South Tasman Rise; CP Cascade
Plateau; DR Dampier Ridge; LHR Lord Howe Rise; NR Norfolk Ridge; MP Marion Plateau; QP Queensland
Plateau.

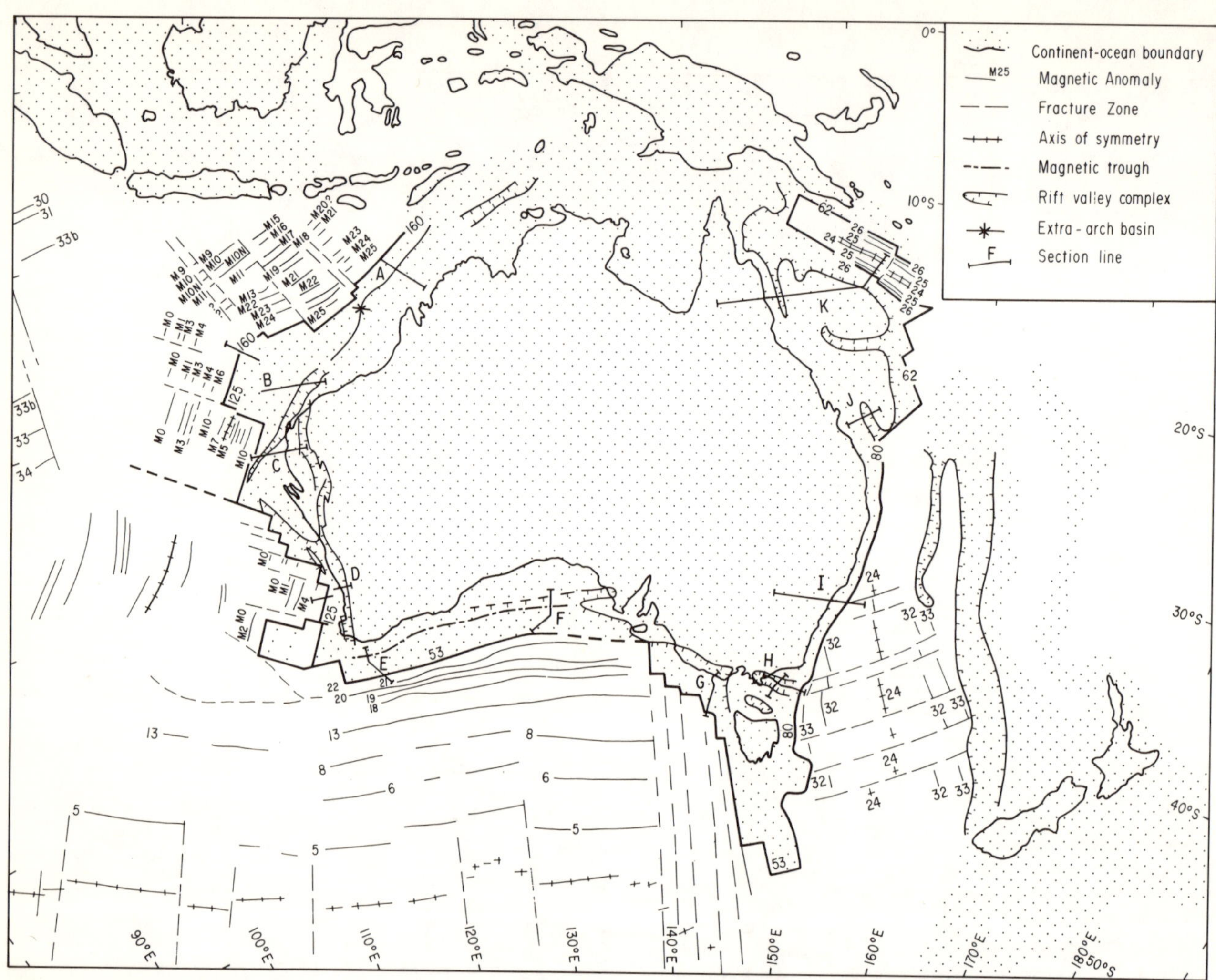

Fig. 2. Distribution of continental (stippled) and oceanic lithosphere, showing seafloor spreading magnetic anomalies seaward of continent-ocean boundary (COB-dated in M.Y.), modified on the west from Veevers and Cotterill [1978], on the south from Talwani et al.[1978, 1979], and on the east grossly approximated by the 4 km isobath and by somewhat shallower contours on the Lord Howe Rise. Magnetic anomalies (anticlockwise from the northwest) from Heirtzler et al. [1978], Larson et al [1969], Johnson et al. [1980], Markl [1978a,b], Talwani et al. [1979], Weissel, Hayes, and Herron [1977], Weissel and Hayes [1977], and Weissel and Watts [1979]. Details of rift valley complexes on the Australian margin from Laws and Kraus [1974], Balke and Burt [1976], Veevers and Cotterill [1976, 1978], Allen et al. [1978], Symonds and Cameron [1977], Talwani et al. [1978, 1979], Boeuf and Doust [1975], James and Evans [1971], Jongsma and Mutter [1978], and Taylor and Falvey [1977]. Lines A to K show the location of sections given in Figures 5 to 9.

Comparative Structure of Australia and
East Africa

The Neogene rift system of East Africa (Fig. 3B) is developed across the Ethiopian and Central Plateaus (outlined by the 1000 m contour) and is underlain by late Proterozoic fold belts wrapped round Archean and early Proterozoic blocks. Except the Albert and Albert Nile rifts, the Great Ruaha Valley, Lake Eyasi, and the western part of the Kavirondo Rift, which constitute only 20% of its length, the system is restricted to (and follows the grain of) the later Proterozoic fold belts, and, without exception, so are all the volcanics of the system [McConnell, 1974]. Within the Proterozoic fold belts, the volcanics predominate in the Ethiopian and eastern (Gregory) rifts that overlie the Mozambiquian Fold Belt, and are absent, outside the Lake Kivu and Rungwe areas, in the western, Great Ruaha, and Lake Nyasa rifts, most of which are in the older Ubendian Fold Belt. The preference of the

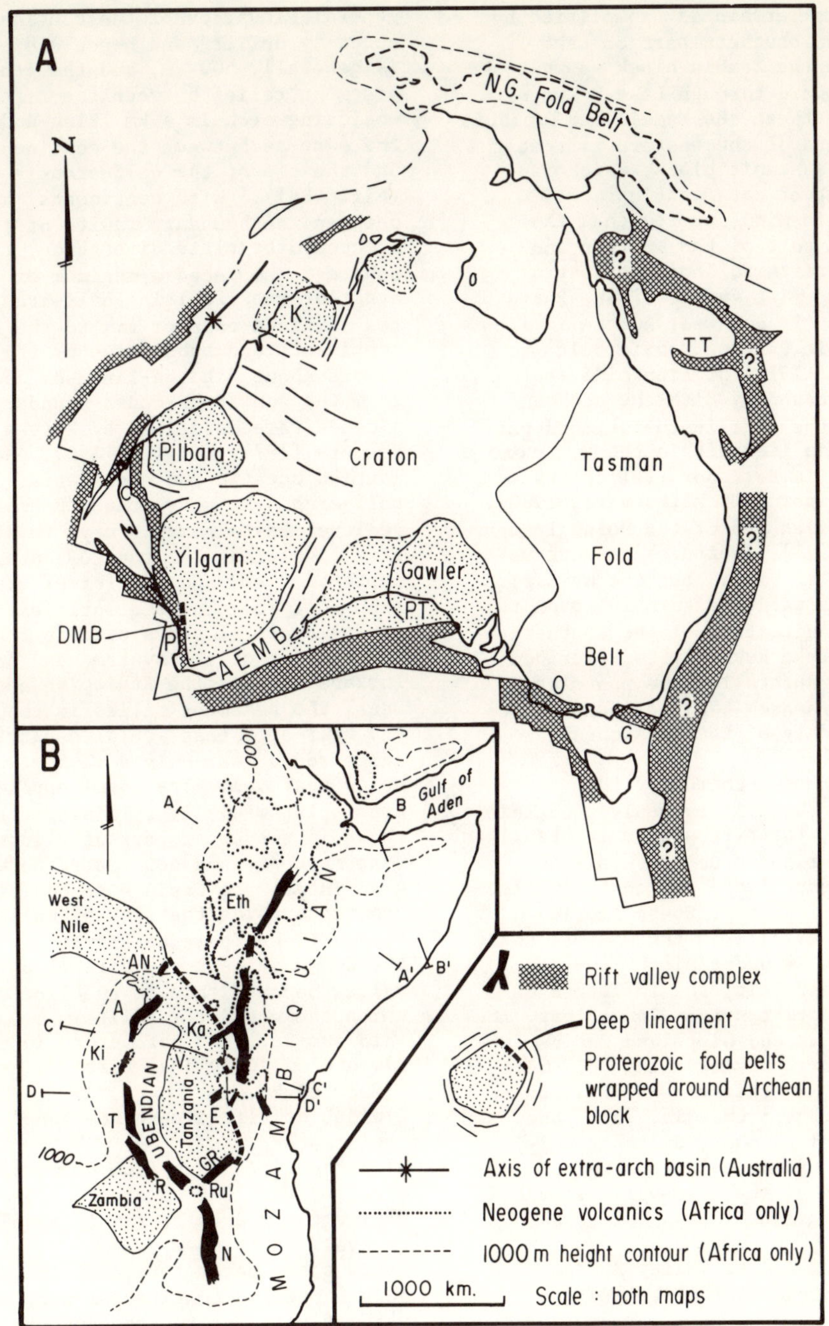

Fig. 3. Rift valley systems and deep structure: Mesozoic to early Tertiary integrated palinspastic reconstructions of Australia (A) and Neogene rift system in East Africa (B). Dotted lines in New Guinea indicate present features only. Australian structure from Plumb [1979] and Wellman [1976], rift systems from Veevers and Cotterill [1976, 1978], Symonds and Cameron [1977], Talwani et al. [1978, 1979], Jongsma and Mutter [1978], and Taylor and Falvey [1977]. African structure from Kröner [1977], and International Tectonic Map of Africa [1968]. In A, features (anticlockwise from northwest) are K Kimberley Block; C Carnarvon Basin; DMB Darling Mobile Belt; P Perth Basin; AEMB Albany-Esperance Mobile Zone; PT Polda Trough; O Otway Basin; G Gippsland Basin; TT Townsville Trough. In B, features (anticlockwise from the north) are Eth Ethiopia; AN Albert Nile rift; A Albert rift; KI Lake Kivu; T Lake Tanganyika; R Lake Rukwe; N Lake Nyasa; RU Rungwe; GR Great Ruaha Valley; E Lake Eyasi; V Lake Victoria; Ka Kavirondo rift. Lines show the location of the sections of Fig.4.

rifts for Proterozoic terrain is exemplified by
the deflection of the southern part of Lake
Tanganyika away from the Zambia block so that the
rift continues southward through Lake Rukwe within the narrow zone between the Zambia and Tanzania
blocks, and by the end of the eastern rift at the
eastern edge of the Tanzania block, which is
marked by a lineament of cataclastic rocks.

With this pattern in mind, we see that the
location of at least part of the western and
southern Australian rifts follows the East
African pattern: the rift system of the Perth
and Carnarvon Basins of southwest Australia lies
within the Proterozoic Darling Mobile Belt
[Glikson and Lambert, 1976] or alongside its
locally mylonitized boundary with the Archean
Yilgarn block; and the rift in the western part
of the southern margin lies within the Proterozoic
Albany-Esperance Mobile Belt parallel to its
boundary with the Yilgarn Block [Veevers, 1980].
Farther east, the failed arm of the Polda Trough
[Geol. Soc. Aust., 1971] penetrates a short way
into the Archean Gawler block, but the main rift
is deflected southeastward, presumably round the
block, recalling the situation of the southern
Lake Tanganyika and Lake Rukwe rifts described
above. In northwest Australia, the postulated
rift and related extra-arch basin trend parallel
to the presumed boundary of the (?)Archean
Kimberley block.

Thus, in the western two-thirds of the
Australian continent that is obviously underlain
by Precambrian rocks, the rift system is clearly
influenced by Precambrian structure, as round
much of Africa [Kennedy, 1965] and on the reflected Atlantic margin of part of South America
[Fyfe and Leonardos, 1973]. In the eastern third
of the continent that is underlain by the
Phanerozoic Tasman Fold Belt, the influence of
deep structure on the rift system in unclear. The
easterly trending Otway and Gippsland Basins cut
sharply across the northerly trending fold belt
but each is a failed arm that ends blindly on
either side of Melbourne; that is, each penetrates across the grain a short distance only.
The postulated rifts of the eastern margin,
except the failed arm of the Townsville Trough,
grossly parallel the grain.

Actualistic Model of a Rift Valley System
and its Succeeding Rifted Continental Margin

An actualistic model of the geodynamic development of a rift valley system and its succeeding
rifted continental margin, derived from East
Africa and the Red Sea and Gulf of Aden, can be
assembled from elements given by Kinsman [1975],
Rabinowitz and La Brecque [1977], Veevers [1977a,
b] and Veevers and Cotterill [1976, 1978], and
elaborated by Veevers [in prep.] (Figure 4). The
rift valleys are some 60 km wide, and rift divergence zones at least 300 km wide; the half-width
of the rifted arch, and the width of the succeeding half-arch out to an elevation of 0.5 km - the

general elevation of the continental crust unaffected by uplift, and hence a hinge after breakup -
is generally 500 km, and the general maximum
basement relief between the rift valley and the
enclosing arch is 4 km [King and Williams, 1976],
the same as between the continent-ocean boundary
and the top of the half-arch in the Gulf of Aden
(Fig. 4BB'). With continental breakup, located
between the boundary faults of the rift valley,
succeeded by plate divergence, the oceanic lithosphere is emplaced a maximum of 4 km below the
crest of the resulting half-arch, which constitutes a raised edge or rim to the newly defined
continent, and subsidence by thermal contraction
pivots about a hinge-line some 500 km distant
from the continent-ocean boundary. With a subsidence rate described by curves derived from
Veevers [1977b, figs. 5 & 6], and an assumed
coupled oceanic and continental lithosphere, the
half-arch 4 km above the COB emplaced at or about
sealevel takes some 40 m.y. to sink below sea-
level; ultimately, the COB subsides to a final
depth of 6 km, and the former crust of the half-
arch to 2 km. These quantities apply to the
crestal part of the system in Ethiopia; at the
lowest part of the system, in the saddle of Lake
Turkana between the Ethiopian and Central Plateaus, the basement relief is about 1.5 km, and
the half-arch that would result from plate divergence would last only some 5 m.y. before sinking
below sealevel. The model applies to an original
arch split medially by a single rift valley, as
shown in the upper part of Figure 4. Multiple
systems contain a depocentre between the arches
- an inter-arch basin - in addition to the depocentres outside the arches, called extra-arch
basins (lower part of Figure 4). Furthermore,
the COB must mark the site of a former rift
valley because this is the line of weakness along
which the continent splits.

In summary, the parameters in this model are
1)width of individual rift valleys: 30-90 km,
   mode 60 km.
2)width of rift divergence zone: several hundred
   kilometers.
3)distance of hinge from COB: mode of 500 km.
4)relief from COB to crest of ½-arch: 1.5 to 4 km.
5)time for ½-arch or rim to subside below sea-
   level: 5 to 40 m.y.
All these parameters are, in principle, measurable on rifted continental margins. In the next
part of this review, the Australian margins are
scrutinized for values of these parameters. In
particular, the postulated existence of a continental rim as a successor of the half-arch, and
its duration above sealevel, should be reflected
in the stratigraphical relations of the post-
breakup sediment to the surface, shaped by events
up to and including breakup, on which it rests.

As suggested in Figure 4 (BB' and related sections), the main relation at the rim will be the
lateral termination of strata at their depositional limit, called lapout. The kind of lapout,
where the deposited strata are initially horizon-

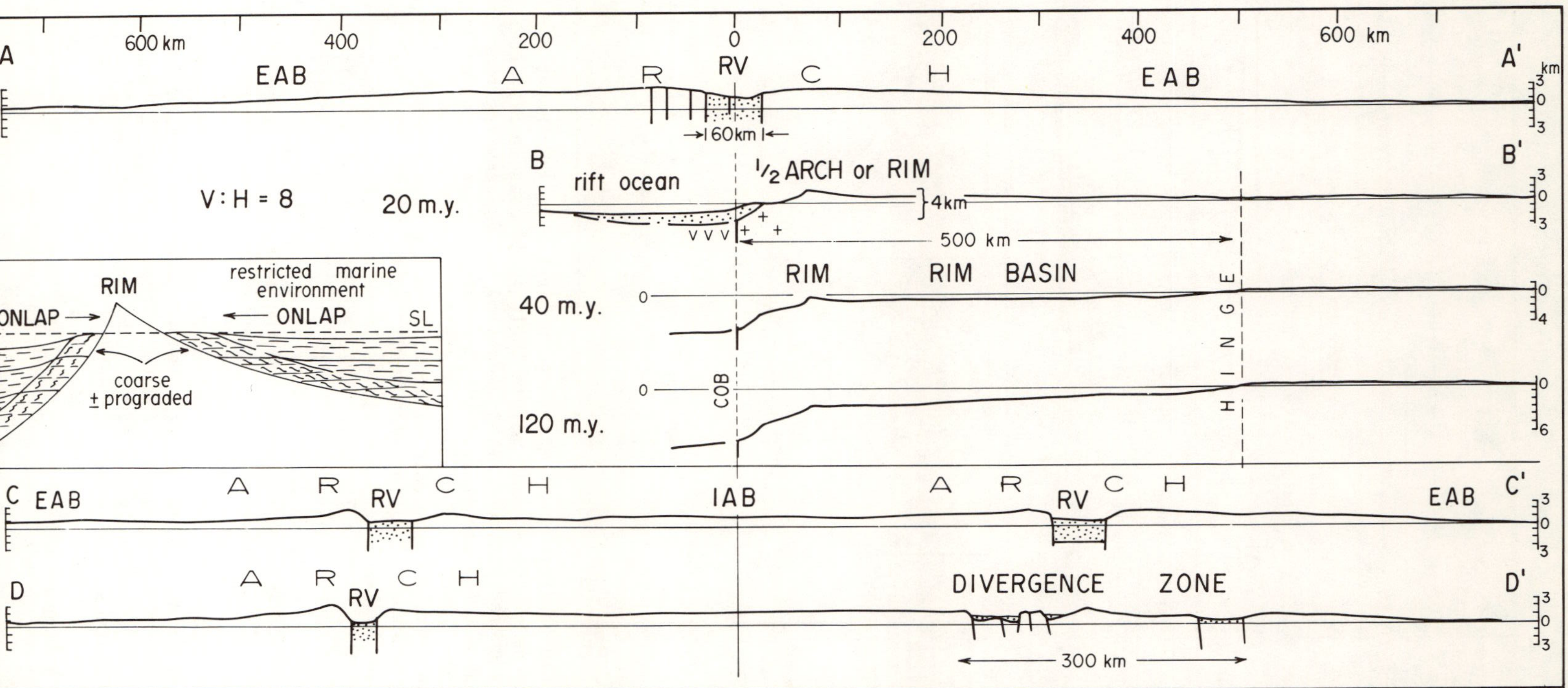

Fig. 4. Cross-sections of East Africa, located in Fig. 3, and future projections, arranged to show the geodynamic development of a rift valley system and its succeeding rifted continental margin. AA': Ethiopian Plateau split medianly by rift valley (RV) and flanked by extra-arch basins (EAB). Topographic information from Map of the World [1968]. BB': Gulf of Aden and Somalia, today, 20 m.y. after continental breakup, with half-arch or continental rim 4 km above the COB, which lies about 500 km distant from the 0.5 km height contour. The two sections below are projections 40 m.y. and 120 m.y. after breakup, and were derived from BB' by subsidence about a hinge 500 km from the COB to depths, at the COB, of 4 and 6 km. CC': Central Plateau, comprising the Western Rift, Lake Victorian basin – an inter-arch basin (IAB) – and Eastern Rift. DD': Central Plateau, comprising the Western Rift, Tanzania Plateau, and rift divergence zone. Inset shows the facies developed at a subsiding rim.

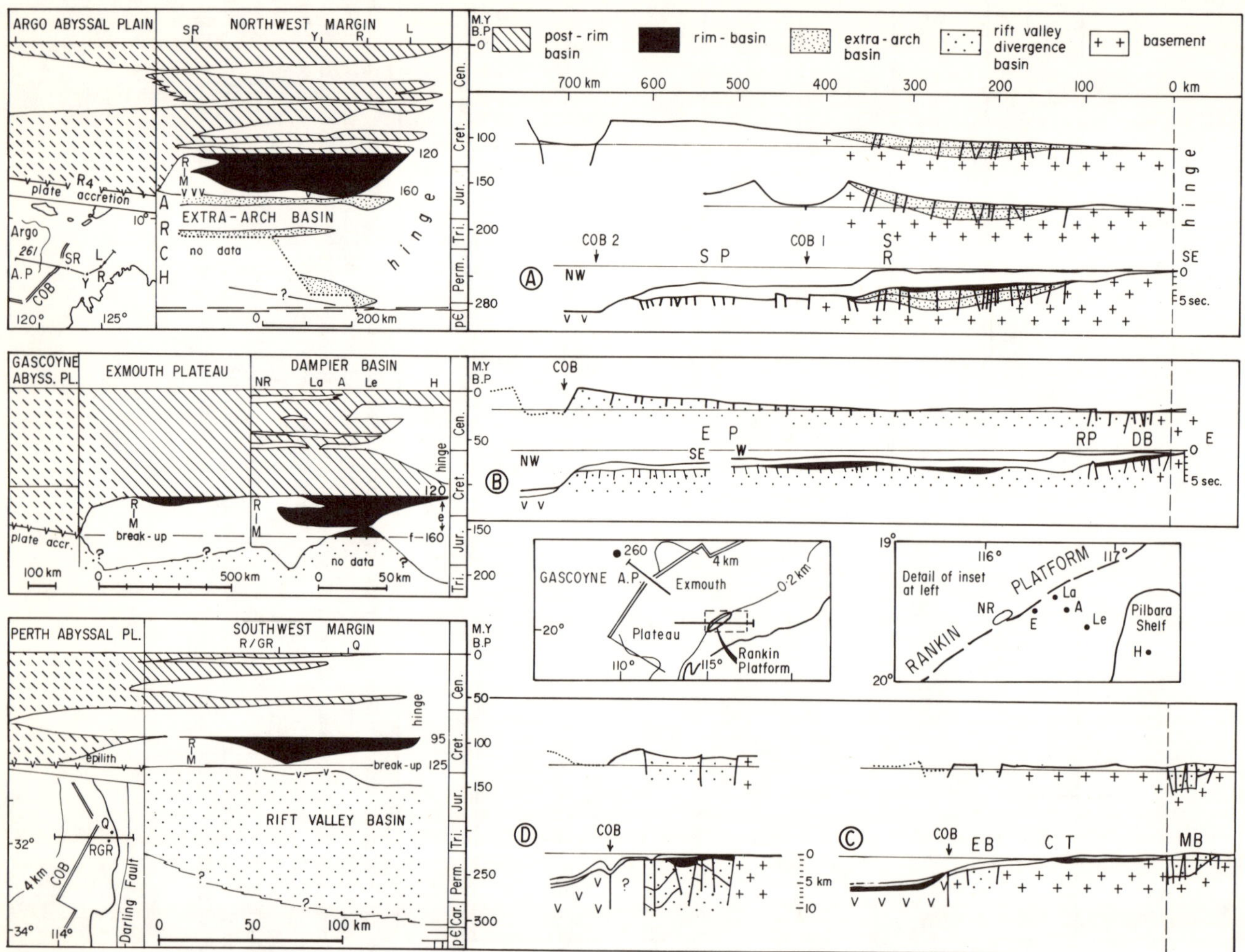

Fig. 5. Cross-sections and time-space diagrams of the western margin. A: northwest margin: on the right, seismic time (2-way) sections of the present margin (below), located in Fig. 2, and modified from Powell [1976, fig. 3], and restored sections at breakup (160 M.Y.), assuming that most of the Scott Plateau (SP) is a post-breakup volcanogenic structure of oceanic crust or an epilith (middle) or is continental crust (top). SR - Scott Reef-1 Well. COB 1 and 2: alternative positions of the continent-ocean boundary. On the left, a time-space diagram, assuming that the Scott Plateau is an epilith, modified from Veevers and Cotterill [1978, fig. 16A]. Drill sites: SR Scott Reef; Y Yampi; R Rob Roy; L Londonderry. Oblique broken lines indicate sediment deposition oceanward of the continental rim. B: Exmouth Plateau (EP), Rankin Platform (RP), and Dampier Basin (DB): on the right, seismic time section of the present margin (below) located in Fig. 2 and in insets, and restored section (above) at breakup (160 M.Y.), modified from Veevers et al. [1974, fig. 2, line SR] on west, and from Veevers and Cotterill [1978, fig. 10B] on east; on the left, a time-space diagram of the margin, from Veevers and Cotterill [1978, fig. 16B], with location in insets on right. Drill sites: NR North Rankin; E Egret; La Lambert; A Angel; Le Legendre; H Hauy. C: depth section of the Carnarvon Terrace (CT) and Merlinleigh Basin (MB), located in Fig. 2, modified from Veevers and Cotterill [1978, fig. 11] and Symonds and Cameron [1977]; above, restored section at breakup [125 M.Y.]. D: depth section of the southwest margin (below), located in Fig. 2, modified from Veevers and Cotterill [1978, fig. 12], and restored section (above) at breakup (125 M.Y.): and time-space diagram (left), modified from Veevers and Cotterill [1978, fig. 16C]. Drill sites: R-GR Roe/Gage Roads; Q Quinns Rock. The rift valley stage consists of diverging individual rift valleys, called a diverging rift valley basin.

tal, and the surface on which they are deposited is inclined, is called onlap; the other form of basal lapout, called downlap, comes about when initially inclined strata terminate downdip against an initially horizontal or inclined surface. Vail et al. [1977, p.57,58], who discuss these terms, note that "Onlap or downlap usually can be readily identified. However, later structural movement may necessitate the reconstruction of depositional surfaces. In areas of great structural complication, the discrimination between onlap and downlap may be practically impossible, and the worker may be able to determine only that the strata are in a baselap relation". This kind of uncertainty is inherent in Figure 4 (BB' and related sections). Whereas, during the first 40 m.y., the strata are deposited, on either side of the rim, essentially horizontally over inclined surfaces, and so express onlap, subsequently, by the continued subsidence of what was the rim by rotation about a hinge, these strata may become inclined while correspondingly the surface on which they were deposited becomes horizontal, to simulate the relation of downlap. This is the kind of situation described above by Vail et al. [1977], in which the geometry alone indicates nothing more than baselap or the termination of strata at the lower boundary of a depositional sequence. This ambiguity can be resolved by seeking facies trends along a shoreline that indicates the local derivation of strata from the postulated rim. In the case of onlap (Fig. 4, inset), the sediments will coarsen towards the shoreline, and, if deposition was in deltas, may dip, by progradation, away from it. In the case of downlap, neither feature will occur. Another kind of test will be whether or not the basin fill was deposited in an open or a restricted marine environment. An open environment would tend to suggest the absence of a rim, whereas a restricted environment could point to the presence of a rim, but not uniquely because the juvenile ocean of which the sea over the margin is only a part may itself have a restricted circulation.

Western Margin

A generally clear set of magnetic anomalies off the western margin (Fig. 2) [Heirtzler et al., 1978; Larson et al., 1979; Johnson et al., 1980; Markl, 1978a,b] indicates that the margin broke up in two stages: the northwest a short time before M-25 (=153 M.Y.), by extrapolation to the COB calculated to be 160 M.Y., in the Jurassic; and the southwest a short time before M-10, again by extrapolation to the COB calculated to be 125 M.Y., in the Cretaceous. Larson et al. [1979] show that the boundary between the Jurassic and Cretaceous inception of spreading, placed by Veevers and Cotterill [1978] at the southern flank of the Exmouth Plateau, lies farther north, in the vicinity of DSDP Site 260, along the western flank of the Exmouth Plateau. Regardless of the precise

time of breakup, the entire margin seems to have started the preliminary stage of rift valley formation at the same time in the Late Carboniferous and Permian, and faulting, as the chief expression of structure, likewise started at the same time in the Triassic [Johnstone et al., 1973; Powell, 1976]. The rift system of the western margin thus involves Late Carboniferous to Middle Jurassic (300-160 M.Y.) sediments in the north, and Late Carboniferous to Early Cretaceous (300-125 M.Y.) sediments in the south. A second difference is that the rift system in the north was simple, with the COB indicating the preceding line of the simple rift system, whereas that in the south was multiple, with the COB indicating the outer part, and the Perth and Carnarvon Basins representing the inner part. The rift system extends to the Timor Sea between Timor and Australia [Laws and Kraus, 1974; Balke and Burt, 1976] but its outer part is obscured by the modern diastrophism that accompanied plate convergence in the Timor area.

The precise position of the COB along the northwestern margin is poorly known. The COB is generally drawn along the lower part of the continental slope [Stagg, 1978; Allen et al., 1978; Exon and Willcox, 1978] but Veevers and Cotterill [1978] suggested that the outer part of the Scott Plateau and the northwestern tip of the Exmouth Plateau are post-breakup volcanic excrescences (or epiliths) that lie seaward of the COB, as shown in Figure 2, a point discussed by Stagg and Exon [1979] and Veevers [1979]. The reconstructions in Figure 5A show both interpretations.

The northwesterly baselap of the Permian to Jurassic fluviodeltaic to shallow marine sediments in the Scott Reef area (stippled in Fig. 5A), and dips in these sediments that suggest a northwest source [Allen et al., 1978] indicate onlap against a broad uplift on the west. The main faults were initiated in the latest Triassic to Early Jurassic, were rejuvenated near the end of the Middle Jurassic, and, together with the widespread eruption of basalt, mark continental breakup, in harmony with the age of the oldest adjacent seafloor in the Argo Abyssal Plain. Several dredge hauls of volcanic rocks from the slope of the Scott Plateau [Hinz et al., 1978] represent flows either over postulated continental crust or are parts of the postulated epilith. Thereafter the half-arch subsided so that the part of it sampled by Scott Reef-1 Well was covered by the sea in the latest Jurassic (140 M.Y.) and then, after a short hiatus, completely in the Late Neocomian (120 M.Y.). All these events are consistent with the interpretation of the pre-breakup sediments as an extra-arch basin succeeded by the restricted marine sediments of a rim basin until open marine sediments covered the whole area in the Late Cretaceous, as shown in the cross-sections and time-space diagram of Figure 5A. The precise duration of the rim basin is ambiguous. On the one hand, the overlap of the Scott Reef area in the latest Jurassic (140 M.Y.) would suggest that

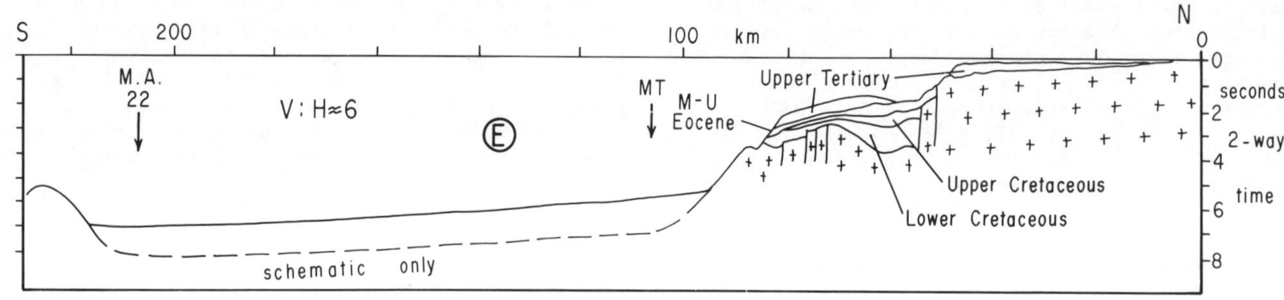

Fig. 6. Time (seismic) section of western part of southern margin, located in Fig. 2, and modified from Cooney et al. [1975] and Talwani et al. [1979]. MT: magnetic trough.

the rim basin lasted only 20 m.y.; on the other hand, the retreat of the sea, from 140 to 120 M.Y., possibly due to eustatic fall - the highest stand in the Jurassic, according to Vail et al., [1977, p.91], was at 140 M.Y. - and the continuation east of Scott Reef of "a restricted marine environment...through the Early Cretaceous until the Scott Plateau had subsided sufficiently to no longer form an effective barrier to the basin" [Allen et al., 1978, p.31] suggest that the rim basin lasted at least until the definitive marine onlap of Scott Reef at 120 M.Y. This second possibility is shown in the time-space diagram of Figure 5A, and agrees with the 40 m.y. maximum duration of the rim in the actualistic model. The distance of the COB from the hinge - 425 km if most of the Scott Plateau is an epilith, 670 km if it is continental - likewise corresponds with that of the model. Later depositional episodes of this and other parts of the margin are described by Quilty [1977] and Apthorpe [1979].

The broad margin of the Exmouth Plateau-Rankin Platform-Dampier Basin (Fig. 5B) has been affected by both the Jurassic and Cretaceous stages of spreading. By extrapolation from the Rankin Platform and Dampier Basin, Willcox and Exon [1976], Exon and Willcox [1978], and Wright and Wheatley [1979] forecast the Permian and younger stratigraphy of the Exmouth Plateau, which is now being tested by the drill. A mid-Jurassic unconformity separates fault blocks below from essentially unfaulted sediments above, and is interpreted as reflecting continental breakup [Powell, 1976]. The thick pre-breakup fill of the Dampier Basin is interpreted by Veevers and Cotterill [1976] as a rift valley or rift divergence zone fill - its structure resembles that of the rift divergence zone of northeast Tanzania. Lenses of sediment immediately above the unconformity are interpreted as rim basins, and the one beneath the crest of the plateau was deposited by progradation mainly from the south, probably including a local source in the uplift of the southern flank of the plateau during transform fault motion from 125 to 120 M.Y. [Veevers and Powell, 1979] during the Cretaceous stage of spreading. The pre-breakup reconstruction (Fig. 5B) shows that the entire margin was sub-

jected to diffuse rifting, with some concentration in the Rankin Platform and Dampier Basin; uplift of the western flank produced a broad inter-arch basin over the entire plateau, which, after breakup, then became a rim basin separated from that of the Dampier Basin by the Rankin Platform. Of measurable parameters, the duration of the inner rim basin of 40 m.y., from 160 M.Y. to 120 M.Y., when the Rankin Platform became covered by marine sediment [Powell, 1976], again matches the maximum value of the model precisely. As yet, direct evidence of a sediment source from the northwest is lacking. Facies trends in the Middle Triassic sequence 'suggest a possible northwesterly [sediment] source' [Crostella and Barter, 1980], and 'seismic data correlation over the Exmouth Plateau to the west of the Kangaroo Trough [immediately west of the Rankin Platform] indicates that most of the Jurassic section is missing below the mid-Jurassic unconformity, implying aerial exposure of the plateau during the Late Jurassic. By contrast, the sharply defined nature of most of the fault blocks infers little erosion since their formation. These observations could be explained by two phases of tectonic movement; an initial one in which the plateau was elevated and erosion occurred and a later one at which time the present main fault block structures were formed. By analogy with the Dampier Sub-basin, the first phase, which probably included block-faulting, would have occurred in the late Middle Jurassic and the second phase at the end of the Jurassic prior to the Cretaceous deposition' [Wright and Wheatley, 1979, p.25].

South of the Exmouth Plateau, the Carnarvon Terrace (Fig. 5C) lies between two rift valleys: the Permian Merlinleigh Basin on the east, and the southern part of the Dampier Basin (the Exmouth Sub-basin) on the west [Symonds and Cameron, 1977], and faces the Cuvier Abyssal Plain, which started being generated 125 M.Y. [Larson et al., 1979; Roots et al., 1979; Johnson et al., 1980]. Drilling on the shelf shows that the oldest post-breakup sediment that overlaps Early Carboniferous basement is Aptian (~110 M.Y.), from which it is inferred that the rim underlain by the Exmouth Sub-basin, which was onlapped also from the west, did not become submerged until some time later in the Late Cretaceous, perhaps 30 to 50 m.y. after breakup. A late structural event in this area was broad folding associated with reverse faults in the Late Cenozoic [Geol. Survey WA, 1975; Van de Graaff et al., 1975]. Farther south, in the Perth Basin (Fig. 5D), oil exploration wells show that the rim basin there lasted from breakup at 125 M.Y. to 95 M.Y., or for 30 m.y. The COB in this area was thought earlier to be dated by the oldest sea-floor spreading magnetic anomaly, which Markl [1974] identified as M-11 (=125.5 M.Y.) but this anomaly is now thought to be due to the edge effect [Markl, 1978a; Johnson et al., 1980]. The date of 125 M.Y. for the COB remains nevertheless because extrapolation from the oldest adjacent anomaly (M-8), and the age of the breakup unconformity both point to this date. A geomorphic analysis of the southwestern Australian region, in the light of rift tectonics, is given by Fairbridge and Finkl [1978] and Finkl and Fairbridge [1979].

The crustal structure of the western margin has been interpreted from gravity anomalies by Branson [1974], Symonds and Willcox [1976], and Willcox [1977], and this work indicates that the continental crust thins across the margin from 30 km at an average distance of 200 km from the COB to 18-22 km at the COB itself. A detailed gravity study of the Cuvier Abyssal Plain-Carnarvon Terrace area by Roots et al. [1979] suggests that the continental and oceanic crusts are about 17 km thick at the COB, located at a water depth of 3 km, and that the oceanic crust thins seaward from the COB for a distance of 80 km before reaching its normal thickness of 7 km; this thick oceanic crust is taken to be responsible for the elevated position of new oceanic crust in young ocean basins, as described by Veevers [1977b]. Larson et al. [1979] made four refraction probes of the Cuvier Abyssal Plain to within 70 km of the COB, and found that it contains oceanic crust of normal thickness; unfortunately, they did not probe closer to the COB in the area crucial to Roots et al.'s [1979] model.

The only other determinations of crustal structure along the western margin are refraction probes by Hawkins et al. [1965a] in the area off Perth, but they failed to reach the top of the mantle. Onshore, between Perth and the Yilgarn block, Mathur [1974] has interpreted from seismic and gravity data an anomalously dense (ρ = 3.10 gm cm^{-3}) lowermost continental crustal layer that thickens over a distance of 600 km from zero beneath Kalgoorlie to a maximum of 30 km near the Darling Fault. Glikson and Lambert [1976] point to the possibility that this layer "formed by accumulation of basic magma along the upper mantle-lowermost crust boundary during the breakdown of Gondwanaland in the Late Palaeozoic and Early Mesozoic", and the width of this layer corresponds with that of the half-arch in the actualistic model.

The crustal structure of the Wallaby Plateau/ Zenith Seamount and the Naturaliste Plateau is obscure. Direct evidence from dredging indicates that at least the uppermost 2 km of the Wallaby Plateau consist of volcanogenic rocks [von

Stackelberg et al., 1980], consistent with it being an epilith, as suggested by Veevers and Cotterill [1978]. Deep sea drilling on the Naturaliste Plateau revealed volcanogenic rocks too but a dredge haul of deeply weathered crystalline rocks "of continental affinities" [Heezen and Tharp, 1973] points to the possibility that part if not all of the Naturaliste Plateau is continental - hence the alternative positions of the COB shown for this area in Figure 2.

Vulcanism along the rift system that preceded the western margin is known, so far, from isolated centres of alkaline vulcanism [Veevers and Evans, 1975], as in the western rift of Africa, from rare fragments of silicic volcanics in the Late Jurassic Yarragadee Formation of the Perth Basin [Veevers and Larsen, in prep.], and from extrusion of basalt at the time of breakup. Dredge hauls and cores containing volcanic or volcanogenic rock from the margin and adjacent ocean floor are reported by Cook et al. [1978], Hinz et al. [1978], and von Stackelberg et al. [1980].

Southern Margin

The southern margin has been much less intensively explored than the western margin, and offshore drilling has been done only in the eastern half. In compensation, two complementary surveys, by Boeuf and Doust [1975] and Talwani et al. [1978, 1979], provide excellent data, and these surveys, augmented by work by the Bureau of Mineral Resources [Willcox, 1978; Fraser and Tilbury, 1979] and others [Griffiths, 1971; Falvey, 1974; Deighton et al., 1976; Denham and Brown, 1976] comprise the source of information for this review. The margin extends eastward from the southwestern tip of Australia to longitude 132°E, where it splits off the failed arm of the Polda Trough, then bears southeastward to be offset by transform faults past Tasmania and the South Tasman Rise. Between longitude 115°E and 125°E (Fig. 6), the margin has a narrow shelf, a steep (≤8°), locally terraced, continental slope which passes at 3.5 km into a wide smooth continental rise, and a rough, ridged, 5 km-deep ocean floor, the Diamantina Zone. Farther east, the rise narrows by the buildup of the Ceduna Terrace (Fig. 7,F) [Fraser and Tilbury, 1979], and the Diamantina Zone gives way to isolated seamounts in an abyssal plain at a depth of 5.5 km. East of longitude 135°E, the slope becomes steep and is crossed by deep canyons [von der Borch, 1968], and then widens again off the Otway Basin (Fig. 7,G).

Talwani et al. [1978, 1979] have mapped a magnetic trough along the western half of the margin (Fig. 2), which marks the northern boundary of a smooth magnetic field in the east and a progressively more disturbed, but not aligned, one toward the west, all called a magnetic quiet zone. This zone is bounded southward at the COB by the oldest identified seafloor spreading magnetic anomaly 22 (=53 M.Y.). The quiet zone continues southeastward seaward of the Otway Basin and western side

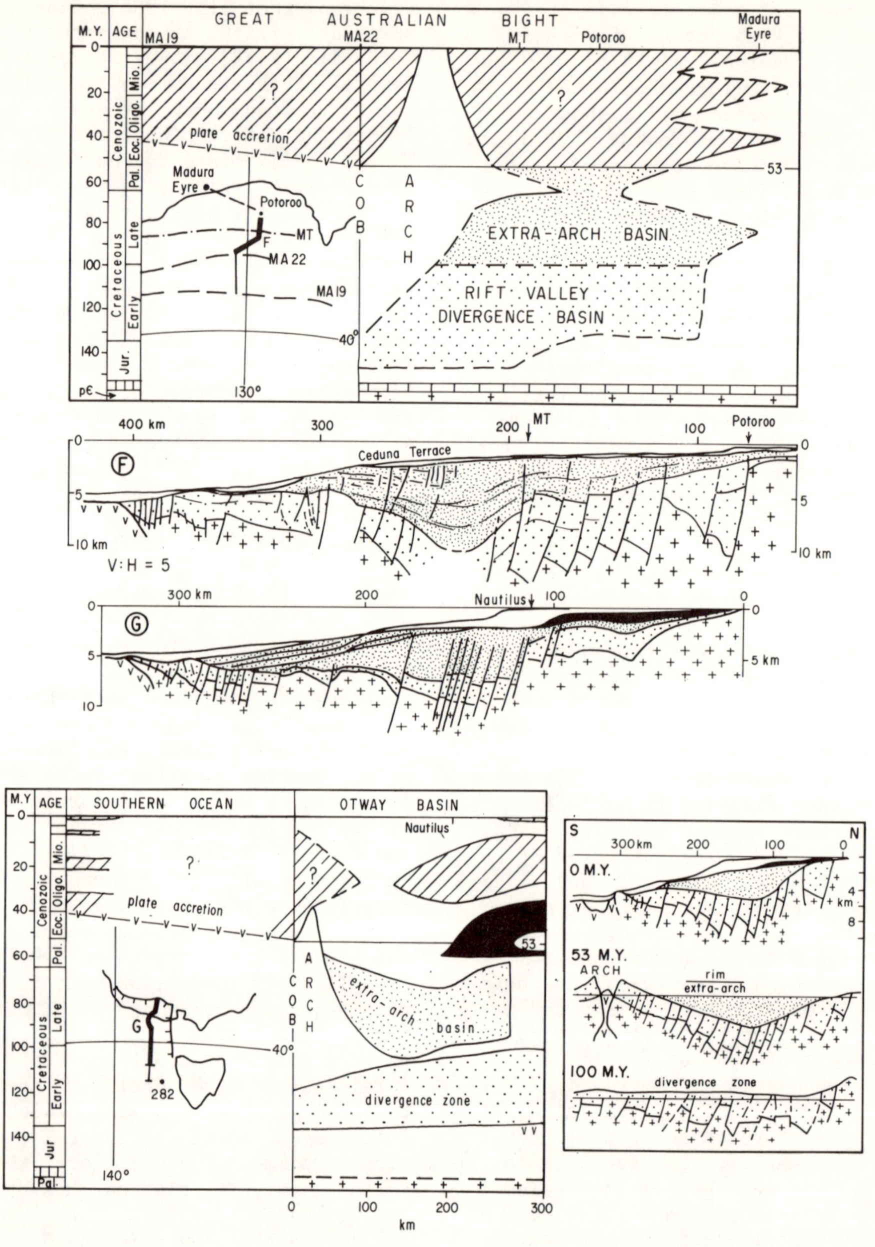

Fig. 7. Southern margin. Above - time-space diagram and location map of Great Australian Bight.
Location map shows also the section line (heavy line) of F, the magnetic trough (MT), and seafloor
spreading magnetic anomalies. Details of time-space diagram from Boeuf and Doust [1975, fig. 5],
Lowry [1976], and Fraser and Tilbury [1979, fig. 7]; information beyond the Potoroo well by extrapolation of seismic profiles. F - Cross-section of the Great Australian Bight Basin to the continent-
ocean boundary, modified from Boeuf and Doust [1975, fig. 8] by the later work of Willcox [1978] and
Fraser and Tilbury [1979] that shows a thinner Cenozoic section and a correspondingly thicker Late
Cretaceous section. Location (heavy line) above. G - Cross-section of the Otway Basin to the continent-ocean boundary, modified from Boeuf and Doust [1975, fig. 8] by the work of Denham and Brown
[1976] that shows, as in the west, the Cenozoic section to be thinner and the Late Cretaceous correspondingly thicker. Location (heavy line) below. Below - Time-space diagram and location map of
Otway Basin. Details out to Nautilus well from Boeuf and Doust [1975, fig. 4]; beyond Nautilus, by
extrapolation of seismic profiles. Side - Schematic cross-sections of the southern margin, modified
from Boeuf and Doust [1975, fig. 9], updated by Denham and Brown [1976] and Fraser and Tilbury [1979],
showing the succession of rift divergence basin, extra-arch basin, rim basin, and post-rim basin.

of Tasmania [Weissel and Hayes, 1972]. On the
basis of seismic refraction probes, including work
by Hawkins et al. [1965b], Talwani et al. find
four different kinds of crust in the area south of
the magnetic trough: oceanic seaward of the quiet
zone, and three groups in the quiet zone: Group I
(mainly in the eastern and central parts), in which
the velocity of the main crustal layer ranges from
6.5 to 7.1 km sec^{-1}; Group II (mainly in the western part), 7.2 to 7.5 km sec^{-1}; and Group III
(central part), 5.8 to 6.2 km sec^{-1}. Those groups,
whose distribution overlaps each other, constitute
an exceptional variety of crustal layers - some
typically continental, others oceanic, though much
thicker than is typical, others still neither continental nor oceanic - which Talwani et al. interpret as indicating a unique, rift crust "comparable with the deep structure of the Rhine Graben"
generated in the 200 km-wide continental rift that
was later split evenly between Australia and
Antarctica. The quiet zone that extends southeastward to the west side of Tasmania is explained by
Roots [1976] as due to narrow compartments of
oceanic crust generated alongside a continental
margin oblique to the spreading direction.

The shallower structure of the west is shown in
Figure 6, and of the central and eastern parts,
augmented by time-space diagrams, in Figure 7. The
region is dominated, between longitude 115° and
135°E, by a broad saddle, seen in the COB and in
the crest of the mid-ocean ridge, that is flanked
by crests at Naturaliste Plateau and south of
Tasmania. In the west, a terrace underlain by
inferred Cretaceous and Cenozoic sediment, lies
landward of the magnetic trough. Section F (Fig.
7), along the head of the Great Australian Bight,
shows a faulted Late Jurassic and Early Cretaceous
sequence unconformably overlain by a thick lens of
Late Cretaceous sediment, draped by Cenozoic sed-
ment. The same elements appear eastward, in the
Otway Basin (Fig. 7,G). Boeuf and Doust's [1975]
interpretation of the development of these sections, which I have labelled in terms of the actualistic features described in the model, with some
changes in the ages required by the later work of
Denham and Brown [1976] and Fraser and Tilbury
[1979], is given in Figure 7 (side). I call the
first stage, to 100 M.Y., a divergence zone in
place of Boeuf and Doust's [1975] rift valley because its width of more than 300 km greatly exceeds that of any individual rift valley but agrees
with that of the divergence zone of Tanzania, as
does its structure. Contemporaneous vulcanism
supplied an appreciable part of the accumulation
of fluvio-lacustrine sediments. During the Late
Cretaceous, what was to become the "Outer Continental Margin Ridge" [Bouef and Doust, 1975, p.40]
rose as part of a rifted arch along which rifting
was concentrated along a single axis in contrast
to the distributed rifting of the previous divergence zone. Shallow to deeper marine and fluviodeltaic sediments accumulated in the extra-arch
basin behind the arch to thicknesses exceeding 5 km
until the breakup of Australia and Antarctica at

53 M.Y. Thereafter the newly formed continental
rim subsided with the cooling of the lithosphere,
and was onlapped by marine detrital sediment until
it sank below sealevel soon after breakup in the
Great Australian Bight Basin, and some 10 to 15
m.y. after breakup in the Otway Basin. The short
duration of the rim stems from the situation of
this part of the margin in a saddle.

This account departs in some important ways from
that of Falvey [1974] and Deighton et al. [1976].
In devising a model of rifted margins, Falvey
[1974, fig. 11] identified the Early Cretaceous
sequence of the Otway Basin as a "pre-rift basin"
and the Late Cretaceous sequence as a "rift valley
basin", so that the unconformity between these
sequences was correspondingly called a "rift onset
unconformity". In company with Boeuf and Doust
[1975], I regard the Early Cretaceous sequence as
no less a part of the rift system than later ones,
with the result that its boundary with basement is
recognised as the rift onset unconformity, and the
no less conspicuous surface between the Early and
Late Cretaceous sequences as the unconformity that
separates the sequence of the divergence zone from
that of the extra-arch basin. In revising part of
Falvey's [1974] model, Deighton et al. [1976,
fig. 7] recognised that "the early 'rift valley'
section would probably have been deposited on the
flank of the initial rift" but showed this flank
to be below sealevel before breakup, thus implying
that the seaward lapout of sediment is due to downlap, and not to onlap against an outer ridge. On
such a crucial point, hard evidence is not likely
to accrue without deep marginal drilling.

In a novel application of the fission-track dating method to the study of rift systems, Gleadow
[1978] and Gleadow and Lovering [1978] measured
the fission-track ages of sphene and apatite from
mid-Palaeozoic granites of King Island, off the
northwest tip of Tasmania, and found that whereas
the sphene ages match K-Ar ages and indicate emplacement at about 350 M.Y., apatite ages are all
younger by about 80 to 200 m.y. Fission tracks
were not fully retained in the apatite, which has
an annealing temperature about 110°C, until the
Cretaceous, the youngest age being 112 M.Y., or
close to the boundary between the Early and Late
Cretaceous, indicating the latest age that these
rocks cooled through 110°C, probably through the
uplift and removal of overlying rock at a rift
shoulder.

Eastern Margin

Except along the Gippsland Basin, at latitude
38°S, the eastern margin is too little known for a
coherent account to be given. And even along the
Gippsland Basin, published descriptions of the outer part of the margin seaward of the shelf are
lacking. The history of the adjacent oceanic
Tasman Sea basin is also incomplete. According to
Weissel and Hayes [1977], the Tasman Sea started
opening a short time before magnetic anomaly 33
(=76.5 M.Y. - La Brecque et al., 1977), say 80 M.Y.,

and continued to magnetic anomaly 24 (=55 M.Y.),
while the Coral Sea basin [Weissel and Watts, 1979]
began opening at about 62 M.Y. (magnetic anomaly
26-27) and continued also to about 55 M.Y. (anomaly
24). In each basin, the single deep sea drill hole
that reached basement confirms the magnetic ages.
Closure of the geometrically simple Coral Sea basin
appears to be straightforward, but closure of the
Tasman Sea basin, bounded as it is on the east by
the complex and poorly known submarine Dampier
Ridge and Lord Howe Rise, is not. The difficulties
are (1) while at least some parts of the Lord Howe
Rise have been shown by refraction probing to be
underlain by continental crust [Shor et al., 1971]
- it is one of the few possible submarine microcontinents to be positively identified as a microcontinent [Scrutton, 1976] - the COB, in particular
on its western side, is poorly defined. Unlike
the southeastern Australian margin, with its close
uniformly spaced 2- and 4-km isobaths, the isobaths on the western side of the Lord Howe Rise
are widely and unevenly spaced, so that much error
is likely to result from adopting a particular
isobath as the COB, as Weissel and Hayes [1977]
adopt the 3-km isobath. In the absence of an
accurately determined COB, attempts at refining
the reconstruction of Australia and the Lord Howe
Rise would seem to be futile; (2) the truncation
of magnetic anomalies by the southeast Australian
margin at an angle of 45° [Roots, 1975] can be
explained only by there being numerous transform
faults each with a short offset to accommodate
this geometry. But, according to Weissel and
Hayes [1977], the lineation of the magnetic anomalies in the compartment between their fracture
zones 6 and 7 trends uniformly at 330°, without
sensible offsetting by transform faults. From
geometrical arguments of this kind, Roots [1975]
suggests that some 15 m.y. production of seafloor
has been subducted under northern New South Wales
so that the mid-ocean ridge was originally parallel to the southeast Australian margin and the
Lord Howe Rise, with the result that New Zealand
would be reconstructed some 2000 km farther north
than all fits except that of Jones [1971]. The
expected volcanic products of the postulated subduction are not found.

Added to these difficulties on the oceanic side
are those from the adjacent continental margin,
which, with its steep slope, locally as steep as
20°, and undiscerned pre-breakup stages, except
the failed arm of the Gippsland Basin, differs
radically from what is regarded as a typical rifted margin. Two attempts at reconciling these differences are (a) that the entire pre-breakup stage
of development by incipient divergence now lies on
the Lord Howe Rise by breakup along the western
boundary fault of a rift valley system [Jongsma
and Mutter, 1978; Mutter and Jongsma, 1978], as
shown in Figure 2; and (b) that strike slip motion
was the dominant development in the northern Tasman
Sea during the first 15 m.y. of spreading [Shaw,
1978].

Further 'anomalous' features in southeast
Australia are (1) the persistent elevation of the
ranges behind the margin, possibly since the Early
Cretaceous, as mentioned earlier; (2) the fairly
continuous history, since about 90 M.Y., of vulcanism [Wellman and McDougall, 1974a,b; S.Y. Wass,
pers. comm.]; and (3) the zone of present seismicity [Sykes, 1978]. Any satisfactory account of
the development of the southeastern margin must
take all these factors into account.

Two parts of the southeastern margin are illustrated here. Burke and Dewey [1973] identified the
Gippsland Basin (Fig. 8) as a failed arm of the
Tasman Sea basin, and Mutter and Jongsma [1978]
detail the possible spreading geometry. The cross-
section AA' in Figure 8 shows an Early and Late
Cretaceous graben or rift valley, with appropriate
width of 60 km, succeeded by a Cenozoic basin that
has expanded across the graben probably by regional isostatic adjustment and flexure of the graben
in the manner outlined by Beaumont and Sweeney
[1978]. The longitudinal section (BB'B") in Figure 8 contains data of varied reliability: the
boxed part of BB' is controlled by drilling and
processed seismic profiles, and what lies beneath
is a simple downward projection; section B'B",
derived from a single-channel seismic profile, is
highly interpretative, and the only secure parts
of this interpretation are the shape of the sea-
floor, the shape of the oceanic basement, and the
dominantly westward dip of the deepest reflections,
which provide the basis for the speculation that
the outermost Gippsland Basin is part of a subsided arch and successor marginal rim to which are
related the Early Cretaceous rift valley fill, Late
Cretaceous extra-arch basin, and Late Cretaceous to
Eocene rim basin (Fig. 8C). In this scheme, the
rim lasted some 30 to 40 m.y., in harmony with the
actualistic model, before the widespread Oligocene
marine transgression of the Lakes Entrance Formation and Gippsland Limestone. In this scheme, the
breakup unconformity lies within the Late Cretaceous
sequence, but is not seen. In a radically different scheme, the unconformity between the Eocene and
older sediments below and the Oligocene sediments
above is recognised as the breakup unconformity,
implying continental breakup in the latest Eocene,
about 40 M.Y. This scheme would lend support to
Jones' [1971] and Jones and Roots' [1974] postulated
later opening of the Tasman Sea. Against later
opening, however, are the magnetic anomaly identi-
fications [Weissel and Hayes, 1977] and the basal
sediment age of 65 M.Y. at DSDP Site 283 [Partridge,
1976]. A reconciliation is provided by a third
scheme [Griffiths, 1971], in which the Gippsland
Basin is affected by movement, from the mid-
Cretaceous to the Eocene, of a Tasmania 'sub-plate'
relative to the main Australian plate, so that the
regional unconformity of the Gippsland Basin is
relatable to adjustments between Australia and
Tasmania during the early stage of opening of
Australia and Antarctica.

The en echelon anticlines in the Gippsland Basin
(that now serve as hydrocarbon reservoirs) were
generated by east-west right lateral shear during

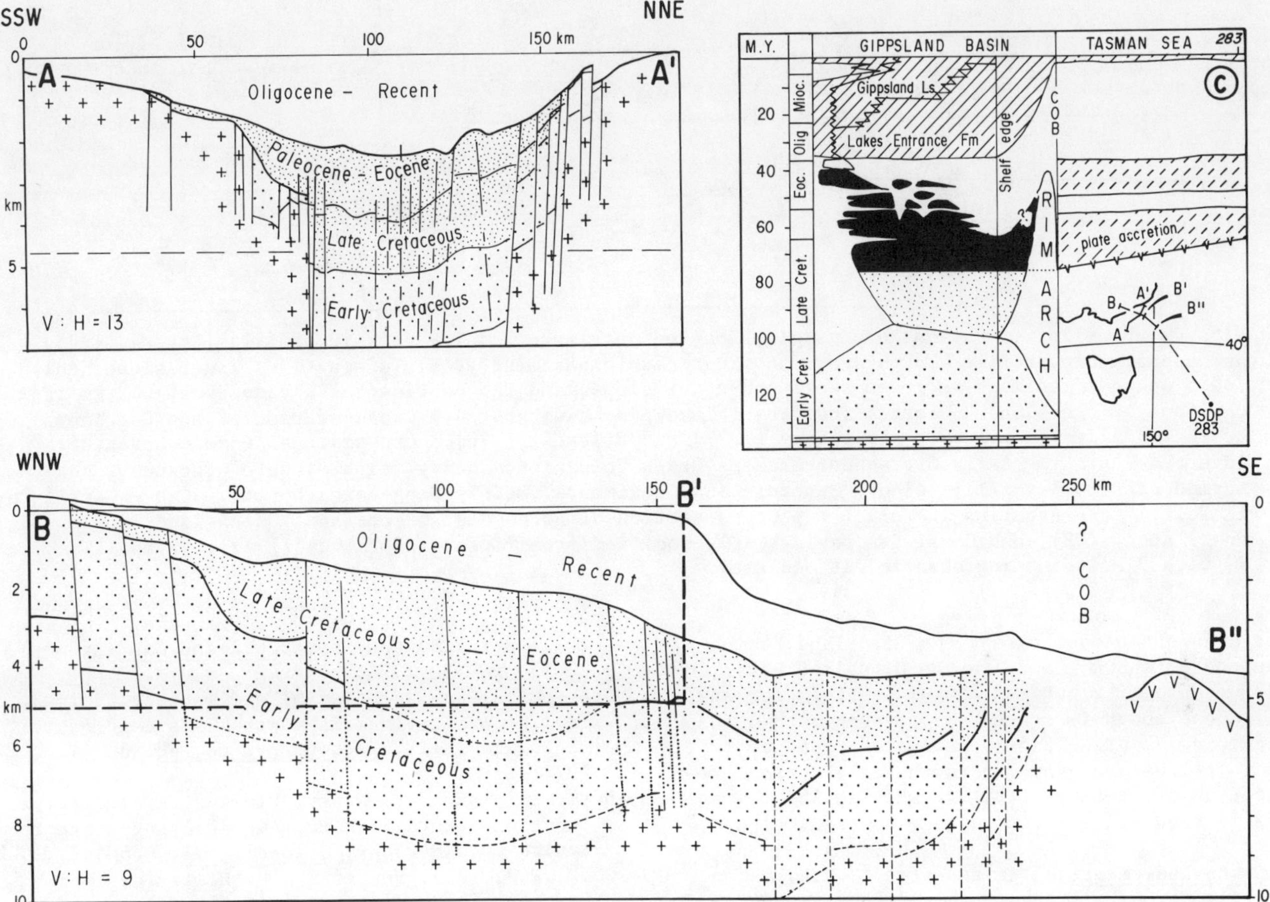

Fig. 8. Southeastern margin, off Gippsland Basin, sections located at H in Figure 2, and in inset of C. AA' and BB': depth sections across and along axis of Gippsland Basin. Top part of sections, above broken line, modified from James and Evans [1971, figs. 4A and 4B]; lower parts projected downward by me. B'B": depth section converted from a single-channel seismic profile made in 1973 by Shell International Petroleum from the MV Petrel, and available under the terms of the Submerged Lands Act. The precise boundary between the extra-arch basin and the succeeding rim basin is not known, and both these basins are shown together by the pattern of light dots. C: Time-space diagram of profile BB" extended to Site 283. Offshore Gippsland Basin out to shelf edge and at DSDP Site 283 modified from Partridge [1976, fig. 8]; interpolation by me. Boundary between extra-arch basin and rim basin shown at about 80 M.Y.

the Late Eocene and Oligocene and again in the Late Miocene [Threlfall et al., 1976].

Representative of the rest of the southeastern margin is the section off Sydney (Fig. 9,I), with a steep lower slope, a thin, presumably Tertiary, sediment wedge beneath a narrow shelf, and elevated land adjacent to the coast.

The northeastern margin widens northward of latitude 24°S into the broad shelf of the Great Barrier Reef and adjacent Marion and Queensland Plateaus. Sections of this margin, incorporating the results of the only marginal drilling in the region, are shown in Figure 9 (J and K), and are modified from Taylor and Falvey [1977]. Section J, across the presumed failed arm of the Capricorn Basin, shows a Late Cretaceous rift valley fill overlain by Cenozoic sediments. In section K,

which has been extended westward to show the high land adjacent to the margin, sediments that occupy the graben of the Queensland Trough (QT) − another presumed failed arm − and the marginal graben of the Queensland Plateau (QP) are inferred to be Late Cretaceous, of equivalent age and tectonic setting as the Capricorn Basin. This interpretation is based on indirect, mainly seismic refraction, evidence, and some other possibilities, including a lack of evidence of any pre-breakup stage, are promoted by Mutter and Karner [1978].

Discussion

The development of the western and southern margins of Australia matches a model derived from the African-Arabian rift systems. The eastern margin

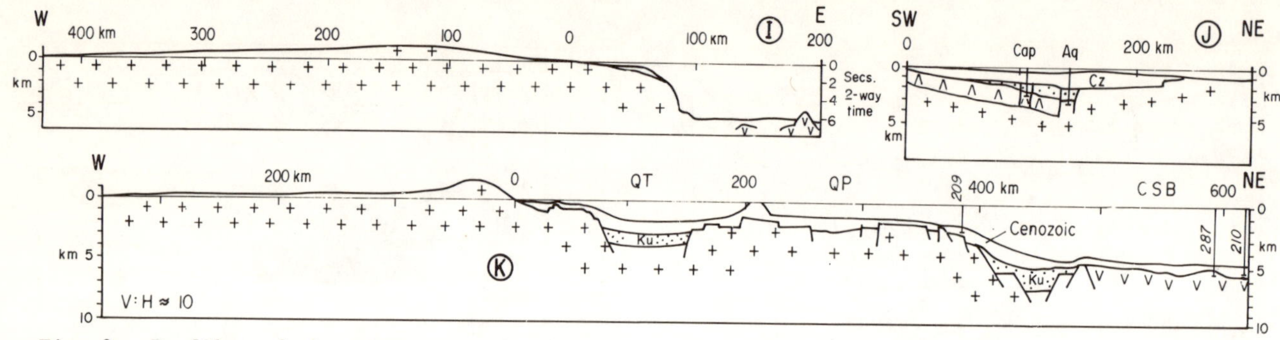

Fig. 9. Profiles of the eastern margin, located on Figure 2. I - Seismic profile [Davies, 1975] and topographic profile near Sydney, showing oceanic basement (V's), steep (15°) lower slope, thin (<0.5 second) sediment wedge on shelf and upper slope, and the continental divide about 200 km from the foot of the slope. Crosses indicate Palaeozoic basement. J - Cross-section of the Capricorn Basin, modified from Taylor and Falvey [1977] and Ericson [1976]. Crosses: Palaeozoic basement; inverted V's: the Early Cretaceous Grahams Creek Volcanics; heavy stipple: Late Cretaceous non-marine rift valley fill; clear: Cenozoic (Cz) sediments, mainly Neogene carbonate. Cap - Capricorn-1A Well; Aq - Aquarius-1 Well. K - Cross-section of north-east Queensland, Queensland Trough (QT) and Plateau (QP), and Coral Sea Basin (CSB), modified from Taylor and Falvey [1977]. Symbols as in J, with V's indicating oceanic lithosphere.

and adjacent ocean, except the Gippsland Basin, are poorly known; and, in the Gippsland Basin, the unconformity between Eocene strata below and Oligocene above is some 40 m.y. younger than the oldest adjacent seafloor, estimated to be 80 M.Y. old, yet the unconformity is the only obvious reflection of breakup. Further work in the Tasman Sea is required to resolve this contradiction.

Acknowledgements. I am indebted to R.L. Larson, M. Talwani, A.R. Fraser, L.A. Tilbury and J.B. Willcox for letting me see work in manuscript before publication, and to J.G. Jones, C.McA. Powell and S.Y. Wass for discussion and criticism. Supported in part by the Australian Research Grants Committee.

References

Allen, G.A., L.G.G. Pearce, and W.E. Gardner, A regional interpretation of the Browse Basin, Aust. Petrol. Expl. Assoc. J., 18, 23-33, 1978.

Apthorpe, M.C., Depositional history of the Upper Cretaceous of the Northwest Shelf, based upon Foraminifera, Aust. Petrol. Expl. Assoc. J., 19, 74-89, 1979.

Balke, B., and D. Burt, Arafura Sea area, Aust. Inst. Min. Metallurgy, Monog. Ser., 7, 209-212, 1976.

Barazangi, M., and J. Dorman, World seismicity maps compiled by ESSA, Coast and Geodetic Survey, Epicenter data, 1961-1967, Bull. Seismol. Soc. Amer., 59, 369-380, 1969.

Beaumont, C., and J.F. Sweeney, Graben generation of major sedimentary basins, Tectonophysics, 50, T19-T23, 1978.

Boeuf, M.G. and H. Doust, Structure and development of the southern margin of Australia, Aust. Petrol. Expl. Assoc. J., 15, 33-43, 1975.

Branson, J.C., Structures of the western margin of the Australian continent, Oil and Gas (Sydney), 20(9), 24-35, 1974.

Branson, J.C., Evolution of sedimentary basins from Mesozoic times in Australia's continental slope and shelf, Tectonophysics, 48, 389-412, 1978.

Burke, K., and J.F. Dewey, Plume-generated triple functions: key indicators in applying plate tectonics to old rocks, J. Geol., 81, 406-433, 1973.

Cook, P.J., J.J. Veevers, J.R. Heirtzler, and P.J. Cameron, The sediments of the Argo Abyssal Plain and adjacent areas, northeast Indian Ocean, BMR J. Aust. Geol. Geophys., 3, 113-124, 1978.

Cooney, P.M., P.R. Evans, and D. Eyles, Southern Ocean and its margins, in Deep Sea Drilling in Australasian Waters, ed. J.J. Veevers, 26-28, Sydney, 1975.

Crostella, A., and T. Barter, Triassic-Jurassic depositional history of the Dampier and Beagle sub-basins, Northwest Shelf of Australia, Aust. Petrol. Expl. Assoc. J., 20, 25-33, 1980.

Davies, P.J., Shallow seismic structure of the continental shelf, southeast Australia, J. geol. Soc. Aust., 22, 345-359, 1975.

Deighton, I., D.A. Falvey, and D.J. Taylor, Depositional environments and geotectonic framework, southern Australian continental margin, Aust. Petrol. Expl. Assoc. J., 16, 25-36, 1976.

Denham, D., G.R. Small, J.R. Cleary, P.J. Gregson, D.J. Sutton, and R. Underwood, Australian earthquakes (1897-1972), Search, 6, 34-37, 1975.

Denham, J.I., and B.R. Brown, A new look at the Otway Basin, Aust. Petrol. Expl. Assoc. J., 16, 91-98, 1976.

Douglas, J.G., and J.A. Ferguson (Eds.), Geology of Victoria, Geol. Soc. Aust. Spec. Publ., 5, 364-371, 1976.

Ericson, E.K., Capricorn Basin, Aust. Inst. Min. Metallurgy, Mongr. Ser., 7, 464-473, 1976.

Exon, N.F., and B.R. Senior, The Cretaceous of the

Eromanga and Surat Basins, <u>BMR J. Geol. Geoph.</u>, <u>1</u>, 33–50, 1976.

Exon, N.F., and J.B. Willcox, The geology and petroleum potential of the Exmouth Plateau area off Western Australia, <u>Bull. Amer. Assoc. Petrol. Geol., 62</u>, 40–72, 1978.

Fairbridge, R.W., and C.W. Finkl, Geomorphic analysis of the rifted cratonic margins of Western Australia, <u>Zeitschr. Geomorph., 22</u>, 369–389, 1978.

Falvey, D.A., The development of continental margins in plate tectonic theory. <u>Aust. Petrol. Expl. Assoc. J., 14</u>, 95–106, 1974.

Finkl, C.W., and R.W. Fairbridge, Paleogeographic evolution of a rifted cratonic margin: S.W. Australia, <u>Palaeogeog., Palaeoclimat., Palaeoecol., 26</u>, 221–252, 1979.

Fraser, A.R., and L.A. Tilbury, Structure and stratigraphy of the Ceduna Terrace region, Great Australian Bight, <u>Aust. Petrol. Expl. Assoc. J., 19</u>, 53–65, 1979.

Fyfe, W.S., and O.H. Leonardos, Ancient metamorphic migmatite belts of the Brazilian and African coasts, <u>Nature, 244</u>, 561–562, 1973.

Geological Society of Australia, Tectonic map of Australia and New Guinea 1:5000000, Sydney, 1971.

Geological Survey of Western Australia, Geology of Western Australia: <u>West. Australia Geol. Surv. Mem., 2</u>, 541pp., 1975.

Gleadow, A.J.W., Fission-track evidence for the evolution of rifted continental margins, <u>4th Int. Conf. Geochron. Cosmochem. Isotope Geology, USGS Open-file Report, 78–701</u>, 146–148, 1978.

Gleadow, A.J.W., and J.F. Lovering, Fission track geochronology of King Island, Bass Strait, Australia: relationship to continental rifting, <u>Earth Planet. Sci. Letters, 37</u>, 429–437, 1978.

Glickson, A.Y., and I.B. Lambert, Vertical zonation and petrogenesis of the Early Precambrian crust in Western Australia, <u>Tectonophysics, 30</u>, 55–89, 1976.

Griffiths, J.R., Continental margin tectonics and the evolution of Southeast Australia, <u>Aust. Petrol. Expl. Assoc. J., 11</u>, 75–79, 1971.

Hawkins, L.V., J.F. Hennion, J.E. Nafe and R.F. Thyer, Geophysical investigations in the area of the Perth Basin, Western Australia, <u>Geophysics, 30</u>, 1026–1052, 1965a.

Hawkins, L.V., J.F. Hennion, J.E. Nafe, and H.A. Doyle, Marine seismic refraction studies on the continental margin to the south of Australia, <u>Deep-Sea Research, 12</u>, 479–495, 1965b.

Heezen, B.C., and M. Tharp, USNS Eltanin cruise 55, <u>Antarctic J. US, 8</u>, 137–141, 1973.

Heirtzler, J.R., P. Cameron, P.J. Cook, T. Powell, H.A. Roeser, S. Suhardi, and J.J. Veevers, The Argo Abyssal Plain, <u>Earth Planet. Sci. Letters, 41</u>, 21–31, 1978.

Hinz, K., H. Beiersdorf, N.F. Exon, H.A. Roeser, H.M.F. Stagg, and U. von Stackelberg, Geoscientific investigations from the Scott Plateau off northwestern Australia to the Java Trench, using R.V. Valdivia, <u>BMR J. Aust. Geol. Geoph., 3</u>, 319–340, 1978.

Holmes, A., <u>Principles of physical geology</u>, 2nd Ed, Nelson, London, 1965.

International Tectonic Map of Africa, <u>Paris, Assoc. Serv. géologiques Africains and UNESCO</u>, 1:5 million, 1968.

James, E.A., and P.R. Evans, The stratigraphy of the off-shore Gippsland Basin, <u>Aust. Petrol. Expl. Assoc. J., 11</u>, 71–74, 1971.

Johnson, B.D., C.McA. Powell, and J.J. Veevers, Early spreading history of the Indian Ocean between India and Australia, <u>Earth Planet. Sci. Letters, 47</u>, 131–143, 1980.

Johnstone, M.H., D.C. Lowry, and P.G. Quilty, The geology of southwestern Australia – a review, <u>J. roy. Soc. W. Aust., 56</u>, 5–15, 1973.

Jones, J.G., Australia's Caenozoic drift, <u>Nature, 230</u>, 237–239, 1971.

Jones, J.G., and W.D. Roots, Evolution of the Tasman Sea, <u>Nature, 252</u>, 613, 1974.

Jongsma, D., and J.C. Mutter, Non-axial breaching of a rift valley: evidence from the Lord Howe Rise and the southeastern Australian margin, <u>Earth Planet. Sci. Letters, 39</u>, 226–234, 1978.

Kennedy, W.Q., The influence of basement structure on the evolution of the coastal (Mesozoic and Tertiary) basins of Africa, in <u>Salt Basins around Africa</u>, edited by D.C. Ion, Inst. Petroleum, London, 7–16, 1965.

King, B.C., and L.A.J. Williams, The East African rift system, in <u>Geodynamics: Progress and Prospects</u>, edited by C.L. Drake, Amer. Geoph. Union, Washington, 63–74, 1976.

Kinsman, D.J.J., Rift valley basins and sedimentary history of trailing continental margins, in <u>Petroleum and global tectonics</u>, edited by A.G. Fischer and S. Judson, Princeton Univ. Press, 83–126, 1975.

Kröner, A., The Precambrian geotectonic evolution of Africa: plate accretion versus plate destruction, <u>Precambrian Research, 4</u>, 163–213, 1977.

Kuenen, P.H., <u>Marine Geology</u>, Wiley, NY, 1950.

La Brecque, J.L., D.V. Kent, and S.C. Cande, Revised magnetic polarity time scale for Late Cretaceous and Cenozoic time. <u>Geology, 5</u>, 330–335, 1977.

Larson, R.L., J.C. Mutter, J.B. Diebold, G.B. Carpenter, and P. Symonds, Cuvier Basin: a product of ocean crust formation by Early Cretaceous rifting off Western Australia, <u>Earth Planet. Sci. Letters, 45</u>, 105–114, 1979.

Laws, R.A., and G.P. Kraus, The regional geology of the Bonaparte Gulf-Timor Sea area, <u>Aust. Petrol. Expl. Assoc. J., 14</u>, 77–84, 1974.

Lowry, D.C., Eucla Basin, <u>Aust. Inst. Min. Metallurgy, Monogr. Ser., 7</u>, 95–98, 1976.

Map of the world, <u>Paris, Inst. géographique national</u>, 1:5 million, sheets 6, 7, 10, 1968.

Markl, R.G., Evidence for the breakup of eastern Gondwanaland by the Early Cretaceous, <u>Nature, 251</u>, 196–199, 1974.

Markl, R.G., Further evidence for the Early Cretaceous breakup of Gondwanaland off southwestern Australia, <u>Marine Geology, 26</u>, 41–48, 1978a.

Markl, R.G., Basement morphology and rift geometry near the former junction of India, Australia and Antarctica, Earth Planet. Sci. Letters, 39, 211-225, 1978b.

Mathur, S.P., Crustal structure in southwestern Australia from seismic and gravity data, Tectonophysics, 24, 151-182, 1974.

McConnell, R.B., Evolution of taphrogenic lineaments in continental platforms, Geologische Rundschau, 63, 389-430, 1974.

Mutter, J.C., and D. Jongsma, The pattern of the pre-Tasman Sea rift system and the geometry of breakup, Aust. Soc. Expl. Geophys. Bull., 9, 70-75, 1978.

Mutter, J.C., and G. Karner, Cretaceous taphrogeny in the Coral Sea, Aust. Soc. Expl. Geophys. Bull., 9, 82-87, 1978.

Partridge, A.D., The geological expression of eustacy in the Early Tertiary of the Gippsland Basin, Aust. Petrol. Expl. Assoc. J., 16, 73-79, 1976.

Plumb, K.A., The tectonic evolution of Australia, Earth-Science Rev., 14, 205-249, 1979.

Powell, D.E., The geological evolution of the continental margin of northwest Australia, Aust. Petrol. Expl. Assoc. J., 16, 13-23, 1976.

Quilty, P.G., Cenozoic sedimentation cycles in Western Australia, Geology, 5, 336-340, 1977.

Rabinowitz, P.D., and J.L. La Brecque, The isostatic gravity anomaly: Key to the evolution of the ocean-continent boundary at passive continental margins, Earth Planet. Sci. Letters, 35, 145-150, 1977.

Roots, W.D., A better fit for Tasman Sea magnetics and reappraisal of Tasman Sea opening, Aust. Soc. Expl. Geophys. Bull., 6, 42, 1975.

Roots, W.D., Magnetic smooth zones and slope anomalies: a mechanism to explain both, Earth Planet. Sci. Letters, 31, 113-118, 1976.

Roots, W.D., J.J. Veevers, and D.F. Clowes, Lithospheric model with thick oceanic crust at the continental boundary: a mechanism for shallow spreading ridges in young oceans, Earth Planet. Sci. Letters, 43, 417-433, 1979.

Scrutton, R.A., Microcontinents and their signnificance, in Geodynamics: progress and prospects, edited by C.L. Drake, 177-189, AGU, Washington, 1976.

Shaw, R.D., Sea floor spreading in the Tasman Sea: a Lord Howe Rise-Eastern Australian reconstruction, Aust. Soc. Expl. Geophys. Bull., 9, 75-81, 1978.

Shor, G.G., H.K. Kirk, and H.W. Menard, Crustal structure of the Melanesian Sea, J. Geophys. Res., 76, 2562-2586, 1971.

Stagg, H.M.J., The geology and evolution of the Scott Plateau, Aust. Petrol. Expl. Assoc. J., 18, 34-43, 1978.

Stagg, H.M.J., and N.F. Exon, Western margin of Australia: evolution of a rifted arch system: Discussion, Geol. Soc. Amer. Bull., 90, 795-797, 1979.

Stephenson, P.J., T.J. Griffin, and F.L. Sutherland, Cainozoic volcanism in north east-

ern Australia, 3rd Aust. geol. Convention, Geol. Soc. Aust., Townsville, 14-15, 1978.

Sykes, L.R., Intraplate seismicity, reactivation of pre-existing zones of weakness, alkaline magmatism, and other tectonism postdating continental fragmentation, Rev. Geoph. Space Physics, 16, 621-688, 1978.

Symonds, P.A., and P.J. Cameron, The structure and stratigraphy of the Carnarvon Terrace and Wallaby Plateau, Aust. Petrol. Expl. Assoc. J., 17, 30-41, 1977.

Symonds, P.A., and J.B. Willcox, The gravity field of offshore Australia, BMR J. Geol. Geoph., 1, 303-314, 1976.

Talwani, M., J. Mutter, R. Houtz, and M. König, The margin south of Australia - a continental margin paleorift, in Tectonics and Geophysics of Continental Rifts, edited by I.B. Ramberg and E.R. Neumann, 203-219, Reidel Publ. Co., Holland, 1978.

Talwani, M., J. Mutter, R. Houtz, and M. König, The crustal structure and evolution of the area underlying the magnetic quiet zone on the margin south of Australia, Amer. Assoc. Petrol. Geol. Mem., 29, 151-175, 1979.

Taylor, L., and D. Falvey, Queensland Plateau and Coral Sea Basin: stratigraphy, structure and tectonics, Aust. Petrol. Expl. Assoc. J., 17, 13-29, 1977.

Threlfall, W.F., B.R. Brown, and B.R. Griffith, Gippsland Basin, offshore, Aust. Inst. Min. Metallurgy, Monog. Ser., 7, 41-67, 1976.

Vail, P.R., R.M. Mitchum, and S. Thompson, Global changes of sea level from seismic stratigraphy, Mem. Amer. Assoc. Petrol. Geol., 26, 1977.

Van de Graaff, W.J.E., P.D. Denman, and R.M. Hocking, Emerged Pleistocene marine terraces on Cape Range, Western Australia, West. Australian Geol. Survey Ann. Rep. (1975), 62-70, 1976.

Veevers, J.J., Rifted arch basins and post-breakup rim basins on passive continental margins, Tectonophysics, 41, T1-T5, 1977a.

Veevers, J.J., Paleobathymetry of the crest of spreading ridges related to the age of ocean basins, Earth Planet. Sci. Letters, 34, 100-106, 1977b.

Veevers, J.J., Western margin of Australia: Reply to Discussion, Geol. Soc. Amer. Bull., 90, 797-798, 1979.

Veevers, J.J., Western and northwestern margin of Australia, in Nairn, A.E.M., F.G. Stehli, and M. Churkin, editors, The Ocean Basins and Margins, 7, Plenum, N.Y., 1980 (in press).

Veevers, J.J., and D. Cotterill, Western margin of Australia: a Mesozoic analog of the East African rift system, Geology, 4, 713-717, 1976.

Veevers, J.J., and D. Cotterill, Western margin of Australia: evolution of a rifted arch system, Geol. Soc. Amer. Bull., 89, 337-355, 1978.

Veevers, J.J., and P.R. Evans, Late Palaeozoic and Mesozoic history of Australia, in Campbell, K.S.W. (ed.), Gondwana Geology, ANU Press, Canberra, 579-607, 1975.

Veevers, J.J., D.A. Falvey, L.V. Hawkins, and

W.J. Ludwig, Seismic reflection measurements of northwest Australian margins and adjacent deeps, Amer. Assoc. Petrol. Geol. Bull., 58, 1731-1750, 1974.

Veevers, J.J., and C.McA. Powell, Sedimentary-wedge progradation from transform-faulted continental rim: Southern Exmouth Plateau, Western Australia, Amer. Assoc. Petrol. Geol. Bull., 63, 2088-2096, 1979.

von der Borch, C.C., Southern Australian submarine canyons; their distribution and ages, Marine Geology, 6, 267-279, 1968.

von Stackelberg, U., N.F. Exon, U. von Rad, P. Quilty, S. Shafik, H. Beiersdorf, E. Seibertz, and J.J. Veevers, Geology of the Exmouth and Wallaby Plateaus off NW Australia: sampling of seismic sequences, BMR J. Geol. Geoph., 1980 (in press).

Weissel, J.K. and D.E. Hayes, Magnetic anomalies in the southeast Indian Ocean, in Hayes, D.E. (ed.), Antarctic Oceanology II: the Australian-New Zealand Sector, Antarctic Res. Ser. (AGU), 19, 165-196, 1972.

Weissel, J.K., and D.E. Hayes, Evolution of the Tasman Sea reappraised, Earth Planet. Sci. Letters, 36, 77-84, 1977.

Weissel, J.K., D.E. Hayes, and E.M. Herron, Plate tectonics synthesis: the displacements between Australia, New Zealand, and Antarctica since the Late Cretaceous, Mar. Geol., 25, 231-277, 1977.

Weissel, J.K. and A.B. Watts, Tectonic evolution of the Coral Sea Basin, J. Geophys. Res., 84, 4572-4582, 1979.

Wellman, P., The Gravity field of the Australian basement, BMR J. Geol. Geoph., 1, 287-290, 1976.

Wellman, P., and I. McDougall, Potassium-argon ages on the Cainozoic volcanic rocks of New South Wales, J. geol. Soc. Aust., 21, 247-272, 1974a.

Wellman, P., and I. McDougall, Cainozoic igneous activity in eastern Australia, Tectonophysics, 23, 49-65, 1974b.

Willcox, J.B., Some gravity models of the continental margin in the Australian region, Bull. Aust. Soc. Expl. Geophys., 88, 118-124, 1977.

Willcox, J.B., The Great Australian Bight: a regional interpretation of gravity, magnetic, and seismic data from the continental margin survey, Bur. Min. Resour. Aust. Rep., 201, 111pp (microform), 1978.

Willcox, J.B., and N.F. Exon, The regional geology of the Exmouth Plateau, Aust. Petrol. Expl. Assoc. J., 16, 1-11, 1976.

Wright, A.J., and T.J. Wheatley, Trapping mechanisms and the hydrocarbon potential of the Exmouth Plateau, Western Australia, Aust. Petrol. Expl. Assoc. J., 19, 19-29, 1979.

Postscript. Another review of Australian rifted margins, by D.A. Falvey and J.C. Mutter, seen after this review was made, is scheduled for publication in the BMR Journal of Australian Geology and Geophysics in late 1980. Falvey and Mutter's review relies on essentially the same sources of information as mine but concentrates on the driving mechanisms of subsidence before and after break-up.

GRAVITY MEASUREMENTS BORDERING PASSIVE CONTINENTAL MARGINS

Philip D. Rabinowitz

Lamont-Doherty Geological Observatory of Columbia University, Palisades, N.Y. 10964

Abstract. Gravity measurements in the vicinity of the zone between ascertained oceanic crust and ascertained continental crust are reviewed for many of the world's passive continental margins.

Elongate belts of free-air gravity anomalies paralleling the margins are ubiquitous to the passive margins. These are: free-air gravity highs over the shelf break, free-air lows over the continental slope/rise, and in places, relative free-air gravity highs farther seaward. An isostatic gravity anomaly is generally present when the simple Airy isostatic correction is applied to the free-air anomaly. This anomaly is not necessarily located over any particular bathymetric contour but is shown to be at the continental shelf, as well as landward or seaward of the shelf edge. Bordering some of the passive margins, such as off eastern United States, southern Australia and eastern Brazil (Sao Paulo Plateau) an isostatic gravity high persists near the shelf edge as well as farther seaward.

In many areas a rather pronounced magnetic anomaly, which can be modelled as an edge effect separating oceanic from continental basement, is associated with the continental margin isostatic gravity anomaly. In these areas it has been suggested that the marginal isostatic gravity anomaly is a manifestation of thickened elevated ocean crust adjacent to the continent. The detailed seismic mapping of both the upper crustal and deeper structure is necessary to test the various models for the origin of the marginal gravity anomaly and to learn more with respect to the zone between oceanic and continental basement.

Introduction

The early gravity measurements over passive continental margins were obtained with pendulum apparatus aboard submarines (Vening Meinesz, 1932; Worzel, 1965). Despite the relatively large separations between gravity stations, these measurements provided the primary data for deducing the first crustal models showing the transition from continental to oceanic crustal thicknesses (Worzel and Shurbet, 1955). Off the coast of North America the pendulum data have shown the presence of a prominent free-air gravity high near the shelf edge. This high is bordered on its seaward side by a prominent free-air low. Worzel and Shurbet (1955) have shown that these gravity anomalies, in most part, are accounted for by models which assume thinning of the crustal layers as the continental margin is traversed from continent to ocean. These anomalies have been termed the gravity "edge effect."

Since the initiation of continuous gravity measurements at sea (Worzel, 1959) a large quantity of gravity data has been obtained over the passive continental margins. These measurements have shown us that the shelf edge free-air anomalies described by Worzel and Shurbet (1955) align as long continuous belts of anomalies. A major difficulty in interpreting the shelf edge free-air anomalies is in determining which part of the anomaly arises from the topography and its compensation and which part, if any, results from mass excesses or deficiencies within or below the crust. Along continental margins where the bathymetry is not steep, the edge effect may be small and the free-air anomaly may be a good indicator of crustal inhomogeneities. Steep slopes, however, are encountered bordering most of the continental margins. In these cases, the free-air anomaly may be an artifact of changes in crustal thicknesses and the edge effect may be large. For this reason many investigators have used the isostatic gravity anomaly as an indicator of structure or density contrasts bordering the passive continental margins.

On some of the passive continental margins the isostatic correction reduces but does not completely eliminate the shelf edge free-air gravity high. One example of this is off the east coast of North America, where shelf edge isostatic anomalies suggest that mass excesses are present beneath the shelf edge. In some regions not only does a shelf edge isostatic gravity high persist but a more seaward gravity high is also present. Some examples which will be discussed are off the coast of eastern North America, Norway and parts of southern Australia and Antarctica. In some other regions, however, such as bordering large

parts of the southern South Atlantic, the shelf
edge free-air gravity high is eliminated when the
isostatic correction is applied, and generally,
only a more seaward isostatic gravity anomaly is
observed. In many cases a prominent magnetic a-
nomaly is associated with the shelf edge gravity
high and/or the more seaward gravity high with a
magnetic quiet zone on the landward side of the
gravity high.

In this report, the gravity observations
bordering selected passive continental margins
will be reviewed. Free-air gravity maps, free-
air and isostatic gravity profiles, and magnetic
anomaly profiles will be shown in the selected
regions. The areas include portions of the opposing continental margins of the Central and
South Atlantic Ocean, as well as portions of the
continental margins of Norway, eastern Africa,
western India, Antarctica and southern Australia.
Linear trends of free-air and isostatic gravity
anomalies have been reported for other passive
continental margins not discussed in this report.
For example, Stacey (1968) and Bott and Watts
(1970) discuss the free-air gravity field on the
margin between northern Scotland and Iceland.
Taylor and Falvey (1977) and Falvey (1972) discuss and model gravity anomalies off northern Australia bordering the margins of the Coral Sea
Basin. Markl (1978) shows isostatic profiles for
the passive margin of western Australia. He reports steep isostatic gradients in the vicinity
of the oldest recognizable seafloor spreading
magnetic lineation.

Central Atlantic Ocean

<u>East Coast North America</u>

We show in Figure 1 a free-air gravity map
bordering the continental margin of the east
coast of North America (modified after Keen et
al., 1971; Emery et al., 1970; Rabinowitz, 1973,
1974). The contour interval is 25 mgal. The
most prominent feature of the free-air gravity
map is the gravity high which roughly parallels
the east coast at the seaward edge of the continental shelf. It is a continuous high ranging in
magnitude from values near 0 mgal to values more
positive than 75 mgal. Seaward of the shelf edge
high, a trend of negative anomalies is observed
on the continental slope and rise, with magnitudes
ranging from about -25 mgal to values more negative than -125 mgal in the region of the precipi-
tous Blake escarpment in the south. An important
gradient is observed in the gravity field seaward
of the continental slope/rise low which is rather
broad ($\sim$100 to 150 km) and has small relative
amplitudes ($\sim$5 to 20 mgal).

More recent gravity maps have been published at
smaller contour intervals for large parts of the
eastern margin of the United States, by Grow et
al., (1976, 1979a), and Ewing (in press). The detail in the latter maps allow inferences to be
made with respect to fracture zone intersections

with the coastlines. In particular, the locations
where the shelf edge gravity high changes rather
abruptly in magnitude have been shown by Grow et
al., (1979a) to correspond closely with the
boundaries between platform and basin areas on
the continental shelf.

In Figure 2, projected topographic, free-air
and isostatic gravity and magnetic profiles are
given for the east coast of North America. All of
the profiles are aligned with respect to the shelf
edge. All isostatic profiles shown in this report
are simple Airy models which assume two-dimen-
sionality and density contrasts of 1.57 and 0.47
gm/cm^3 at the water-crust and crust-mantle interfaces, respectively, together with a depth of
compensation of T=30 km.

The isostatic computations (Figure 2) demonstrate that the free-air shelf edge high is not
totally a simple "edge effect" resulting from the
topography and its compensation. In all cases,
an isostatic gravity high is present close to the
shelf edge which strongly suggests that intra-
basement density highs and/or changes in thickness of basement are present.

The very prominent east coast magnetic anomaly
(Figure 2) has been interpreted by some investigators as being a manifestation of the boundary
between oceanic and continental basement (Keen,
1969; Emery et al., 1970; Luyendyk and Bunce,
1973; Klitgord and Behrendt, 1979). In particular, the very detailed magnetic investigations
by Klitgord and Behrendt (1979) have suggested
that the east coast magnetic anomaly arises from
a combination of a magnetic basement high, most
probably an uplifted block of oceanic crust with
a minimum depth of 6 to 8 km below sea level,
and an "edge effect" where the edge is between
the uplifted block and flat-lying sediments to
the west.

Gravity models by Grow et al., (1979a) utilizing extensive multichannel seismic measurements for sediment control (Schlee et al., 1976;
Grow and Schlee, 1976) and refraction measurements (Sheridan et al., 1979) have suggested
that the oceanic crust is $\sim$11 to 18 km in thickness near the ocean/continent boundary (as defined
by the east coast magnetic anomaly). The ocean
crust thins to $\sim$8 km within 20 to 90 km seaward
of the east coast magnetic anomaly. A composite
schematic model, showing crustal structure as well
as magnetic and isostatic gravity anomalies, is
shown in Figure 3 (after Grow et al., 1979b).

It should be noted that although the east coast
magnetic anomaly is modelled as separating ocean
from continent and the shelf edge positive isostatic gravity anomaly is modelled as thickened
ocean crust adjacent to the boundary, the gravity
and magnetic anomalies are not always coincident.
The regions where they do not coincide may represent local disturbances (effects between platform and basin edges on continental shelf, local
intrusions, etc.).

We have noted that an important, broad gravity
gradient is observed seaward of the continental

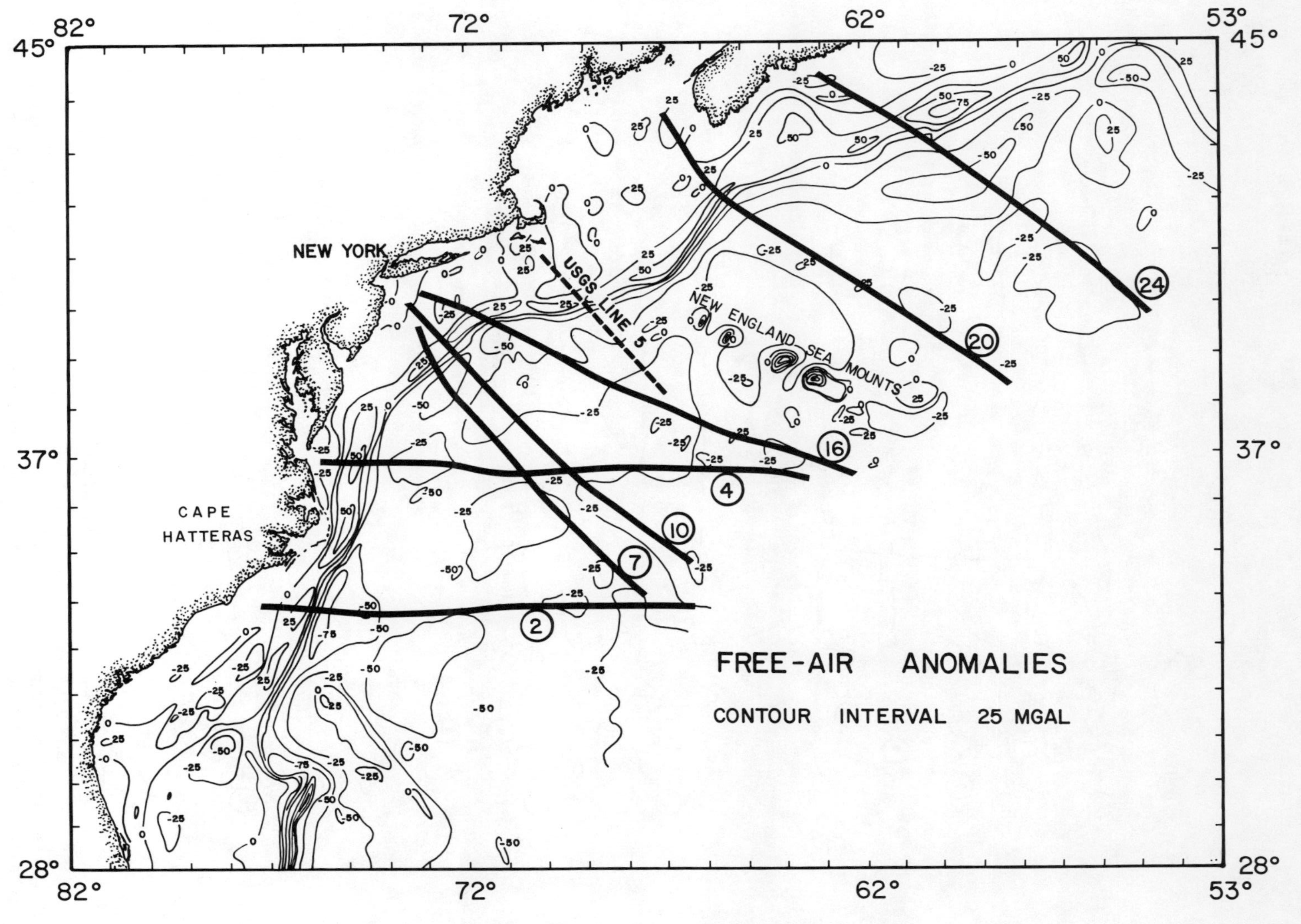

Fig. 1. Free-air gravity anomalies off eastern North America. Modified after Emery et al., (1970), Keen et al., (1971), and Rabinowitz (1973, 1974). Contour interval 25 mgal. Heavy lines across margin show locations of projected gravity magnetics and topography profiles given in Fig. 2. U.S.G.S. line 5 given in Fig. 3.

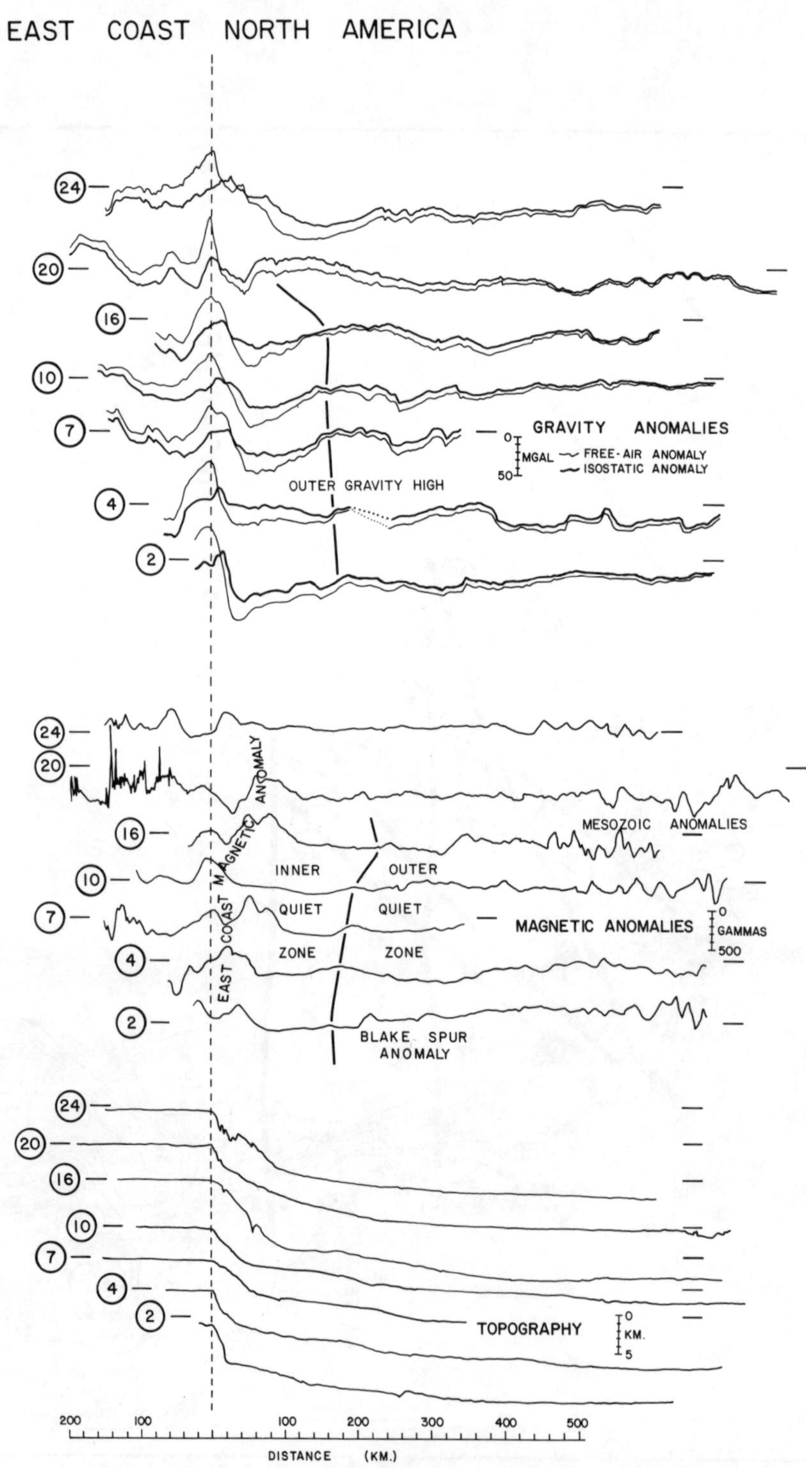

Fig. 2. Projected gravity, magnetics and topography profiles, east coast North America. All profiles are aligned with respect to the shelf edge (modified after Rabinowitz, 1974). Location of profiles in Fig. 1.

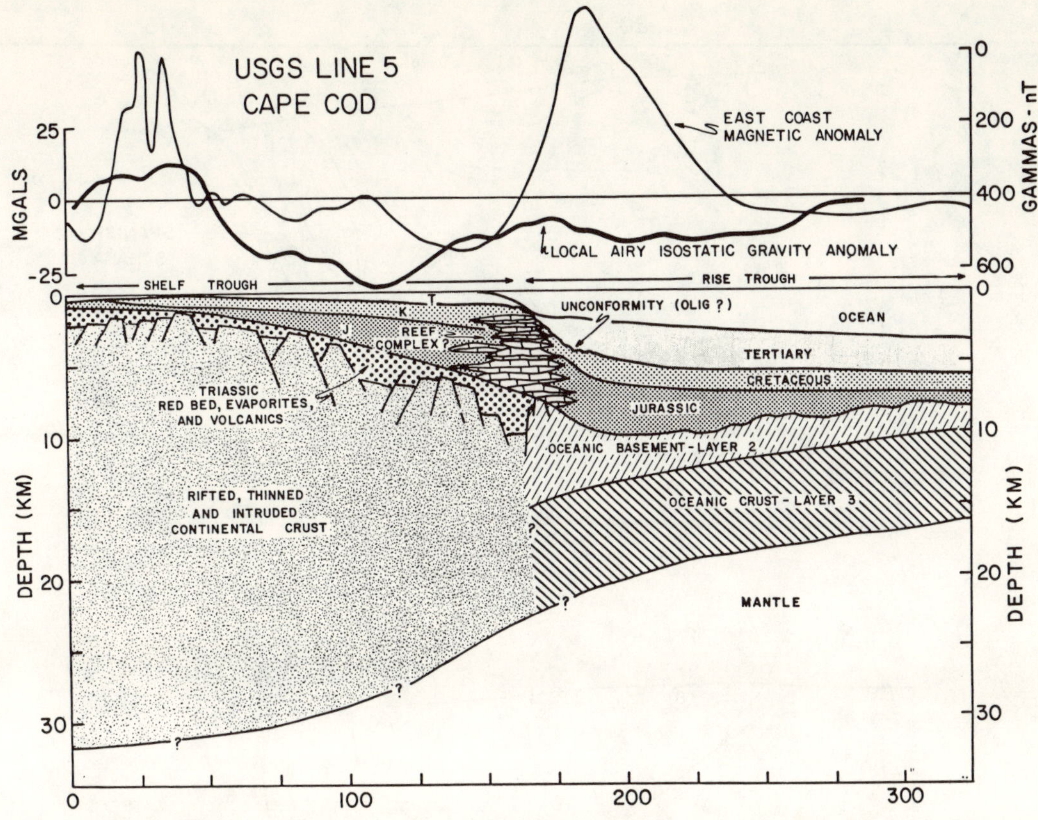

Fig. 3. Schematic cross-section summarizing seismic measurements together with isostatic gravity and magnetic anomalies (modified after Grow et al., 1979b). Location in Fig. 1. This line crosses the eastern end of the Long Island Platform and is described by Grow et al., (1979b) as typical of a narrow transition zone between continental and oceanic crust.

rise/slope low (Figure 2). The location of the seaward part of the gravity gradient (the outer gravity high) separates the magnetic quiet zone (region between the Mesozoic magnetic anomalies and the east coast magnetic anomaly) into an inner landward quiet zone characterized by a very smooth magnetic field and an outer quiet zone characterized by many small amplitude magnetic anomalies. The inner quiet zone has been interpreted as being situated on subsided continental crust (Rabinowitz, 1974). Multichannel seismic measurements (Schlee et al., 1976) appear to trace reflectors associated with oceanic basement across part of the inner quiet zone and are thus in conflict with the above interpretation. The outer gravity high is similar in shape to the observed gravity anomalies bordering some of the other passive continental margins (e.g., southern Australia and Antarctica) and may reflect processes active during the early separation of the continents.

Northwest Africa

The conjugate margin in pre-drift reconstructions to the east coast of North America, is the continental margin of Northwest Africa. In Figure 4 a free-air gravity map is contoured at a 20 mgal interval and shown for a selected portion of this margin. This map is modified, with additional data, from Uchupi et al., (1976). Other gravity studies off northwest Africa include those of Behrendt et al., (1974) and Roeser et al., (1971). The most prominent feature of the map in this area is a continuous free-air gravity high near the shelf edge which obtains values, in places, greater than 60 mgal. Seaward of the shelf-edge free-air gravity high a trend of negative anomalies is observed with magnitudes from about 0 to -40 mgals. In places, a gradient in the gravity field is observed seaward of the continental slope/rise low, which is similar to that observed off the east coast of North America.

There are locations where the shelf-edge gravity high and/or the slope/rise gravity low change rather abruptly in magnitude (e.g., near 17°N, 20°N, 22.5°N, 24.5°N: Figure 4). These locations may correspond with boundaries between platform and basin areas on the continental shelf and/or fracture zone intersections with the coastline in a similar fashion to that suggested by Grow et

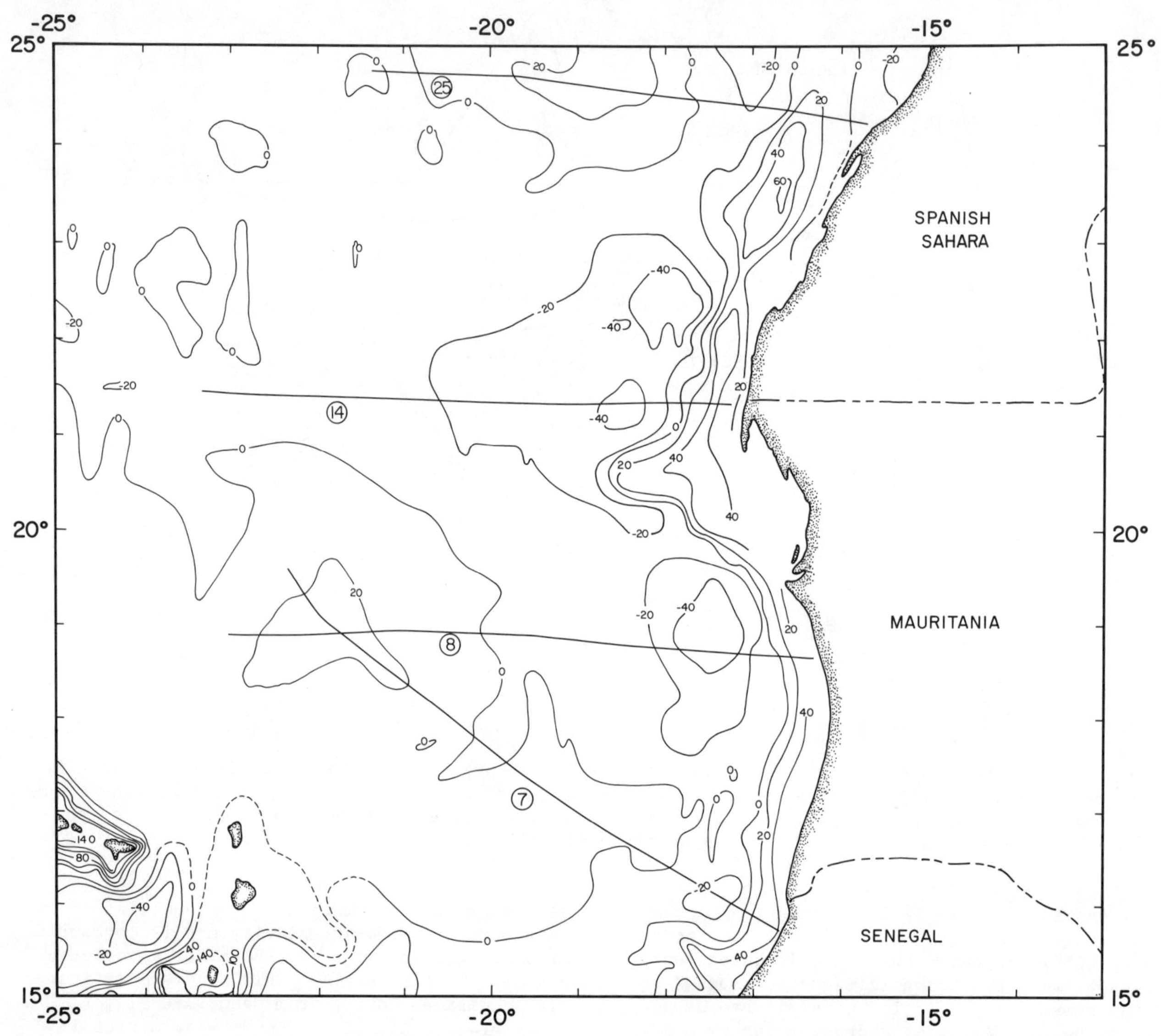

Fig. 4. Free-air gravity anomalies off part of northwest Africa's continental margin. This map is modified after Uchupi et al., (1976) utilizing an extensive Lamont-Doherty Data Library. Lines 7, 8, 14 and 25 are locations of projected magnetic and topography profiles given in Fig. 5.

al., (1979a) and Klitgord and Behrendt (1979) for the conjugate margin off eastern United States.

In Figure 5, projected topographic and magnetic profiles are given for the Northwest Africa margin. A conspicuous magnetic anomaly is observed bordering the continental margin, that may be a conjugate feature to either the east coast magnetic anomaly or the Blake Spur Anomaly (Rona et al., 1970; Hayes and Rabinowitz, 1975; Uchupi et al., 1976; Klitgord and Schouten, 1977). This magnetic anomaly, which is situated in places landward of the shelf edge (e.g., profile 25, Figure 5) may be present, in places, landward of the coastline (e.g., profile 7, Figure 5). Because of its spatial location, Uchupi et al., (1976)

suggest that the magnetic anomaly is a manifestation of intrusions of basic dikes and sills in a broad area of continental basement which may have begun as early as 40 million years before the onset of seafloor spreading. This model, if correct, rules out the conventional seafloor spreading and/or magnetic edge effect interpretation as suggested by many investigators for the east coast magnetic anomaly.

Isostatic and/or other gravity computations are not published in the literature for this part of the margin. We thus do not know if free-air anomalies on this margin are total "edge effects" or if isostatic anomalies are present. Clearly, much more work is necessary

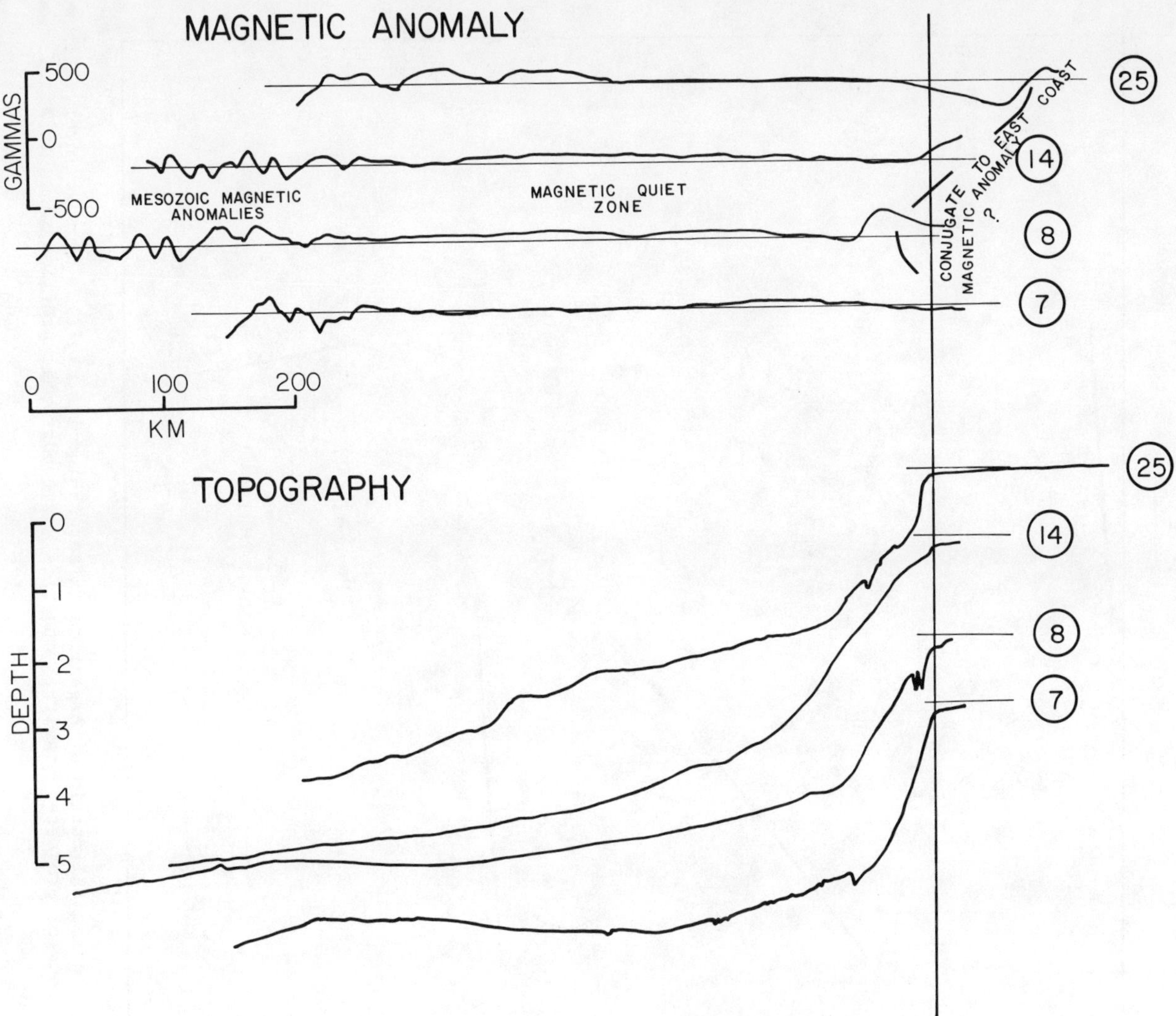

Fig. 5. Projected magnetic and topographic profiles off northwest African margin (after Hayes and Rabinowitz, 1975). All profiles aligned with respect to shelf edge. Locations in Fig. 4.

to ascertain whether or not there are spatially aligned gravity and magnetic lineaments and what their relationship is to the ocean continent boundary.

Margins of the South Atlantic Ocean

The gravity observations bordering the continental margin of the South Atlantic Ocean will be discussed in two parts: those measurements observed south of the Rio Grande Rise - Walvis Ridge lineaments, and those to the north.

South of the Walvis Ridge (Africa) and Rio Grande Rise (South America)

A free-air gravity map of the continental margin of Argentina is shown in Figure 6 (after Rabinowitz, 1977). The area includes the predominantly rifted margin of eastern Argentina and the predominantly sheared margin of the northern Falkland Plateau. In Figure 7 a free-air gravity map is given of western Africa south of the Walvis Ridge and includes part of the Agulhas Fracture Zone (after Emery et al., 1975a). Other gravity maps published for this area include that of Simpson

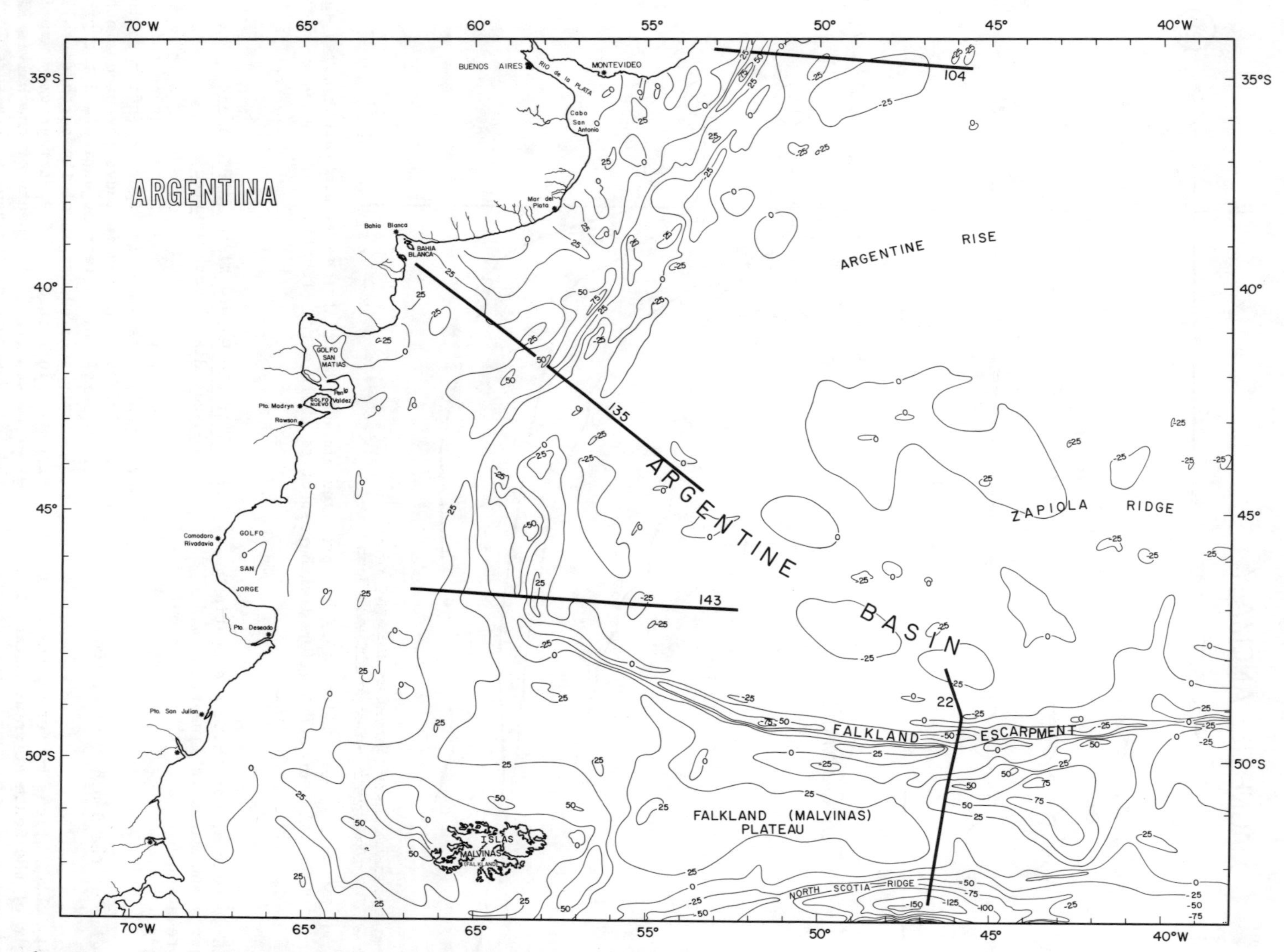

Fig. 6. Free-air gravity anomalies, continental margin of Argentina (after Rabinowitz, 1977). Contour interval 25 mgal. Heavy lines crossing margin are locations of projected gravity, magnetics and topographic profiles shown in Figs. 8 and 9.

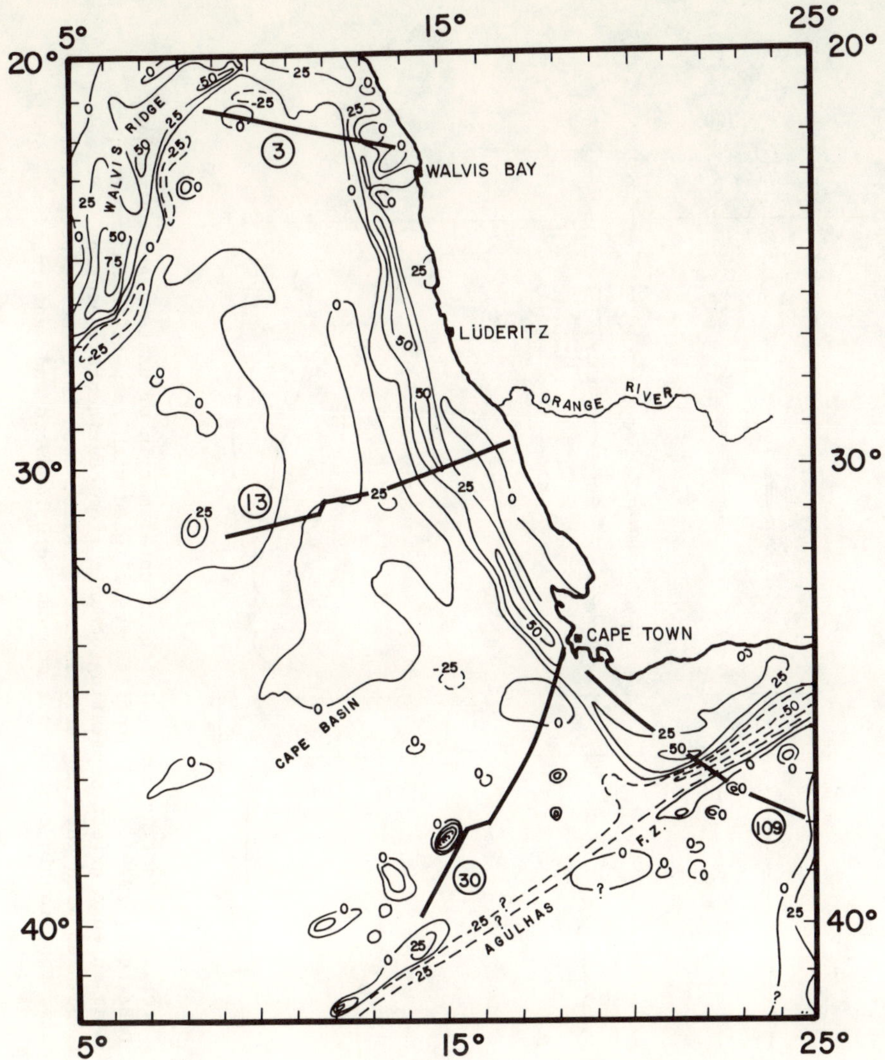

Fig. 7. Free-air gravity anomalies, continental margin of southern Africa (after Emery et al., 1975a). Contour interval 25 mgal. Heavy lines crossing margin are locations of projected gravity, magnetics and topography profiles given in Figs. 8 and 9.

and du Plessis (1968); Graham and Hales (1965); Goslin and Sibuet (1975); and Scrutton (1973, 1976). Selected free air and isostatic gravity, magnetics and topographic profiles are given for the predominantly rifted margins of eastern Argentina and its conjugate location off southwestern Africa (Figure 8) as well as for the predominantly sheared margin of the northern Falkland Plateau and its conjugate location off the Agulhas Fracture Zone (Figure 9).

One of the characteristic features of the gravity field on the Argentine side is the continuous free-air gravity high observed near the continental shelf edge. This high ranges in amplitude from approximately +25 to greater than +75 mgal. North of Rio de la Plata this high is not totally an isostatic edge effect. In particular, note the steep landward gradient and more gentle seaward gradient in the isostatic gravity high in profile 104 (Figure 8). In some regions a more seaward isostatic gravity high is observed. Note the similarity of the seaward isostatic gravity high in profile 135 to the shelf edge isostatic gravity high in profile 104 (Figure 8).

Elongate belts of free-air gravity anomalies, which closely follow the trends of the topographic lineaments, are associated with the northern edge of the Falkland Plateau, the Falkland Escarpment, and the deep ocean floor just north of the escarpment. The free-air gravity high at the northern edge of the Falkland Plateau is considerably reduced when the isostatic correction is applied (Figure 9). Farther

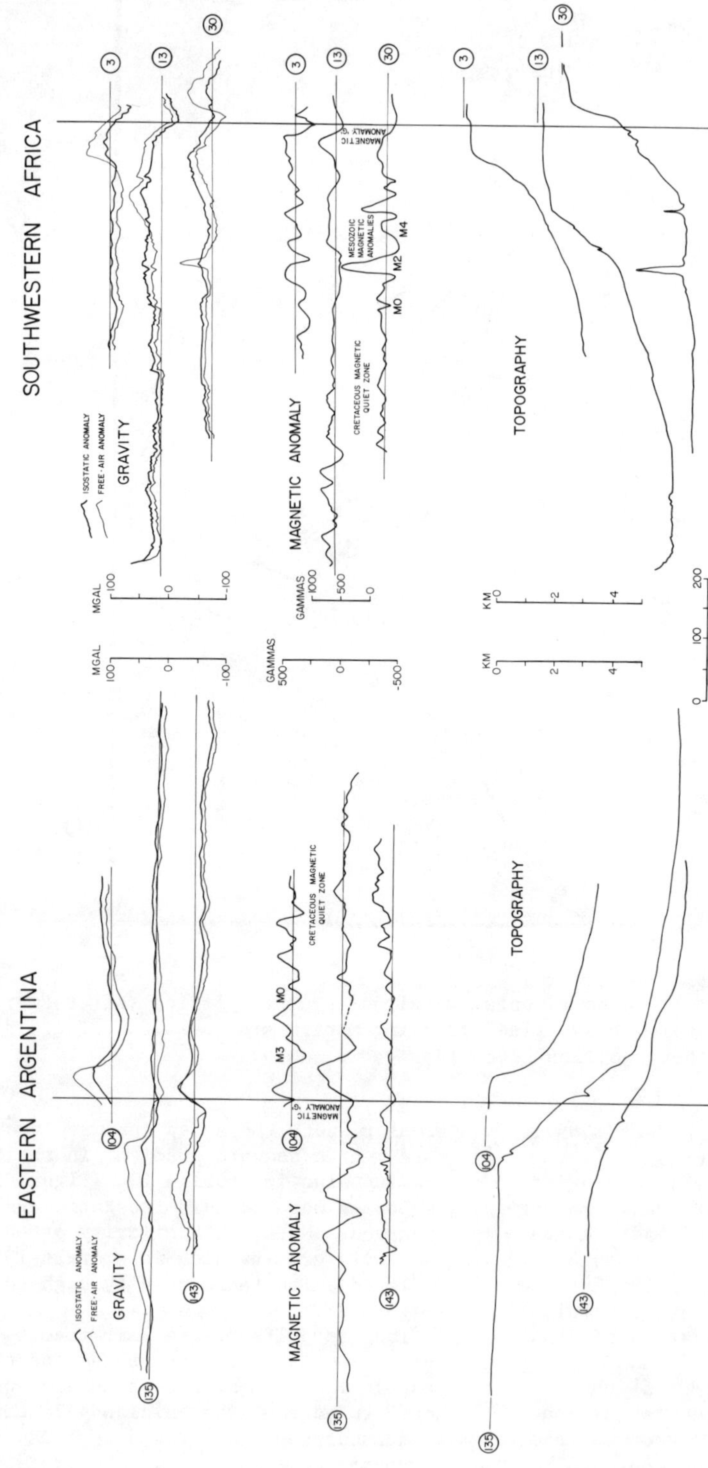

Fig. 8. Gravity, magnetics and topography profiles bordering conjugate continental margin of eastern Argentina and southwestern Africa (after Rabinowitz and LaBrecque, 1979). All profiles are aligned with respect to an important magnetic anomaly which has been interpreted as an edge effect separating oceanic from continental basement. Note a corresponding isostatic gravity gradient associated with this magnetic anomaly. Also, note that the magnetic and gravity anomalies do not follow any particular topographic contour but meander both landward and seaward of the shelf edge.

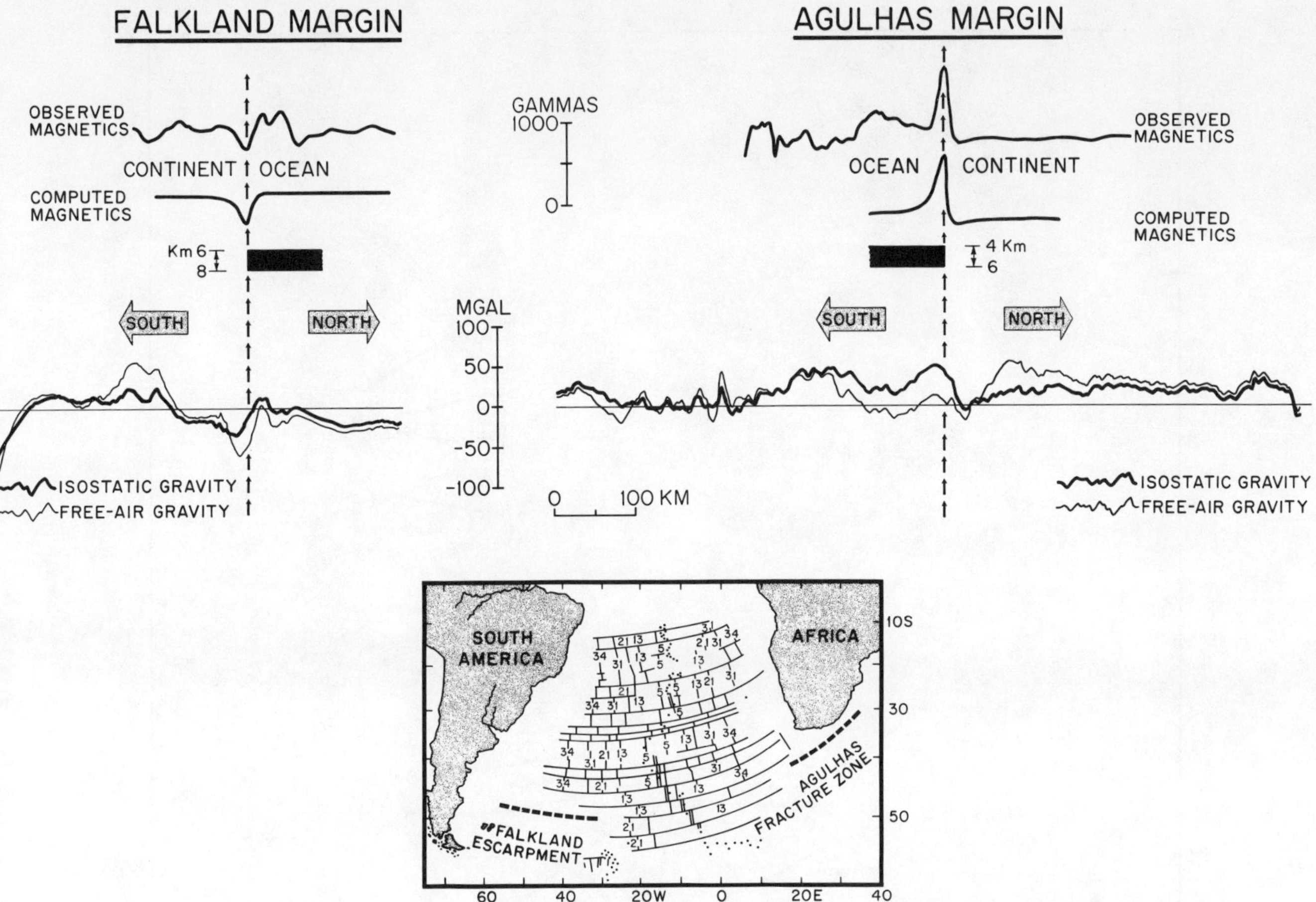

Fig. 9. Model magnetic computations for conjugate Agulhas and Falkland margins (after Rabinowitz and LaBrecque, 1979). Agulhas model assumes semi-infinite, magnetized horizontal slab (oceanic), 2 km thick, lying seaward of the fracture zone and with its edge coincident with the fracture zone. Magnetization J = 0.007 emu. Falkland model is similar to Agulhas with magnetization J = 0.012 emu. Both computations assume remanent magnetization using Mesozoic African pole of McElhinny (1973) 65°S, 82°E. Note that when oceanic basement is located south of ocean-continent boundary, the magnetic anomaly is positive (Agulhas); when oceanic basement is north of boundary, the magnetic anomaly is negative (Falkland). On both margins a gradient in the isostatic gravity anomaly is observed coincident with magnetic anomaly.

north, an important relative high in the free air gravity anomaly is observed, which is _enhanced_ when the isostatic correction is applied (Figure 9). This is true for most profiles across the Falkland Escarpment (Rabinowitz et al., 1976).

Off southwestern Africa one of the characteristic features of the gravity map (after Emery et al., 1975a) is the continuous free-air gravity high near the shelf edge which attains values, in places, greater than 50 mgals. Seaward of this shelf-edge high negative anomalies are observed on the continental rise/slope area. Emery et al., (1975a) noted that in selected regions these negative anomalies are not, in general, as low as those values observed bordering many of the other rifted continental margins, and hence, the shelf-

edge high - continental rise/slope low do not totally constitute an "edge effect." Isostatic computations (Figure 8) indeed show that an isostatic anomaly is present bordering most of the southwestern continental margin of Africa. These anomalies are similar in shape to those observed bordering the Argentine continental margin. In general, the isostatic anomaly shows a sharp landward gradient and a more gentle seaward gradient. Rabinowitz (1976) has shown that the isostatic anomaly is situated at or landward of the shelf break for the region north of Cape Town (Profiles 3 and 13; Figure 8). South of approximately the latitude of Cape Town the isostatic anomalies are situated seaward of the shelf break (Profile 30; Figure 8).

A free-air gravity high which attains values

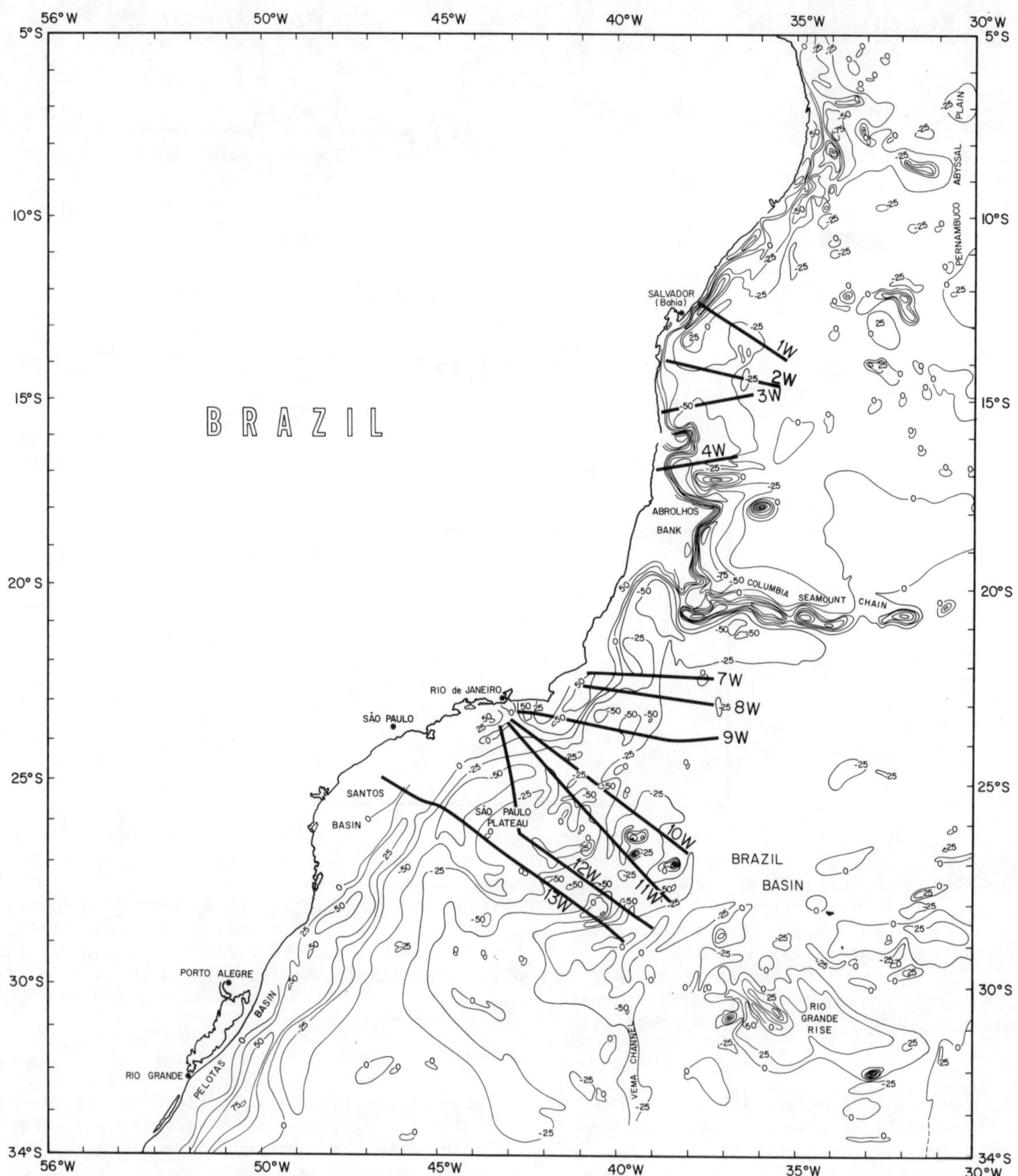

Fig. 10.   Free-air gravity anomalies bordering Brazil (after Rabinowitz and Cochran, 1979).
Contour interval 25 mgal.   Heavy lines crossing margin are locations of projected gravity and
topography profiles given in Fig. 12.

more positive than 75 mgals, is observed near the
shelf edge of southern South Africa (Agulhas
Fracture Zone region).   This high is considerably
reduced in amplitude or eliminated when the isostatic correction is applied (Figure 9) as noted
by Talwani and Eldholm (1973) and Rabinowitz
(1976).   The continental slope in this region is
characterized by anomalies, in places, more negative than -75 mgals.   To the south of this low an
important free-air gravity high is observed,
which reaches values, in places, greater than 100
mgals.   This seaward free-air gravity high is
present and in some places enhanced when the isostatic corrections are applied in all profiles
across the Agulhas Fracture Zone (Figure 9).

Linear magnetic anomalies are observed coinci-

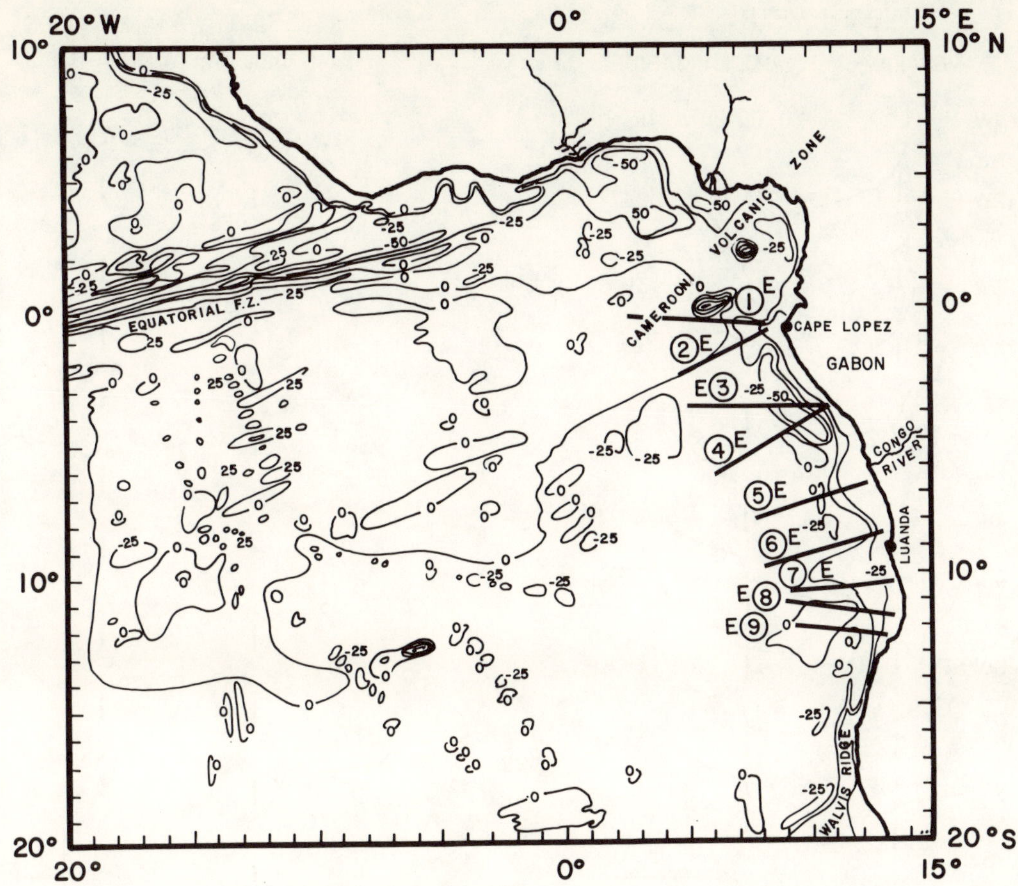

Fig. 11. Free-air gravity anomalies bordering western Africa (after Emery et al., 1975b). Contour interval 25 mgal. Heavy lines are locations of projected gravity and topography profiles given in Fig. 12.

dent in places with isostatic gravity anomalies bordering the conjugate continental margins of southern Africa and South America. The Agulhas Fracture Zone, where bounded to the north by the African continent, is characterized by a nearly linear high amplitude (300-800 gammas) positive magnetic anomaly. The conjugate segment at the base of the Falkland Escarpment is characterized by nearly linear negative anomalies. Magnetic quiet zones are observed on their landward sides. Simple magnetic edge effect computations that assume that magnetized oceanic crust is abutting a less magnetized continental basement, reproduce very nicely the shape of the anomalies for both the Agulhas and Falkland Margins (Figure 9). In both cases these magnetic edge effect anomalies are associated with the seaward isostatic anomalies discussed earlier. In a similar fashion, nearly linear magnetic anomalies, perhaps similar in origin to the celebrated East Coast magnetic anomaly, are observed and are coincident with the isostatic gravity anomalies on both the rifted segment of Argentina and southwestern Africa.

It is important to note that spatially coinci-

dent isostatic gravity and magnetic anomalies, and hence the presumed ocean-continent boundaries, are independent of their location from the shelf break. For example, southwest of Cape Town, off western Africa, both of the anomalies are situated seaward of the shelf break (Profile 30; Figure 9); farther north, at about the latitude of the Orange River, the anomalies are landward of the shelf break (Profile 13; Figure 9).

North of Rio Grande Rise - Walvis Ridge Lineaments

A free-air gravity map of the continental margin of Brazil, is shown in Figure 10 (after Rabinowitz and Cochran, 1979). The free-air gravity anomalies for the opposing continental margin off Angola-Gabon, Africa, is given in Figure 11 (after Emery et al., 1975b). Computed isostatic gravity anomalies for the opposing margins are given in Figure 12 (after Cande and Rabinowitz, 1978).

Off Brazil, a free-air gravity high, ranging in amplitude from ∿25 mgal to more than 125 mgal, is situated near the continental shelf break. This high is not everywhere an isostatic edge effect,

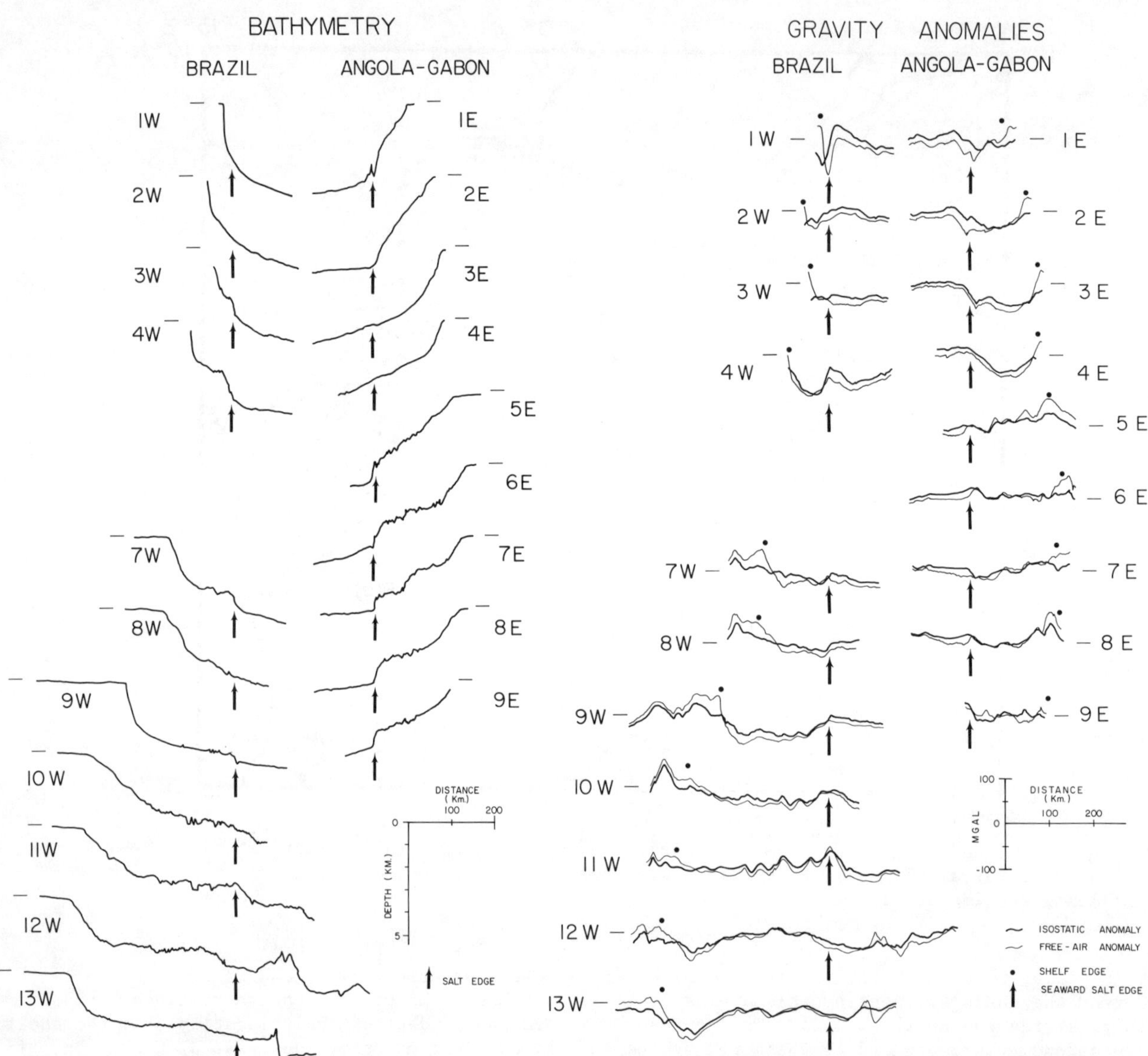

Fig. 12. Projected gravity (free-air and isostatic) and topography profiles off conjugate continental margins of Brazil and Angola-Gabon (after Cande and Rabinowitz, 1978). Locations in Figs. 10 and 11. Note that an isostatic gravity anomaly is generally observed at the seaward salt edge as well as, in places, farther landward.

but is present, in places, in the isostatic anomaly (e.g., profiles 9W, 10W, Figure 12: margin of Sao Paulo Plateau). Near Salvador (profile 1W, Figure 12) application of the isostatic correction greatly reduces the shelf edge free-air high. In this region, a very pronounced outer isostatic gravity high is present.

Diapiric structures of presumed salt origin are observed on the Sao Paulo Plateau, and extend northward to about 11°S (Butler, 1970; Leyden et al., 1976; Mascle and Renard, 1976). The Brazilian salt was probably deposited in the Aptian concurrently with deposition of salt along the conjugate Angolan-Gabon margin (Leyden et al., 1972; Rabinowitz, 1972; von Herzen et al., 1972; Pautot et al., 1973). Negative free-air and isostatic gravity anomalies lie between the continental slope and seaward edge of the salt diapir field between Abrolhos Bank and the Sao Paulo Plateau. A distinct gravity gradient (both isostatic and free-air) is associated with the seaward salt edge.

Off Africa, between the Walvis Ridge and Cape Lope (Gabon), positive free-air anomalies are observed near the continental shelf break (after Rabinowitz, 1972; Emery et al., 1975b: Figure 11).

This positive free-air anomaly is flanked on its
seaward side by a free-air gravity low. Farther
seaward the gravity pattern is irregular, as noted
by Emery et al., (1975b). An important isostatic
gravity gradient, however, is associated with the
seaward edge of the salt diapiric field (Figure
12). This salt diapir field is at the opposite
margin to the one discussed earlier off Brazil.

The isostatic gravity anomalies bordering the
conjugate margins of Brazil and Angola-Gabon,
thus show marked similarities. In most cases an
isostatic gravity gradient is observed nearly
coincident with the seaward salt edges. Isostatic
anomalies are also present, in places, farther
landward. Considerable discussion has been given
in the literature with respect to whether the salt
was deposited upon oceanic or continental basement. If rigid plate tectonics prevails, then
seafloor spreading must have started between Angola and Brazil in the Valanginian, the same
time as spreading south of the Rio Grande-Walvis
Ridge (Larson and Ladd, 1973) and the bulk of the
salt would be deposited on oceanic crust. However, if considerable necking and stretching of
continental crust has occurred, then the salt
could conceivably have been deposited on continental crust or a combination of continental and
oceanic crust. In addition, there is evidence
that strongly suggests major ridge crest migrations during the early opening of the South
Atlantic in the region of the Sao Paulo Plateau,
just to the north of the Rio Grande Rise-Walvis
Ridge complex (Rabinowitz and LaBrecque, 1979).

In view of these problems and inasmuch as isostatic anomalies are present, in places, both at
the seaward edge of the salt as well as farther
landward, the utilization of this anomaly as
diagnostic of the ocean continental boundary as
presented earlier, is further complicated.

Norwegian Continental Margin

The evolution of the Norwegian-Greenland Sea
and the continental margin of Norway, has been
extensively discussed in the literature (e.g.,
Talwani and Eldholm, 1977). In Figure 13, a
free-air gravity map of part of the Norwegian
Margin (Voring Plateau) is given at a 10 mgal
contour interval (Talwani and Eldholm, 1972;
Gronlie and Talwani, 1978). In Figure 14, projected isostatic and free-air gravity profiles,
Magnetic profiles, as well as a composite seismic
refraction section, is shown across the Voring
Plateau Escarpment (after Talwani and Eldholm,
1972).

A free-air gravity high with values, in places,
greater than 100 mgals, is observed near the shelf
edge. This high is also observed when the isostatic correction is applied. Seaward of this
high a gravity low is observed over the Voring
Plateau. Farther seaward over the Voring Plateau
Escarpment a steep landward gradient is observed
in both the isostatic and free-air gravity fields.
Talwani and Eldholm (1972) have shown that oce-

anic basement can be traced landward from the
basins of the Norwegian Sea onto the Voring Plateau where it abruptly terminates at the seaward
edge of a deep sedimentary basin. Magnetic sea-
floor spreading lineations are observed seaward
of the basin; a magnetic quiet zone is observed
over the basin. The magnetic anomaly just seaward of the magnetic quiet zone and associated
with the Voring Plateau Escarpment, was modelled
by Talwani and Eldholm (1972, 1973) as a magnetic edge effect separating oceanic from continental basement. The isostatic anomaly associated with this presumed magnetic edge effect a-
nomaly was termed diagnostic of the boundary between oceanic and continental crust (Talwani and
Eldholm, 1973) and the magnetic quiet zone was
interpreted as being situated over subsided
continental crust.

Continental Margin of Eastern Africa

The continental margin of part of eastern Africa (Kenya and Northern Tanzania) most probably
evolved as a result of transform motion associated with the separation of Madagascar from the
African continent (McElhinny et al., 1976;
Heirtzler and Burroughs, 1971; Bunce and Molnar,
1977; Scrutton, 1978). In Figure 15 a free-air
gravity map is given for this margin. The map
is characterized by alternating belts of highs
and lows. In particular, a relative free-air
gravity high, which trends approximately north-
south, appears to extend onshore near 2.6^{o}S.
This important gravity high lies on a trend
which when extrapolated southwards, connects
with Davie Ridge (Scrutton, 1978). This ridge
is a presumed manifestation of the transform
motion of Madagascar from Africa.

The topography in the region south of 2.6^{o}S
has small gradients and the isostatic edge effect is thus small (Figure 15, after Rabinowitz, 1971). Therefore, the highs and lows observed in the free-air anomaly are not artifacts
of topography and its compensation, but should
reflect structural changes within the crust.
It has been suggested that the free-air (and isostatic) gravity anomaly arises from a basement
elevation which separates continental basement in
the west from oceanic basement to the east (Rabinowitz, 1971).

Continental Margin of Western India

A free-air gravity map of the continental margin of western India is shown at a 25 mgal interval in Figure 16 (after Naini, 1980). The regional field is negative over the Arabian Sea. A
linear belt of positive anomalies with values a-
bout 50 mgals is observed near the shelf break;
seaward of the shelf edge positive, a negative
belt of anomalies is observed on the continental
slope/rise with values more negative than -75
mgals. This latter trend of anomalies is interrupted to the southwest of Goa by a relative positive anomaly associated with the Chagos-Laccadive

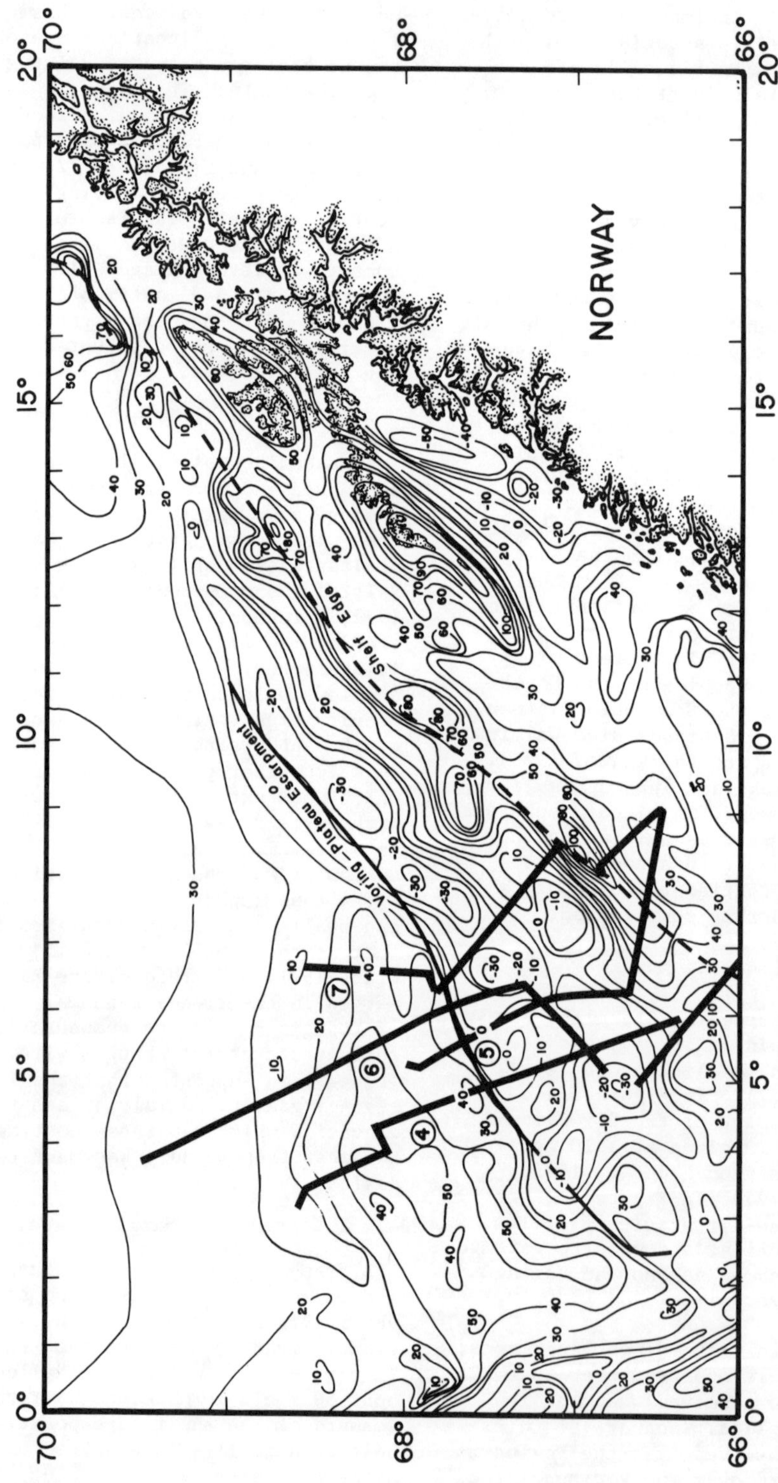

Fig. 13. Free-air gravity anomalies, continental margin of Norway (after Gronlie and Talwani, 1978). Contour interval 10 mgal. Heavy lines across margin show location of projected profiles given in Fig. 14.

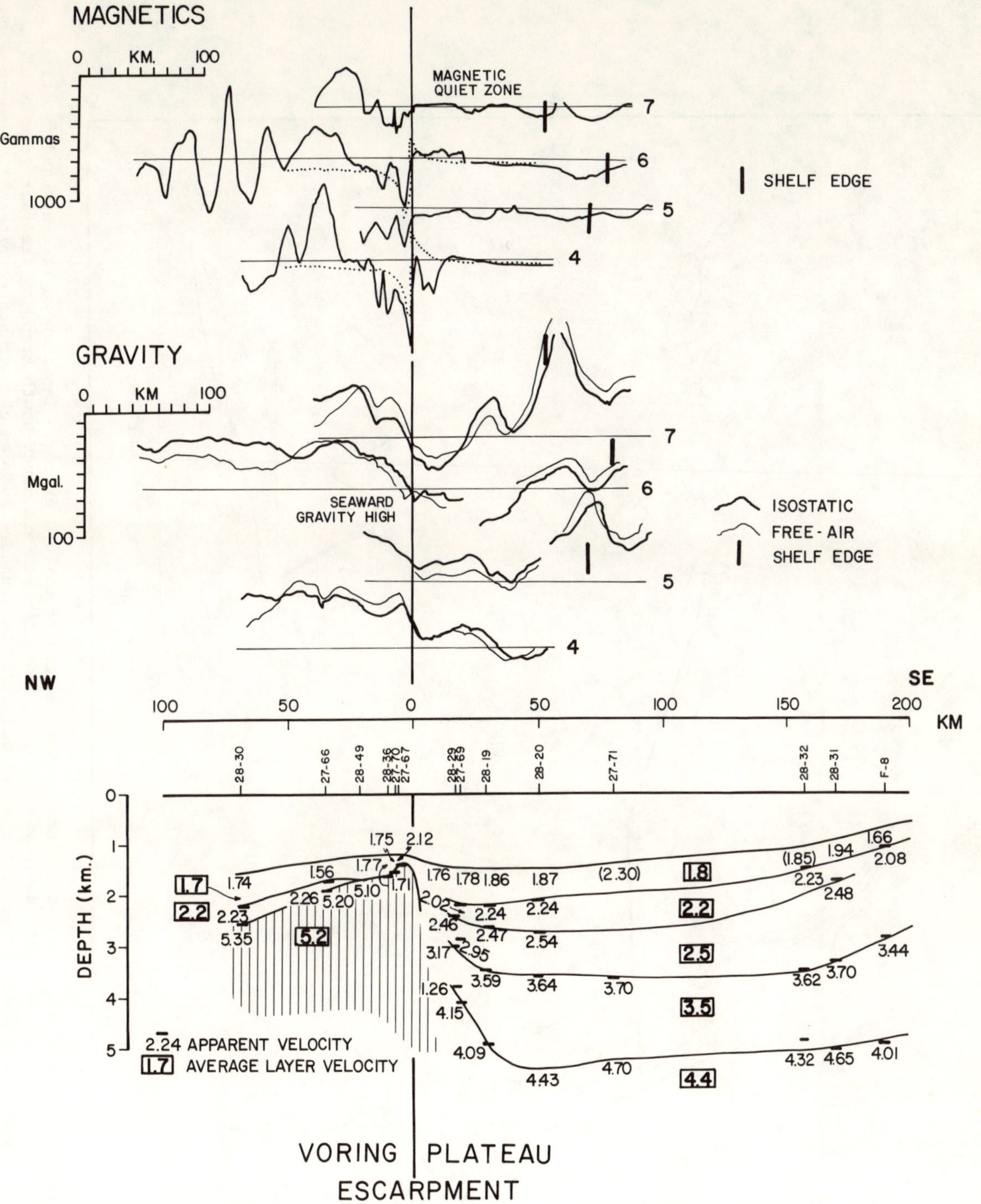

Fig. 14. Projected gravity (free-air/isostatic) and magnetic profiles as well as composite seismic refraction section across Voring Plateau Escarpment (after Talwani and Eldholm, 1972). Location of profiles given in Fig. 13. Note magnetic edge effect anomaly (model shown dotted) over Voring Plateau Escarpment. Landward of this anomaly a magnetic quiet zone is observed; seaward of this anomaly seafloor spreading lineations are observed as well as a seaward isostatic gravity high.

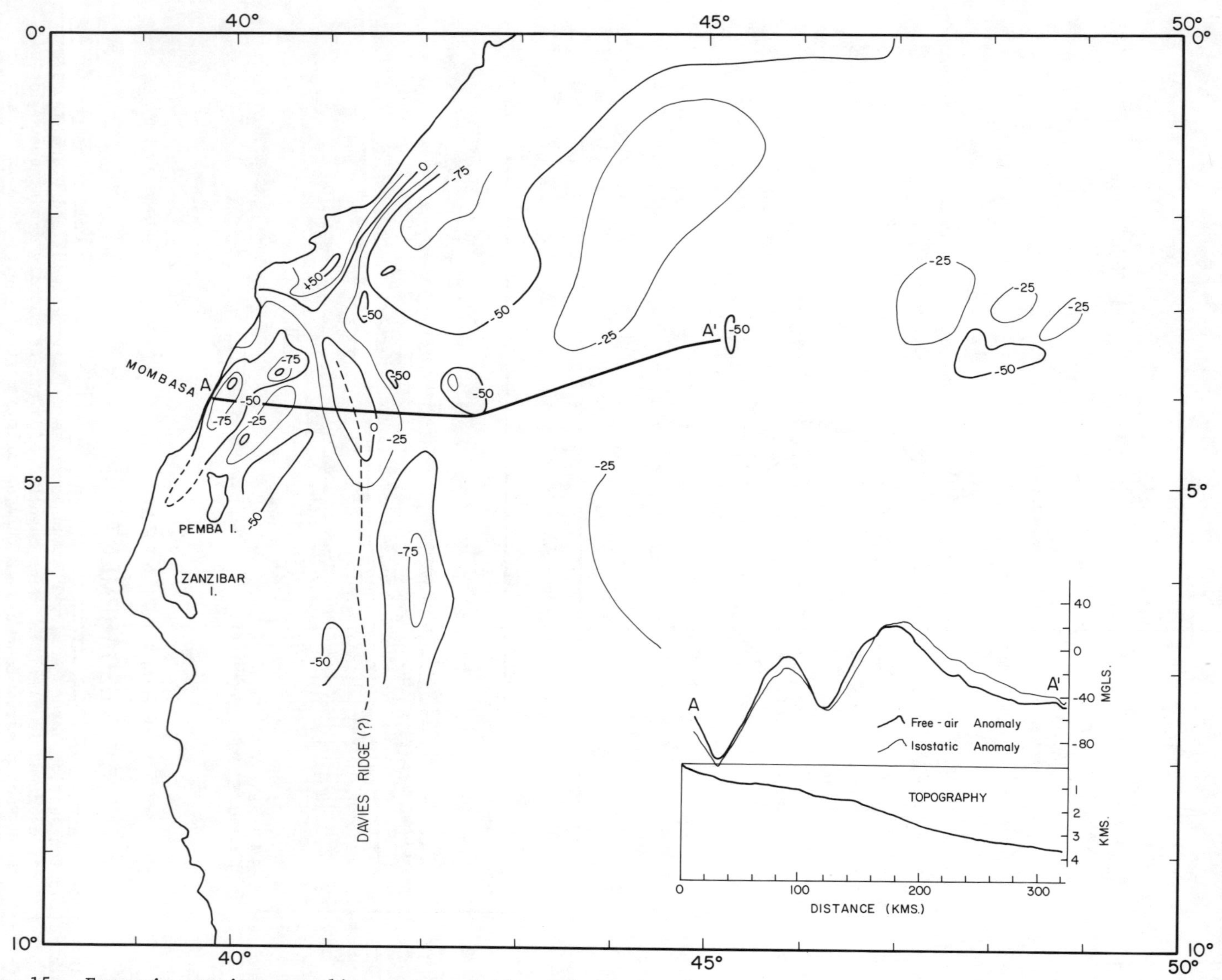

Fig. 15. Free-air gravity anomalies across continental margin of eastern Africa. Inset shows free-air and isostatic gravity and topography profile (after Rabinowitz, 1971). Note approximate north-south trending free-air gravity high. This high may connect with Davie Ridge to the south and mark the boundary between oceanic and continental basement.

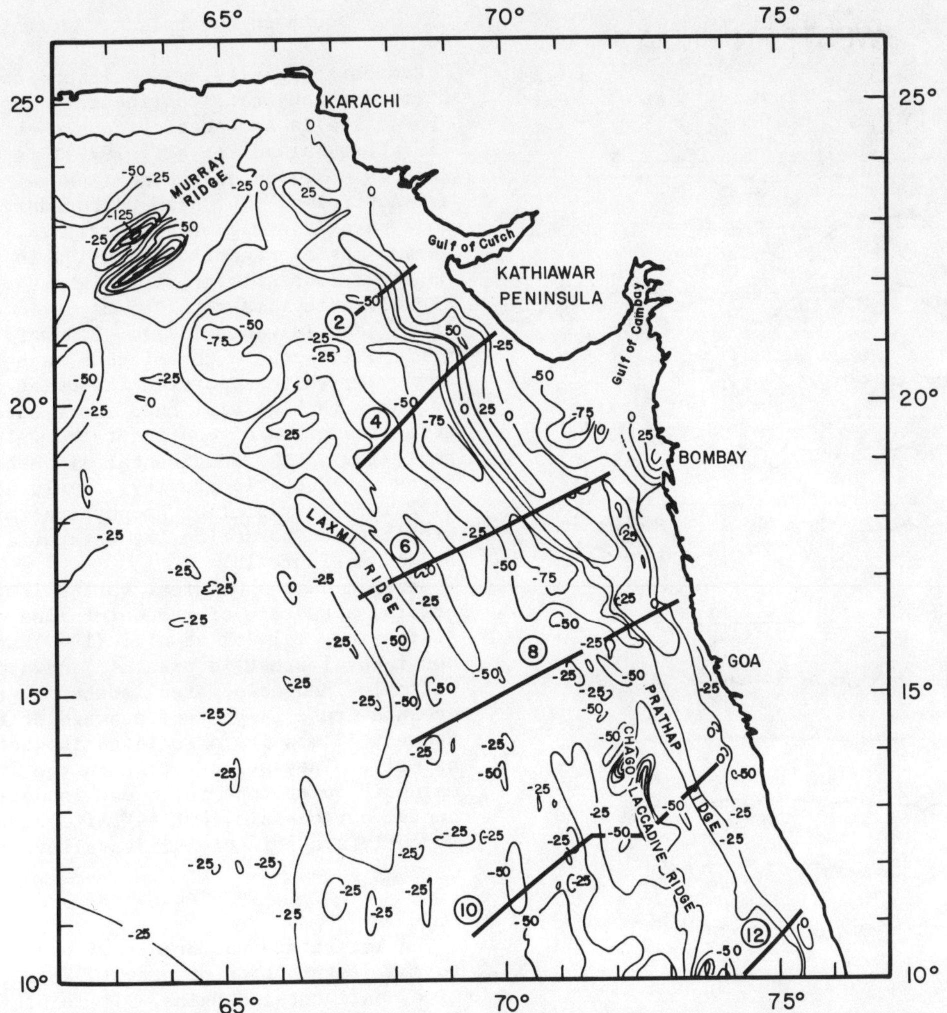

Fig. 16. Free-air gravity anomalies off continental margin of western India (after Naini, 1980). Contour interval 25 mgal. Heavy lines crossing margin are locations of projected gravity, magnetic and topography profiles given in Fig. 17.

Ridge. Between the latitudes of Goa and Kathiawar Peninsula a positive anomaly is observed seaward of the continental rise low, which coincides with a broad structural high to the northeast of Laxmi Ridge.

The isostatic gravity anomalies, as shown in profile form (Figure 17, after Naini, 1980) are more subdued than the free-air anomalies. In the northern profiles (profiles 2 and 4, Figure 17) the isostatic correction reduces but does not eliminate the shelf-edge free-air gravity high; a more seaward outer isostatic gravity high is also present. In the southern profiles (profiles 8, 10, 12, Figure 17) no marked isostatic anomalies are observed. Naini (1980) suggests that the outer gravity high observed on the northern profiles, may be a manifestation of an extinct spreading center. Although Naini (1980) reports that some magnetic anomalies may be traced for short distances over the continental margin, there are no distinct or conspicuous magnetic anomalies associated with the gravity anomalies.

West of the outer gravity high, Naini (1980) reports average seismic velocity structures similar to that observed over typical ocean basin, i.e., a 1.7 km thick 5.5 km/sec layer overlying a 3.0 km thick 6.7 km/sec layer, overlying mantle at average depths of ~11.5 km below sea level with an average velocity of 8.1 km/sec. In the region of the outer gravity high and to the east, the average crustal structure is somewhat different and resembles the velocity structures bordering the other passive margins (Drake and Nafe, 1968). A 1.6 km thick 5.4 km/sec layer overlies a 2.8 km thick 6.3 km/sec layer, which overlies a layer with a velocity of 7.2 km/sec.

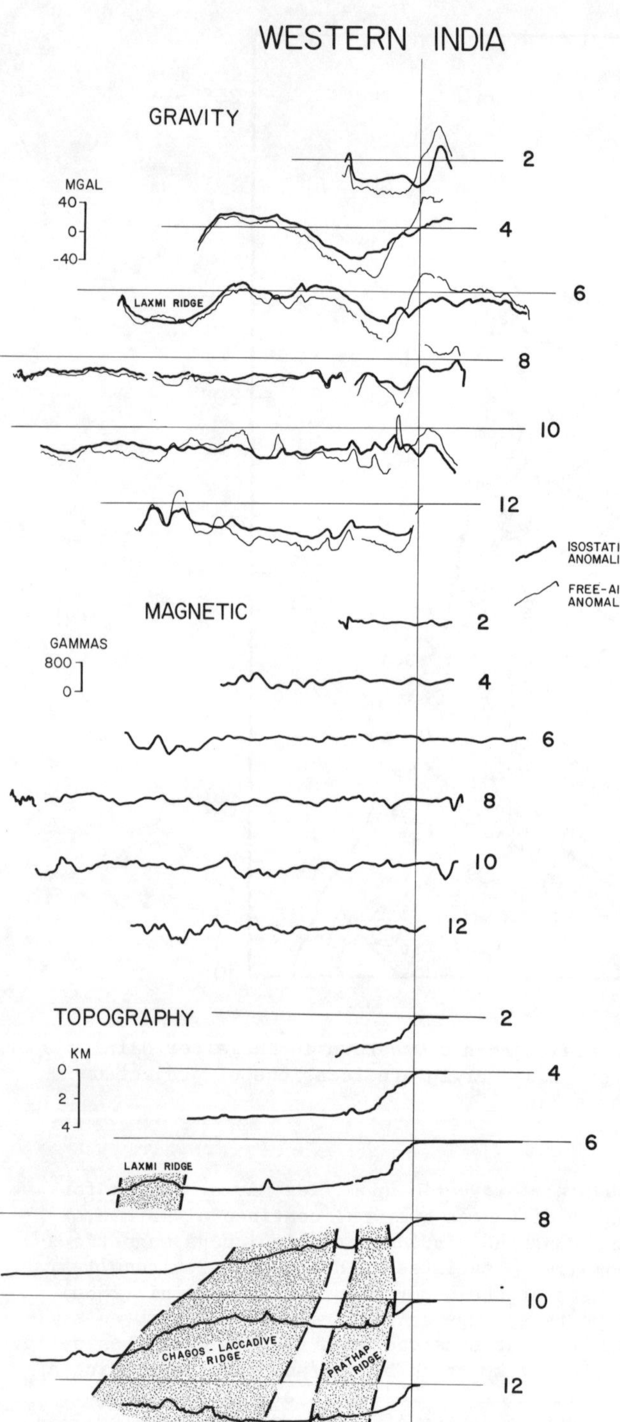

Fig. 17. Projected gravity (free-air and iso-static) magnetic and topography profiles off continental margin of western India (after Naini, 1980).

Composite gravity, magnetic and topographic pro-files for conjugate continental margins of south-ern Australia and Antarctica are shown in Figure 18 (after Talwani et al., 1979). A magnetic quiet zone is observed on the continental rise of south-ern Australia. The seaward boundary of the quiet zone parallels magnetic anomaly 22, the oldest ob-served marine magnetic lineation in this region. An outer isostatic gravity high is observed which closely coincides with the magnetic anomaly 22 (Konig and Talwani, 1977). Landward of the mag-netic quiet zone a conspicuous magnetic low is observed, which separates the quiet zone from a region with high amplitude short wavelength mag-netic anomalies of continental origin. This mag-netic low on the continental slope has in most cases an isostatic gravity anomaly associated with it. Very similar geophysical measurements are observed bordering the Antarctic margin, as shown in Figure 18.

Seismic data show great variability in the crustal structure of the quiet zone off southern Australia. Talwani et al., (1979) suggest that continental crust is present landward of the mag-netic low and associated isostatic gradient, and oceanic crust is present seaward of the magnetic anomaly 22 and its associated isostatic gravity anomaly. They suggest that in the intervening magnetic quiet zone the crust is neither conti-nental nor oceanic, but a "rift crust" which was formerly a continental rift valley.

Discussion

The literature has been reviewed with respect to the distribution of gravity anomalies on pas-sive continental margins. It is clear that e-longate belts of free-air gravity anomalies paralleling the margins, are indeed ubiquitous to the passive continental margins. A free-air gravity high is observed near the shelf break, a free-air gravity low is observed on the conti-nental rise/slope, and in places, a relative free-air gravity high is observed farther sea-ward.

In general, an isostatic gravity anomaly is present when the simple Airy isostatic correc-tion is applied to the free-air gravity anomaly. This anomaly is not necessarily dependent on its location with respect to the shelf edge. In par-ticular, off southwestern Africa and Argentina, an isostatic gravity anomaly has been shown to be, in places, at the continental shelf edge as well as, in places, either landward or seaward of the shelf edge.

In many areas a rather pronounced magnetic a-nomaly is associated with the continental margin isostatic gravity anomaly. This magnetic anoma-ly can, in many regions, be modelled as a mag-netic edge effect separating oceanic from conti-nental basement. Further, this magnetic anomaly, when used as an ocean continent boundary indicator,

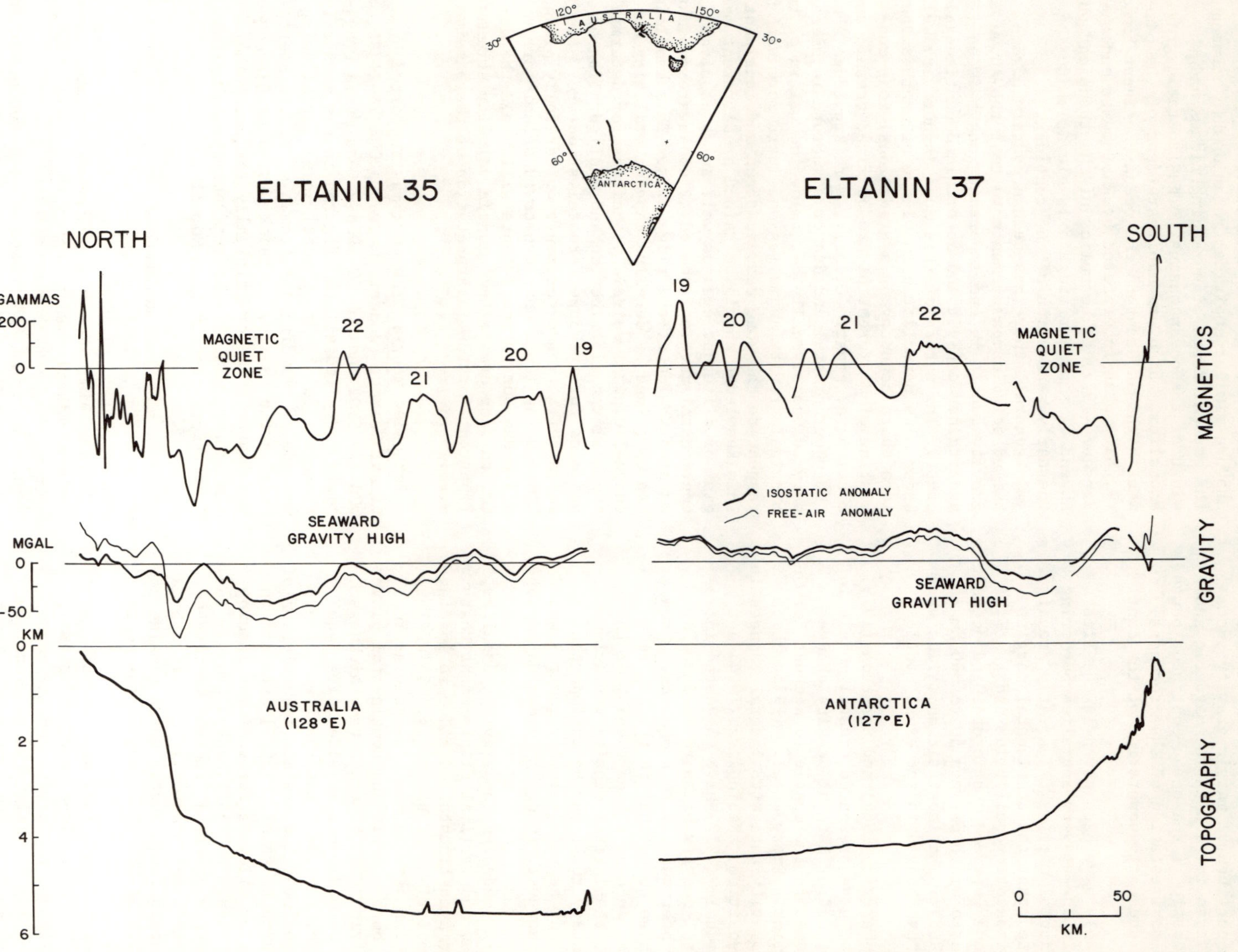

Fig. 18. Projected gravity, magnetic and topographic profiles for conjugate continental margins of Australia and Antarctica (after Talwani et al., 1979). Note marked similarities in both profiles. Magnetic quiet zones are observed over slope/rise areas. Landward of the quiet zones magnetic negatives are observed associated with isostatic gravity anomalies. Seaward of the magnetic quiet zones magnetic seafloor spreading lineations are observed as well as seaward isostatic gravity highs.

matches very well on opposing sides of the mid-ocean ridge axis in paleo-reconstructions, e.g.,
Rabinowitz and LaBrecque, 1979; Klitgord and
Behrendt, 1979. These observations strongly suggest that the continental margin gravity anomalies that are associated with the magnetic anomalies are diagnostic of the ocean-continent boundary and are a manifestation of the processes
active during the early separation of the continents.

Numerous models have been given in the literature to explain the gravity anomalies bordering
passive continental margins. The early gravity
models suggested that the continental margins
were in near isostatic equilibrium and that the
"transition zone" between continental and oceanic
thicknesses has a width between 50 to 300 km (e.
g., Worzel, 1968). The isostatic anomalies were
accounted for by changes in the crust mantle
interface.

Keen and Loncarevic (1966) and Scrutton (1979)
amongst others, have presented crustal models in
which vertical and/or lateral density changes
have been assumed in the mantle to account for the
gravity anomalies. Emery et al., (1975a) have
modelled the gravity data with lateral changes in
the crustal densities. Rabinowitz (1974) and
Talwani and Eldholm (1972) have suggested intra-
basement crustal density highs to account for
parts of the observed gravity anomalies. Grow et
al., (1979a) have noted that in thickly sedi-
mented margins, such as the east coast of the
United States, the lateral density variations
within sedimentary rocks may be an important effect. Sufficient seismic velocity control would
be necessary, however, in order to evaluate prop-
erly this effect.

Walcott (1972) modelled the gravity anomalies,
flexural stress and vertical displacement that can
be expected at a continental edge due to loading
by large accumulations of sediments. This flexural model, which assumes homogeneous elastic
constants for both oceanic and continental plates
and which explains the shape and the thickness
(up to 18 km) of sediments on wide delta type
passive margins, requires that the free air gravity positive lies over the maximum accumulation of
sediments. Turcotte et al., (1977) extended the
flexural model by allowing for sediment loading
of decoupled oceanic and continental crust. They
concluded that the observed free-air gravity a-
nomalies were most satisfactorily explained by a
continental margin fault system which remains active during much of the evolution of the continental margin.

The continental margin isostatic anomaly, when
associated with a magnetic edge effect anomaly,
has been modelled by Rabinowitz and LaBrecque
(1977) as resulting from elevated thickened o-
ceanic crust adjacent to the continent. These
highs, which are observed in many regions where
deep seismic measurements are available, have been
attributed to oceanic material being injected at
higher elevations during the initial rifting

stage. Once the continental edges are separated
by ∿300 km, oceanic material is extruded at elevations typical of the ridge axes in well developed basins (∿2500 m). Veevers (1977) has
shown a similar empirical relationship between
the crestal depth of spreading ridges and the
width of the ocean basins. The isostatic gravity
anomalies, which are manifestations of the basement highs adjacent to the ocean-continent boundary, have been interpreted by Rabinowitz and La-
Brecque (1977) to be relics of the transient phenomena of a higher ridge axis elevation during
the early separation of the continents.

In general, there has been a paucity of detailed seismic measurements to test the above
gravity models in the important zone between
ascertained oceanic and ascertained continental
crust. The nature of this zone is not very well
known. We have little knowledge of whether there
is a sharp boundary or a gradational zone consisting of different blocks of mixed oceanic and
continental crust. Clearly, more knowledge about
the upper crustal and deeper structure is necessary to unequivocally determine the origin of the
marginal gravity anomalies and learn more with
respect to the zone between ocean and continent.
This knowledge should be gained at different margins of different ages as well as at passive
continental margins located at conjugate positions
on either side of the ridge axis. This knowledge
can be gained by detailed mapping utilizing multi-
channel digital seismic techniques for determining
the sediment distribution and configuration of
basement for the upper crustal structure. Long
range refraction experiments on carefully planned
lines using large aperture towed seismic arrays
and ocean bottom instruments will yield the detailed deep crustal and mantle velocity structures
important for the interpretation of the passive
margin gravity anomalies.

Acknowledgments. The work was supported by National Science Foundation grants OCE 76-21786,
OCE 77-25992 and OCE 79-19389. Ana Maria Draga-
novic redrafted all the figures in the manuscript.
I am grateful to Dr. Bhoopal Naini for allowing
the use of two diagrams from his Ph.D. dissertation in the review. I thank Drs. John LaBrecque
and Steve Cande for critically reviewing the
manuscript. This is Lamont-Doherty Geological
Observatory contribution No. 3163.

References

Behrendt, J.C., J. Schlee, J.M. Robb and M.K.
Silverstein, Structure of the continental mar-
gin of Liberia, West Africa, Bull. Geol. Soc.
Amer., 85, pp. 1143-1158, 1974.
Bott, M.H.P. and A.B. Watts, Deep structure of the
continental margin adjacent to the British
Isles, in Delany, F.M. (Ed.) The geology of the
east Atlantic continental margin: Cambridge
ISCU/SCOR Working Party 31 Symposium, Report
#70/16, pp. 90-109, 1971.

Bunce, E.T. and P. Molnar, Seismic reflection profiling and basement topography in the Somali Basin: Possible fracture zones between Madagascar and Africa, J. Geophys. Res., 82/33, pp. 5305-5311, 1977.

Butler, L.W., Shallow structure of the continental margin, southern Brazil and Uruguay, Bull. Geol. Soc. Amer., 18, pp. 1079-1096, 1970.

Cande, S. and P.D. Rabinowitz, Mesozoic seafloor spreading bordering conjugate continental margins of Angola and Brazil, Proceedings of Offshore Technical Conference, Houston, Texas, Report OTC 3268, pp. 1869-1876, 1978.

Drake, C.L. and J.E. Nafe, The transition from ocean to continent from seismic refraction data, in The Crust and Mantle of the Pacific Area, Geophys. Monograph #12, L. Knopoff, C.L. Drake, P.J. Hart (Eds.) Amer. Geophys. Union, pp. 174-186, 1968.

Emery, K.O., J. Phillips, C. Bowin and J. Mascle, Continental Margin off Western Africa: Angola to Sierra Leone, Amer. Ass. Petrol. Geol. Bull., 59, pp. 2209-2265, 1975b.

Emery, K.O., E. Uchupi, J.D. Phillips, C. Bowin, E. Bunce and S. Knott, Continental rise off eastern North America, Bull. Am. Ass. Petrol. Geol., 54/1, pp. 44-108, 1970.

Emery, K. O., E. Uchupi, C. Bowin, J. Phillips and E.S.W. Simpson, Continental margin off western Africa: Cape St. Francis (South Africa) to Walvis Ridge (South-West Africa), Am. Ass. Petrol. Geol., 59/1, pp. 3-59, 1975a.

Ewing, V., Free-air gravity anomaly map of the east coast of the United States, Amer. Assoc. Petrol. Geol. Bull, in press.

Falvey, D.A., The nature and origin of marginal plateaux and adjacent ocean basins off northern Australia, Ph.D. dissertation, University of New South Wales, Australia, 239 pages, 1972.

Goslin, J. and J.C. Sibuet, Geophysical study of the easternmost Walvis Ridge, South Atlantic: Deep Structure, Bull. Geol. Soc. Amer., 86, pp. 1713-1724, 1975.

Graham, K.W.T. and A.L. Hales, Surface-ship gravity measurements in the Agulhas Bank area south of Africa, J. Geophys. Res., 70, pg. 4005, 1965.

Gronlie, G. and M. Talwani, Geophysical Atlas of the Norwegian-Greenland Sea, VEMA Research Series IV, Lamont-Doherty Geological Observatory, Palisades, New York, 1978.

Grow, J.A., C. Bowin and D.R. Hutchinson, The gravity field of the U.S. continental margin, Tectonophysics, 59, pp. 27-52, 1979a.

Grow, J.A., C. Bowin, D.R. Hutchinson, K.M. Kent, Preliminary free-air gravity anomaly map along the Atlantic continental margin between Virginia and Georges Bank, U.S. Geol. Surv. Misc. Field Ser., Map #MF-795, 1976.

Grow, J.A., R.E. Mattick and J.S. Schlee, Multichannel seismic depth sections and interval velocities over outer continental shelf and upper continental slope between Cape Hatteras and Cape Cod, in Geological and Geophysical Investigations of Continental Margins, J.S. Watkins, L. Montadert and P. Dickerson (Eds.), Am. Ass. Petrol. Geol. Memoir 29, pp. 65-83, 1979b.

Grow, J.A. and J.S. Schlee, Interpretation and velocity analysis of U.S. Geological Survey multichannel reflection profile 4, 5 and 6, Atlantic continental margin, U.S. Geol. Surv. Misc. Field Ser., Map #MF-808,1976.

Hayes, D.E. and P.D. Rabinowitz, Mesozoic magnetic lineations and the magnetic quiet zone off Northwest Africa, Earth and Plan. Sci. Letters, 28, pp. 105-115, 1975.

Heirtzler, J.R. and R.H. Burroughs, Madagascar's paleoposition: New data from the Mozambique Channel, Science, 174, pp. 488-490, 1971

Keen, M.J., Possible edge effect to explain magnetic anomalies off the eastern seaboard of the U.S., Nature, 222, pp. 72-74, 1969.

Keen, M.J., B.D. Loncarevic and G.N. Ewing, Continental margin of eastern Canada: Georges Bank to Kane Basin, in The Sea, vol. 4, A. Maxwell (Ed.) pp. 251-291, 1971.

Keen, C. and B.D. Loncarevic, Crustal structure on the eastern seaboard of Canada: Studies on the continental margin, Canadian Journal of Earth Sciences, 3, pp. 65-76, 1966.

Klitgord, K.D. and J.C. Behrendt, Basin structure of the U.S. Atlantic margin, in Geological and Geophysical Investigations of Continental Margins, J.S. Watkins, L. Montadert and P. Dickerson (Eds.), Am. Ass. Petrol. Geol. Memoir 29, pp. 85-112, 1979.

Klitgord, K.D. and H. Schouten, The onset of seafloor spreading from magnetic anomalies, in Symposium on the geological development of the New York Bight, Palisades, New York, Lamont-Doherty Geological Observatory, pp. 12-13, 1977.

Konig, M. and M. Talwani, A geophysical study of the southern continental margin of Australia: Great Australian Bight and western sections, Bull. Geo. Soc. Amer., 88, pp. 1000-1014, 1977.

Larson, R.L. and J.W. Ladd, Evidence for the opening of the South Atlantic in the Early Cretaceous, Nature, 246/5430, pp. 209-212, 1973.

Leyden, R., H. Asmus and B.G. Zembruscki, South Atlantic diapiric structures, Am. Ass. Petrol. Geol. Bull., 60/2, pp. 196-212, 1976.

Leyden, R., G. Bryan and M. Ewing, Geophysical reconnaissance on the African shelf, 2, Margin sediments from Gulf of Guinea to the Walvis Ridge, Am. Ass. Petrol. Bull., 56, pp. 682-693, 1972.

Luyendyk, B.P. and E.T. Bunce, Geophysical study of the Northwest African margin off Morocco, Deep-Sea Research, 20, pp. 537-549, 1973.

Markl, R.G., Basement morphology and rift geometry near the former junction of India, Australia and Antarctica, Earth and Plan. Sci. Letters, 39, pp. 211-225, 1978.

Mascle, J. and V. Renard, The marginal Sao Paulo Plateau, comparison with the southern Angolan margin, An. Acad. Bras. Cienc., 48 (supplemento) pp. 179-190, 1976.

Mc Elhinny, M.W., <u>Paleomagnetism and Plate Tectonics</u>, Cambridge University Press, New York, 358 pages, 1973.

Mc Elhinny, M.W., B.J. Embleton, L. Daly and J.P. Pozzi, Paleomagnetic evidence for the location of Madagascar in Gondwanaland, <u>Geology, 4</u>, pp. 455-457, 1976.

Naini, B.R., A geological and geophysical study of the continental margin of western India, and the adjoining Arabian Sea including the Indus Cone, Ph.D. dissertation, Columbia University, New York, 173 pages, 1973.

Pautot, G., V. Renard, J. Daniel and J. Dupont, Morphology limits, origin and age of salt layer along South Atlantic African margin, <u>Am. Ass. Petrol. Geol. Bull.</u>, 57, pp. 1658-1671, 1973.

Rabinowitz, P.D., Gravity anomalies across the East African continental margin, <u>J. Geophys. Res., 76/29</u>, pp. 7107-7117, 1971.

Rabinowitz, P.D., Gravity anomalies on the continental margin of Angola, Africa, <u>J. Geophys. Res., 77</u>, pp. 6327-6347, 1972.

Rabinowitz, P.D., The continental margin of the Northwest Atlantic Ocean: A Geophysical Study, Ph.D. dissertation, Columbia University, New York, 181 pages, 1973.

Rabinowitz, P.D., The boundary between oceanic and continental crust in the western North Atlantic, in <u>The Geology of Continental Margins</u>, C.A. Burk and C.L. Drake (Eds.), Springer-Verlag, New York, pp. 67-84, 1974.

Rabinowitz, P.D., Geophysical study of the continental margin of southern Africa, <u>Bull. Geo. Soc. Amer., 87</u>, pp. 1643-1653, 1976.

Rabinowitz, P.D., S.C. Cande and J. LaBrecque, The Falkland Escarpment and Agulhas Fracture Zone: the boundary between oceanic and continental basement at conjugate continental margins, <u>Continental Margins of the Atlantic Type</u>, An. Acad. Bras. Cienc., 48 (supplemento), pp. 241-251, 1976.

Rabinowitz, P.D., Gravity anomalies bordering the continental margin of Argentina, color map with text, <u>Am. Ass. Petrol. Geol. Bull., Tulsa, Oklahoma, Catalog 825</u>, 1977.

Rabinowitz, P.D. and J. LaBrecque, The isostatic gravity anomaly: a key to the evolution of the ocean-continent boundary, <u>Earth and Plan. Sci. Letters, 35</u>, pp. 145-150, 1977.

Rabinowitz, P.D. and J. Cochran, Gravity anomalies of the continental margin of Brazil, <u>Am. Ass. Petrol. Geol. Bull., Tulsa, Oklahoma, Catalog 829</u>, color map, 1979.

Rabinowitz, P.D. and J. LaBrecque, The Mesozoic South Atlantic Ocean and evolution of its continental margins, <u>J. Geophys. Res., 84</u>, pp. 5973-6002, 1979.

Roeser, H.A., K. Hinz and S. Plaumann, Continental margin structure in the Canaries, in <u>The Geology of the East Atlantic Continental Margin, 2</u>, F.M. Delaney (Ed.) ICSU/SCOR Working Party 31 Symposium, Africa Report #70/16, Inst. Geol. Sci., pp. 28-36, 1970.

Rona, P.A., J. Brakl and J.R. Heirtzler, Magnetic anomalies in the Northeast Atlantic between the Canary and Cape Verde Islands, <u>J. Geophys. Res., 75</u>, pp. 7412-7420, 1970.

Schlee, J., J.C. Behrendt, J.A. Grow, J.M. Robb R.E. Mattick, P.T. Taylor and B.J. Lawson, Regional geologic framework off northeastern United States, <u>Am. Ass. Petrol. Geol. Bull., 60/6</u>, pp. 926-951, 1976.

Scrutton, R.A., Gravity results from the continental margin of southwestern Africa, <u>Marine Geophys. Res., 2</u>, pp. 11-21, 1973.

Scrutton, R.A., Crustal structure of the continental margin south of South Africa, <u>Geophys. Journal., 44/3</u>, pp. 601-624, 1976.

Scrutton, R.A., Davie Fracture Zone and the movement of Madagascar, <u>Earth and Plan. Sci. Letters, 39</u>, pp. 84-88, 1978.

Scrutton, R.A., Structure of the crust and upper mantle at Goban Spur southwest of the British Isles - some implications for margin studies, <u>Tectonophysics, 59</u>, pp. 201-215, 1979.

Sheridan, R.E., J.A. Grow, J.C. Behrendt and K.C. Bayer, Seismic refraction study of the continental edge off the eastern United States, in <u>Crustal Properties across Passive Margins</u>, C.E. Keen (Ed.) <u>Tectonophysics, 59</u>, pp. 1-26, 1979.

Simpson, E.S.W. and A. du Plessis, Bathymetric magnetic and gravity data from the continental margin of southwestern Africa, <u>Can. J. Earth Sci., 5</u>, pp. 1119-1123, 1968.

Stacey, A.P., Interpretation of gravity and magnetic anomalies in the Northeast Atlantic, Ph. D. Dissertation, University of Durham, England, 1968.

Talwani, M. and O. Eldholm, Continental margin off Norway: A geophysical study, <u>Bull. Geol. Soc. Amer., 83</u>, pp. 3575-3606, 1972.

Talwani, M. and O. Eldholm, The boundary between continental and oceanic basement at the margin of rifted continents, <u>Nature, 241</u>, pp. 325-330, 1973.

Talwani, M. and O. Eldholm, Evolution of the Norwegian-Greenland Sea, <u>Bull. Geol. Soc. Amer., 88</u>, pp. 969-999, 1977.

Talwani, M., J. Mutter, R. Houtz and M. Koning, The crustal structure and evolution of the area underlying the magnetic quiet zone on the margin south of Australia, in <u>Geological and Geophysical Investigations of continental margins</u>, J. Watkins, L. Montadert and P. Dickerson (Eds.) Am. Ass. Petrol. Geol. Memoir 29, pp. 151-175, 1979.

Taylor, L. and D. Falvey, Queensland Plateau and Coral Sea Basin: Stratigraphy, Structure and Tectonics, <u>APEA Journal</u>, pp. 1-17, 1977.

Turcotte, D.L., J.L. Ahern and J.M. Bird, The state of stress at continental margins, <u>Tectonophysics, 42</u>, pp. 1-28, 1977.

Uchupi, E., L.O. Emery, C. Bowin and J.D. Phillips, Continental margins off western Africa: Senegal to Portugal, <u>Am. Ass. Petrol. Geol., 60</u>, pp. 489-497, 1976.

Veevers, J.J., Paleobathymetry of the crest of spreading ridges related to the age of ocean basins, Earth and Plan. Sci. Letters, 34, pp. 100-106, 1977.

Vening Meinesz, F.A., Gravity Expeditions at Sea, 1923-1930, Vol. 1, Publ. Lands Geodetic Commission, Delft, 110 pages, 1932.

Von Herzen, R.P., H. Hoskins and Tj. Van Andel, Geophysical studies in the Angola diapir field, Bull. Geol. Soc. Amer., 83, pp. 1901-1910, 1972.

Walcott, R.I., Gravity, flexure and growth of sedimentary basins at a continental edge, Bull. Geol. Soc. Amer., 83, pp. 1845-1848, 1972.

Worzel, J.L., Continuous gravity measurements on a surface ship with the Graf sea gravimeter, J. Geophys. Res., 64, pp. 1299-1315, 1959.

Worzel, J.L., Pendulum gravity measurements at sea, John Wiley and Sons, New York, 422 pages, 1965.

Worzel, J.L., Advances in marine geophysical research of continental margins, Can. J. Earth Sci., 5, pp. 963-983, 1968.

Worzel, J.L. and G.L. Shurbet, Gravity interpretations from standard oceanic and continental crustal sections, Geol. Soc. Amer. Spec. Paper 62, pp. 87-100, 1955.

EVAPORITES AT PASSIVE MARGINS

Peter A. Rona

Atlantic Oceanographic and Meteorological Laboratories
National Oceanic and Atmospheric Administration
4301 Rickenbacker Causeway
Miami, Florida   33149

Abstract. The distribution of Atlantic evaporites in space and time is synthesized and examined as a documented example of the association
between evaporites and passive continental margins of a major opening ocean basin. Evaporites
at passive continental margins occupy provinces
that are controlled by the tectonic setting created by rifting and early opening that acts to
restrict oceanic circulation, climatic factors
that favor evaporation in excess of precipitation
and inflow, and eustatic sea level changes that
enforce barriers to circulation. Atlantic evaporites broadly occupy three provinces that correspond to the independently determined history of
opening of the Northeast Atlantic, Central North
Atlantic, and South Atlantic ocean basins and
exemplify relations between evaporites and passive margins.

Introduction

The understanding of evaporites at passive continental margins has undergone a conceptual
breakthrough in conjunction with the development
of the theory of plate tectonics. When continental drift was first advocated, emphasis was
placed on evidence from the continents including
their morphological fit, structural provinces,
and records of paleoclimate and of paleomagnetic
poles (Wegener, 1924; du Toit, 1937; Blackett,
1961; Runcorn, 1962; Holmes, 1965; King, 1967).
Introduction of the concept of seafloor spreading
(Hess, 1962; Dietz, 1961) shifted the locus of
evidence from the continents to the ocean basins
focusing on the history of magnetic reversals in
the polarity reversal time scale (Vine and
Matthews, 1963) and the paleo-oceanographic
record of a widening and deepening ocean environment. Incorporation of the concepts of continental drift and seafloor spreading into the theory
of plate tectonics (Morgan, 1968; LePichon, 1968)
placed opening ocean basins into a global framework of plate motions. An association between
the deposition of evaporites and conditions of
restricted circulation created by the tectonic

settings of rifting and early opening of an ocean
basin was first hypothesized based on geophysical
evidence of possible deep-sea salt diapirs in the
Central North Atlantic Ocean basin, using the
Dead Sea and the Red Sea as actualistic models
(Rona, 1969, 1970; Schneider and Johnson, 1970).
Geological, geophysical and geochemical data subsequently collected during the past decade by
academic and industrial scientists involved in
research cruises, offshore drilling for hydrocarbons, and the Deep Sea Drilling Project have
broadly delineated evaporite deposits of the continental margins of the North Atlantic and South
Atlantic and have confirmed the hypothesized
association of these deposits with rifting and
early opening stages. This paper synthesizes and
examines the distribution of Atlantic evaporites
in space and time as a documented example of the
association between evaporites and passive continental margins of a major opening ocean basin.

Distribution of Atlantic Evaporites

Methods of Determination

The distribution of evaporite deposits of the
Atlantic is illustrated in Figure 1. The location of each evaporite deposit, its tectonic setting, association of minerals present, mode of
occurrence (beds or diapirs), maximum known
thickness (beds), age, and sources of information
are specified in Table 1. Beds refers to strata
that are undeformed or partially deformed. The
maximum thickness of evaporites present is given
where known from physical evidence (drilling,
outcrop, seismic reflection and/or refraction).
The thickness given is that of bedded deposits
and not of diapirs. Theoretical computations are
not used as evidence for thickness of an evaporite deposit. However, it is useful to recall
that theoretical studies indicate that rock salt
(halite) thicknesses of the order of hundreds of
meters beneath a sedimentary overburden of at
least 600 m are considered necessary to produce
diapirism (Nettleton, 1934; Parker and McDowell,

1955). The term "salt" is used to mean rock salt (halite), in distinction to "evaporite" which includes the suite of evaporite minerals.

Evidence for the occurrence of the evaporite deposits described in Table 1 derives from the distinctive physical and chemical properties given in Table 2. Drilling and outcrops furnish direct evidence of the presence of evaporites. Indirect evidence of their presence is furnished by the following methods:

1. Seismic reflection and refraction measurements based on density-velocity contrasts between evaporites and surrounding sediment and delination of characteristic structural features of evaporites (Lohmann, 1979).
2. Magnetic measurements based on the amagnetic properties of evaporites relative to surrounding sediment.
3. Gravity measurements based on the density differential between evaporites and surrounding sediment. In practice, the density of evaporite deposits varies widely depending on the mass of associated caprock and other factors.
4. Thermal gradient and heat flow measurements based on the high conductivity of anhydrite and halite relative to surrounding sediment.
5. Salinity and chlorinity gradient measurements of interstitial water in unconsolidated sediment over evaporite deposits based on the high solubility of certain salts, especially halite, in water.

Seismic, magnetic, and gravity measurements alone may be inadequate to unambiguously distinguish diapirs of salt from diapirs of mud or igneous rock. Thermal gradient and salinity measurements used in conjunction with the other geophysical methods can distinguish salt diapirs from those of other materials.

Salinity gradients in interstial water of unconsolidated sediment have been effectively used by the Deep Sea Drilling Project to indicate the presence of both salt diapirs and beds beneath the seafloor (Manheim et al., 1973). As a consequence of the solubility of halite and rapid rates of ionic diffusion, vertical salinity gradients may develop over halite deposits through several kilometers of water-saturated unconsolidated sediment overburden (Manheim, 1970). The horizontal salinity gradients that develop are approximately equal to the vertical gradients, so that the distribution of vertical salinity gradients effectively delineates the horizontal extent of an underlying salt deposit (Manheim and Bischoff, 1969). The presence or absence of salt based on vertical salinity gradients measured in interstitial pore waters in sediment cores recovered at sites in the Atlantic by the Deep Sea Drilling Project is shown on Figure 1; measured values of chlorinity and salinity are presented in Table 3.

The ages of the evaporite deposits listed in Table 1 are based on various criteria, in order of increasing reliability:

1. Stratigraphic relations: The stratigraphic position of evaporite beds or the base of salt diapirs in a known stratigraphic sequence.
2. Associated strata: Paleontologic or radiogenic dating of strata intercalated with bedded salt or incorporated in salt diapirs.
3. Caprock: Palynologic dating of the caprock of a salt diapir.
4. Salt: Palynologic dating of the salt of diapirs or beds.

Major Features of the Distribution

The information presented in Figure 1 and Table 1 broadly conveys the distribution of Atlantic evaporites, but is incomplete in detail. Evaporites are generally detected in their most spectacular manifestation as salt diapirs. Extensive areas of relatively thick bedded evaporites may remain undetected beneath Atlantic continental margins and the adjacent ocean basin. For example, layers of competent materials such as carbonates and basalt flows and sills may suppress diapirism and mask underlying salt beds. However, major features of the distribution of Atlantic evaporites are deduced, as follows (Figure 1, Table 1):

I. Evaporites are present along those rifted portions of the continental margins of North America, South America, Africa, and Eurasia that trend nearly perpendicular to fracture zones of the Atlantic Ocean basin.
II. Evaporites are absent along those sheared portions of the equatorial continental margins of South America and Africa that trend nearly parallel to fracture zones of the Atlantic Ocean basin.
III. Evaporites are absent in the South Atlantic south of the Rio Grande Rise and the Walvis Ridge.
IV. Evaporite deposits of rifted continental margins may extend continuously seaward in basins that open seaward beneath the continental margin and the adjacent deep ocean basin over both continental and oceanic crust. Examples where evaporites are inferred to extend seaward over oceanic crust are seaward of the Grand Banks, south of Newfoundland, beneath the continental rise off northwest Africa and beneath the São Paulo Plateau off Brazil.
V. The farthest seaward known extent of salt deposits in the Atlantic is beneath the lower continental rise off northwest Africa at least 450 km from the coast, as predicted from geophysical measurements (Rona, 1969, 1970) and confirmed by measurement of salinity gradients (DSDP sites 139, 140; Waterman et al., 1972). Salt deposits beneath deposits beneath the São Paulo Plateau extent 700 km seaward from the coast (Leyden et al., 1978; Kumar and

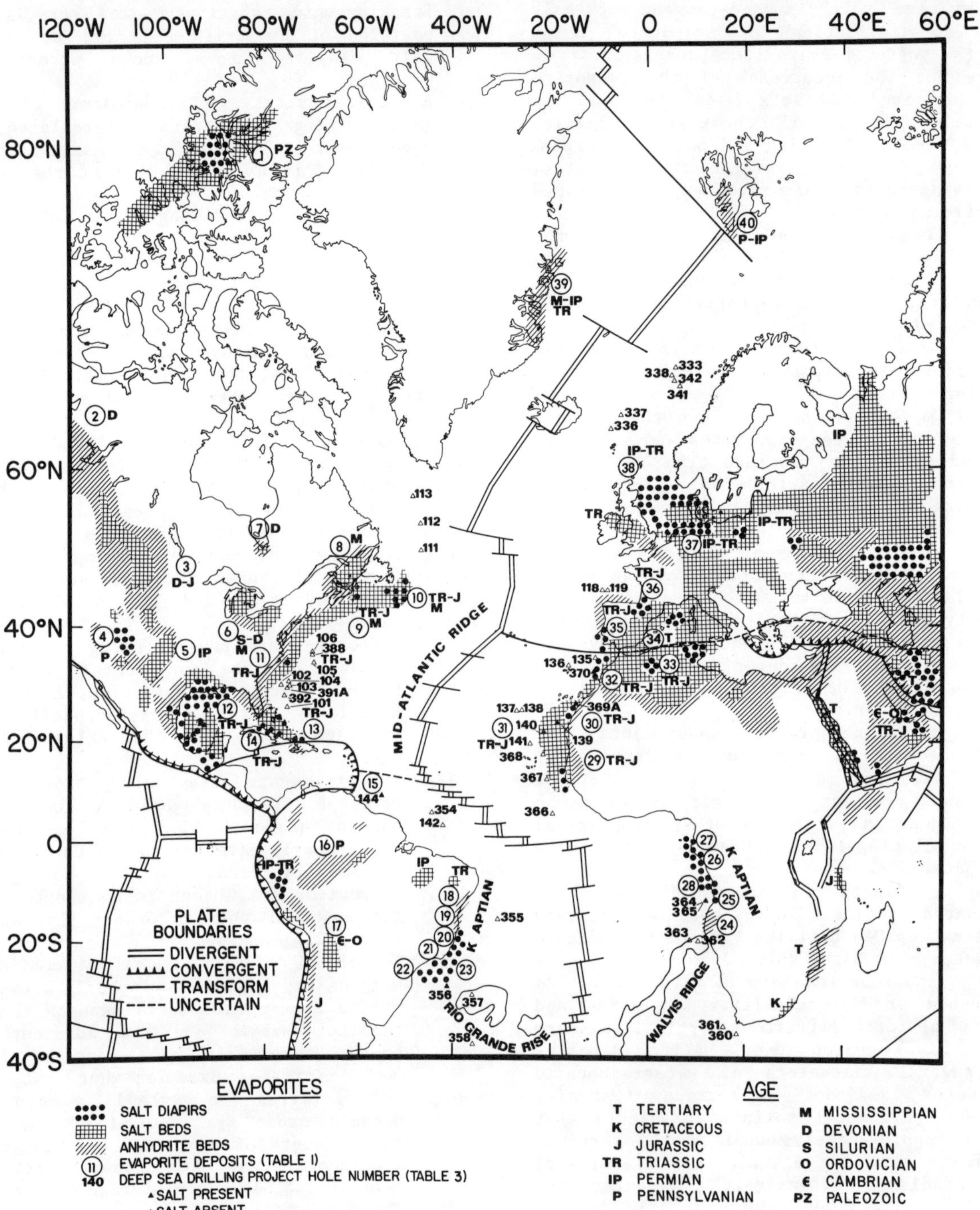

Fig. 1. Generalized distribution of evaporites at passive continental margins and on cratons around the Atlantic Ocean. Information on evaporite deposits at the numbered sites is presented in Table 1.

Gamboa, 1979). According to continental drift reconstructions of the Mesozoic opening of the Atlantic (Dietz and Holden, 1970), the extent of salt deposits off northwest Africa represents a half-width of opening. The extent of salt deposits of the São Paulo Plateau represents a full-width of opening.

TABLE 1. Distribution of Evaporites in Basins Around the Atlantic (Fig. 1) (continued)

Site on Figure 1	Location	Tectonic Setting	Evaporite	Occurrence	Thickness (m)	Evidence	Age	Evidence	References
30	Northwest Africa	Aaiun basin: to continental shelf break	Halite	Beds and diapirs	≤ 1000	Drilling; seismic	Late Triassic and possibly early Jurassic	Stratigraphic relations	Querol, 1966; Reyre, 1966a, b; Choubert and Hottinger, 1971
31	Northwest Africa	Senegal and Aaiun basins: seaward of continental shelf	Halite	Beds and possible diapirs	Unknown	Seismic; salinity gradient	Late Triassic and Early Jurassic	Regional geology; stratigraphic relations	Rona, 1969, 1970; Schneider and Johnson, 1970; Waterman et al., 1972; Uchupi et al., 1976
32	Northwest Africa	Essaouira basin	Halite, anhydrite	Beds and diapirs	≤ 1200	Drilling; seismic	Late Triassic (Keuper) and early Jurassic (Lias)	Stratigraphic relations	Société Cherifienne des Pétroles, 1966; Reyre, 1966a, b; Pautot et al., 1970; Beck and Lehner, 1974; Uchupi et al., 1976; Van Houten, 1977
32	North Africa	Moroccan Meseta basins	Halite, anhydrite, gypsum	Beds and diapirs	≤ 2000	Drilling; outcrop	Late Triassic (Keuper) and early Jurassic (Lias)	Stratigraphic relations	LaCoste, 1934; Tortochaux, 1968; Salvan, 1972; Van Houten, 1977
33	Algerian Sahara	Saharan shelf	Halite, anhydrite, gypsum	Beds and diapirs	≤ 1600	Drilling; outcrop; seismic; gravity	Late Triassic (Keuper) and early Jurassic (Lias)	Stratigraphic relations; brachiopod fauna in carbonates interbedded with salt	Burollet and Dumestre, 1952; Gill, 1965; Demaison, 1965; Tortochaux, 1968
34	Western Mediterranean	Mediterranean basin	Halite, anhydrite	Beds and diapirs	≤ 1800	Drilling; seismic	Upper Miocene (Messinian)	Stratigraphic relations; nannofossils and microfossils in strata overlying and interbedded with anhydrite and gypsum	Auzende et al., 1971; Mauffret et al., 1973; Biju-Duval et al., 1974
35	Portugal	Portugal basin	Halite	Beds and diapirs	Unknown	Drilling; seismic	Triassic and and early Jurassic	Stratigraphic relations	Gill, 1965; Montadert et al., 1974
36	Spain	Pyrenee, Cantabrian, Anadalusian Mountains	Halite, gypsum, anhydrite, potash salts	Diapirs	≤ 500	Drilling; seismic	Late Triassic (Keuper) and early Jurassic (Lias); Eocene; Miocene	Stratigraphic relations	Gill, 1965; Brinkman and Logters, 1968; Rios, 1968
36	Southeastern France	Aquitaine basin	Halite, anhydrite	Beds and diapirs	≤ 800 (Keuper)	Drilling; seismic	Late Triassic (Keuper) and early Jurassic (lower Lias)	Stratigraphic relations	Dupouy-Camet, 1953; Bonnard et al., 1958; Winnock, 1973
37	Germany	North European (Hannover basin)	Halite, potash salts	Beds and diapirs	≤ 1000 (Zechstein)	Drilling; seismic	Late Permian (Zechstein); Triassic (especially Keuper)	Stratigraphic relations	Bentz, 1958; Sannemann, 1968

TABLE 1. Distribution of Evaporites in Basins Around the Atlantic (Fig. 1) (continued)

Site on Figure 1	Location	Tectonic Setting	Evaporite	Occurrence	Thickness (m)	Evidence	Age	Evidence	References
12	Gulf of Mexico	Gulf of Mexico basin	Halite, minor gypsum, anhydrite	Beds and diapirs	≤ 3000 (Louann)	Drilling; seismic; gravity; salinity gradients	Late Triassic through Middle Jurassic (Louann)	Palynology of salt and cap-rock; associated stratigraphy	Jux, 1961; Murray, 1968; Burk et al., 1969; Manheim and Sales, 1970, 1974; Kirkland and Gerhard, 1971; Manheim et al., 1973; Antoine et al., 1974
13	Bahama Banks	Cuban saline basin	Halite, anhydrite, gypsum	Beds and diapirs	Unknown	Drilling; seismic	Late Triassic and Jurassic	Regional geology	Ball et al., 1971, 1974; Tator and Hatfield, 1975
14	Cuba	Cuban saline basin	Halite, anhydrite, gypsum	Beds and diapirs	≤ 5000 (Punta Allegre formation)	Drilling; seismic; gravity	Late Triassic (Rhaetian) to Late Jurassic (Callovian)	Palynology of inter-bedded shale	Meyerhoff and Hatten, 1968, 1974
15	Demerara Rise	Marginal basin	Halite	Beds	Unknown	Salinity gradient	Late Triassic and Jurassic?	Regional geology; stratigraphic relations	Watermann et al., 1972
16–17	South America	Interior basins	Halite, anhydrite, gypsum	Beds and diapirs	≤ 2000	Drilling; seismic; outcrop	Paleozoic; Triassic	Straigraphic relations	Benavides, 1968; Kozary et al., 1968
18–22	South America	Sergipe-Alagoas, Reconcavo, Espírito Santo, Campos, and Santos basins	Halite, anhydrite, gypsum	Beds and diapirs	≤ 800	Drilling; seismic	Early Cretaceous (Aptian)	Ostracod fauna in associated strata	Krömmelbein and Wenger, 1966; Viana, 1966, 1971; Fonseca, 1966; Butler, 1970; Campos et al., 1974; Asmus and Ponte, 1973; Ponte and Asmus, 1976
23	São Paulo Plateau	Offshore portion of Santos basin	Halite	Beds and diapirs	≤ 2000	Drilling; seismic; thermal gradient; salinity gradient	Early Cretaceous (Aptian)	Regional geology; stratigraphic relations	Butler, 1970; Leyden et al., 1978; Kumar and Gamboa, 1979
24	Angola	Moçâmedes basin	Gypsum, anhydrite	Beds	≤ 100	Drilling; seismic	Early Cretaceous (Aptian)	Stratigraphic relations	Reyre, 1966a, b
25–27	Southwest Africa	Cuanza, Lower Congo, and Gabon basins: onshore portions	Halite, carnallite, sylvinite, anhydrite, gypsum	Beds and diapirs	Cuanza: ≤ 750 Lower Congo: ≤ 1000 Gabon: ≤ 2000	Drilling; seismic	Early Cretaceous (Aptian)	Ostracod fauna in associated strata	Belmonte et al., 1965; Reyre, 1966a, b; Brognon and Verrier, 1966; Krömmelbein and Wenger, 1966; Franks and Nairn, 1973; Brink, 1974
28	Southwest Africa	Cuanza, Lower Congo, and Gabon basins: offshore portions	Halite	Beds and diapirs	≤ 3000	Seismic; thermal gradient; salinity gradient	Early Cretaceous	Regional geology; stratigraphic relations	Baumgartner and van Andel, 1971; Von Herzen et al., 1972; Leyden et al., 1972; Pautot et al., 1973; Emery et al., 1975; Bolli, Ryan et al., 1978
29	Northwest Africa	Senegal basin: to continental shelf break	Halite, minor anhydrite, gypsum	Diapirs	≤ 3000	Drilling; seismic	Late Triassic (Keuper) and early Jurassic (Lias)	Palynology of salt	Aymé, 1965; Templeton, 1971

TABLE 1. Distribution of Evaporites in Basins Around the Atlantic (Fig. 1)

Site on Figure 1	Location	Tectonic Setting	Evaporite	Occurrence	Thickness (m)	Evidence	Age	Evidence	References
1	Canadian Arctic Islands	Sverdrup basin	Halite, anhydrite, gypsum	Beds and diapirs	≤ 4000	Drilling; outcrop; seismic; gravity	Ordovician; Late Mississipian, Pennsylvanian, Permian	Stratigraphic relations	Gould and de Mille, 1968; Davies, 1974; Davies and Nassichuk, 1975
2	Western Canada	Interior basins	Halite, anhydrite, gypsum, potash salts	Beds	≤ 375	Drilling; outcrop; seismic	Devonian	Stratigraphic relations	Gorrell and Alderman, 1968; Pierce and Rich, 1972
3	Williston Basin	Interior basin	Halite, anhydrite, gypsum, potash salts	Beds	≤ 275	Drilling; outcrop; seismic	Devonian; Mississipian; Permain; Jurassic		Pierce and Rich, 1972
4	Paradox Basin	Interior basin	Halite, anhydrite, gypsum	Beds and diapirs	≤ 3000	Drilling; outcrop; seismic	Pennsylvanian		Hite, 1968; Pierce and Rich, 1972
5	Permian Basin	Interior basin	Halite, anhydrite, gypsum, potash salts	Beds	≤ 850	Drilling; outcrop; seismic	Permian		Mear, 1968; Pierce and Rich, 1972
6	Michigan and Appalachian Basins	Interior basins	Halite, anhydrite, gypsum	Beds	≤ 550	Drilling; seismic	Silurian; Devonian Mississipian		Landes, 1945; Landes et al., 1945; Withington, 1963; Pearson, 1963; Pierce and Rich, 1972
7	James Bay	Interior basin	Halite, anhydrite, gypsum			Drilling; seismic	Devonian		Kozary et al., 1968
8	Maritime Provinces	Sydney basin; Fundy epieugeosyncline	Halite, anhydrite, minor sylvite, carnallite	Beds and diapirs	≤ 1200	Drilling; outcrop; seismic	Mississippian (Windsor Group)	Stratigraphic relations	Gussow, 1953; Bell, 1958; Howie and Cummings, 1963; Belt, 1965; Bidgood, 1969; Evans, 1970; Howie and Barss, 1974
9	Scotian Shelf	Scotian basin	Halite, anhydrite	Beds and diapirs	≤ 813 (Windsor Group) ≤ 1800 (Argo formation)	Drilling; seismic	Mississippian (Windsor Group); Late Triassic and Early Jurassic (Argo formation)	Stratigraphic relations; palynology of caprock	King and McLean, 1970; Keen, 1970; Permenter, 1971; McIver, 1972; Webb, 1973; Jansa and Wade, 1974; Offshore Exploration Staffs, 1974
10	Grand Banks	Scotian basin; East Newfoundland basin	Halite, anhydrite	Beds and diapirs	≤ 813 (Windsor Group; Scotian basin ≤ 1800 (Argo formation)	Drilling; seismic	Mississippian (Windsor Group); Late Triassic and Early Jurassic (Argo formation)	Stratigraphic relations; palynology of caprock	Watson and Johnson, 1970; Bartlett and Smith, 1971; Offshore Exploration Staffs, 1974; Jansa and Wade, 1974; Walton and Berti, 1976; Jansa et al., 1980
11	Eastern North America, Central Atlantic margin	Marginal basins	Halite, anhydrite, gypsum	Beds and diapirs	0.3 m feather edge anhydrite at Cape Hatteras to unknown thickness seaward	Drilling; seismic; gravity; salinity gradients	Late Triassic and Jurassic	Regional geology; stratigraphic relations	Swain, 1952; Manheim and Horn, 1968; Rona, 1970; Emery et al., 1970; Maher and Applin, 1971; Sheridan, 1974; Mattick et al., 1974; Olson, 1974; Sheridan and Osburn, 1975; Poag, 1978; Grow et al., 1979

TABLE 1. Distribution of Evaporites in Basins Around the Atlantic (Fig. 1) (continued)

Site on Figure 1	Location	Tectonic Setting	Evaporite	Occurrence	Thickness (m)	Evidence	Age	Evidence	References
38	North Sea	Interior basins	Halite, anhydrite, gypsum	Beds and diapirs	≤ 1500 (Zechstein)	Drilling; seismic; gravity	Late Permian (Zechstein); Late Triassic (Keuper)	Stratigraphic relations	Kent, 1969; Ziegler, 1975; Watson and Swanson, 1975; Naylor and Mounteney, 1975
38	British Isles	Rift basins	Halite, anhydrite, gypsum	Beds and possible diapirs	≤ 1000	Drilling; seismic; gravity	Late Permian (Zechstein); Late Triassic (Keuper)	Stratigraphic relations	Audley-Charles, 1970; Naylor and Mounteney, 1975
39	Greenland	Marginal basin	Gypsum	Beds	Unknown	Outcrop	Carboniferous, Permian (Zechstein); Triassic	Stratigraphic relations	Maync, 1961; Kozary et al., 1968; Birkelund et al., 1974
40	Spitsbergen	Marginal basin	Gypsum	Beds	Unknown	Outcrop	Mid-Carboniferous; Permian	Stratigraphic relations	Harland, 1967, 1969; Kozary et al., 1968

VI. Evaporite deposits of two different ages separated by intervals of non-evaporitic sediments are superposed in at least two regions of the North Atlantic:
1. Mississippian and Late Triassic through Jurassic salt deposits are superposed in the Scotian shelf – Grand Banks region.
2. Late Permian and Late Triassic salts are superposed in the North European basin – North Sea region.

VII. Atlantic evaporites exhibit a systematic distribution in time and space, as follows:
1. Ordovician period: Canadian Arctic islands.
2. Late Silurian period: Eastern North America including the Michigan and Appalachian basins.
3. Devonian period: Interior basins of eastern and western North America.
4. Missippian period: Northwestern Atlantic including the Maritime Provinces, Scotian shelf and Grand Banks; Canadian Arctic islands, and interior basins of eastern and western North America.
5. Pennsylvanian, Permian, and Triassic periods: Northeastern Atlantic including Greenland, Spitzbergen, North European basin, the North Sea, and the British Isles; Canadian Arctic islands; the Paradox and Permian basins of western North America; certain interior basins of South America.
6. Late Triassic and Jurassic periods:
a. Central North Atlantic continental margins of eastern North America, northwestern Africa, and southeastern Europe, including the Grand Banks, Scotian shelf, Cuba, Bahama Banks, Senegal basin, Aaiun basin, Essaouira basin, Moroccan Meseta basins, Portugal basin, Aquitaine basin.
b. Gulf of Mexico.
c. Mediterranean region including basins in the areas of the Atlas Mountains and the Algerian Sahara.
7. Aptian stage of the Cretaceous period: South Atlantic including the southeastern continental margin of South America (Sergipe/Alagoas, Recôncavo, Espírito Santo, Campos, and Santos basins), and the southwestern continental margin of Africa (Moçâmedes, Cuanza, Lower Congo, and Gabon basins).
8. Miocene epoch of the Tertiary period: Western Mediterranean Sea.
9. The presence of salt deposits beneath the Demerara Rise off the northeastern continental margin of South America is inferred from a salinity gradient in interstitial pore water recovered from

TABLE 2. Physical and Chemical Properties of Evaporites (Hodgman et al., 1963; Clark, 1966; Odé, 1968)

Material	Composition	Density (gm cm^{-3})	V_p (km sec^{-1})	Viscosity (poises)	Solubility (gm 100 mℓ^{-1} cold water)	Thermal Conductivity (10^{-3} cal cm^{-1} sec^{-1} °C)	Magnetic Susceptibility (cgs emu cm^{-3})
Halite	NaCl	2.16	4.3 - 4.4	$10^{12} - 10^{18}$	35.7	12 - 17	0
Anhydrite	CaSO$_4$	2.9	4.1 - 5.0	$10^{14} - 10^{18}$	0.209	12 - 13	0
Gypsum	CaSO$_4 \cdot$ 2H$_2$O	2.32	2.0 - 3.5		0.241	3	0
Sylvite	KCl	1.99	4.4 - 6.5	$10^{11} - 10^{12}$	34.7	---	0
Carnallite	KMgCl$_3 \cdot$ 6H$_2$O	1.61	4.4 - 6.5		64.5	---	0
Unconsolidated sediment (wet)	variable	1.5	1.5 - 2.0		insoluble	1.9 - 2.7	$10^{-5} - 1$

overlying sediments (Fig. 1). A Late Triassic-Jurassic age is most likely for this salt.

VIII.   The distribution in space and time of evaporites at passive continental margins of the Atlantic broadly occupy three provinces which successively developed from north to south as follows:

1.   Northeast Atlantic province comprising evaporites of Pennsylvanian, Permian and Triassic age off eastern Greenland, Spitsbergen, the North Sea and the North European basin.

2.   Central North Atlantic province comprising evaporites of Late Triassic and Jurassic age off eastern North America, in the Gulf of Mexico, and off northwestern Africa.

3.   South Atlantic province including evaporites of Aptian age off eastern South America and western Africa.

Evaporite deposition within any province was probably not entirely synchronous but involved asynchronous events within the overall time interval, as well as progressive time transgression. Evaporite deposits are expected to transgress time seaward from beneath continental shelves to beneath continental rises and longitudinally along continental margins controlled by the geometry of opening about poles of rotation of adjacent plates with irregular plate boundaries, and by facies migration of lateral fractionation sequences produced by precipitation of evaporite minerals with different solubilities from solution. The dating resolution of the evaporite deposits is presently inadequate to demonstrate these effects, but they are anticipated.

IX.   The distribution of Atlantic evaporites in space and time (Fig. 1, Table 1) demonstrates a genetic relation between the deposition of evaporites and the creation of tectonic settings by rifting and early opening of an ocean basin. The distribution conforms to the independently determined history of incipient rifting and formation of epeiric seas in the Northeast Atlantic during the Pennsylvanian, Permian and Triassic periods, and rifting and early opening of the Central North Atlantic in the Late Triassic and Jurassic periods (Rona, 1969, 1970, 1976; Schneider and Johnson, 1970; Pautot et al., 1970; Olson and Leyden, 1973; Sclater et al., 1977; Evans, 1978), and of the South Atlantic in the Early Cretaceous (Belmonte et al., 1965; Campos et al., 1974).

X.   A high incidence of evaporite deposits in basins both at passive margins and within cratons during portions of the Pennsylvanian, Permian and Triassic periods, indicates that global factors including cli-

TABLE 3. Chlorinity and Salinity of Interstitial Pore Water at Deep Sea Drilling Project (DSDP) Atlantic Sites

Region	DSDP Hole	Depth in Hole (m)	Chlorinity ($^o/oo$)	Salinity ($^o/oo$)	Depth Below Sea Level (m)	Position Latitude	Longitude	Location	Reference
Gulf of Mexico	2 2	25 110	22 38		3572	23°27.3'N	92°35.2'W	Challenger Knoll (diapiric structure)	Manheim and Sayles, 1970
	3 3	34 619	20.0 33.3	33 55	3746	23°01.8'N	92°02.6'W	Abyssal Plain (no diapiric structures present; salt source underlies at least 4 km of sediment)	Manheim and Sayles, 1970
	85 85	33 111	19.8 22.7	34.2 38.0	3749	22°50.5'N	91°25.4'W	Seaward of Campeche Scarps (no diapiric structures present)	Manheim et al., 1973
	88 88	5 135	19.7 27.2	35.2 44.3	2532	21°22.9'N	94°00.2'W	Sigsbee Knolls area (diapiric structure)	Manheim et al., 1973
	89 89	3 224	19.4 30.8	34.8 50.6	3067	20°53.4'N	90°06.7'W	Bay of Campeche (no diapiric structures present)	Manheim et al., 1973
	92 92	35 233	20.2 102.2	35.5 170.5	2573	25°50.7'N	91°49.3'W	Edge of Sigsbee Scarp (diapiric structure)	Manheim et al., 1973; Amery, 1969
Western North Atlantic	98 98	8 223	19.5 19.8	36.0 35.8	2750	25°23.0'N	77°18.7'W	Northeast Providence Channel	Sayles et al., 1972
	101 101	38 70	19.5 19.5	34.1 33.6	4773	25°11.9'N	74°26.3'W	Blake Bahama Outer Ridge	Sayles et al., 1972
	102 102	6 660	19.4 18.6	33.8 30.8	3414	30°43.6'N	74°27.1'W	Blake Bahama Outer Ridge	Sayles et al., 1972
	103 103	6 250	20.0 18.7	-- 30.8	3992	30°27.1'N	74°35.0'W	Blake Bahama Outer Ridge	Sayles et al., 1972
	104 104	42 403	20.2 18.4	36.3 31.1	3833	30°49.6'N	74°19.6'W	Blake Bahama Outer Ridge	Sayles et al., 1972
	105 105	33 325	19.5 19.1	34.9 33.7	5245	34°53.7'N	69°10.4'W	Lower continental rise southeast of New York	Sayles et al., 1972
	106 106	2 115	20.1 19.0	36.2 31.1	4492	36°26.0'N	69°27.7'W	Lower continental rise southeast of New York	Sayles et al., 1972
	111 111	4 158	18.1 18.9	33.0 34.0	1811	50°25.6'N	46°22.0'W	Orphan Knoll	Manheim et al., 1972
	112 112	35, 384	19.5 19.5	35.1 34.0	3667	54°01.0'N	46°36.2'W	Labrador Sea	Manheim et al., 1972
	113 113	55 550	19.1 19.6	34.7 34.2	3629	56°47.4'N	48°19.9'W	Mid-Labrador Sea Ridge	Manheim et al., 1972
	142 142	115 587	19.3 19.7	31.4 34.4	4372	03°22.2'N	42°23.5'W	Ceara Abyssal Plain	Waterman et al., 1972
	144 144	8 270	18.9 27.9	34.1 45.9	2957	09°27.2'N	54°20.5'W	Northern flank of Demerara Rise	Waterman et al., 1972
	354 354	2 854	-- --	35.5 35.2	4052	05°53.9'N	44°11.8'W	Ceara Rise	Supko, P. R., K. Perch-Nielson, et al., 1977
	388 388	40 319	19.7 19.7	34.1 34.1	4919	35°31.3'N	69°23.8'W	Lower continental rise hills	Benson, W. E., R. E. Sheridan, et al., 1978

TABLE 3. Chlorinity and Salinity of Interstitial Pore Water at Deep Sea Drilling Project (DSDP) Atlantic Sites (continued)

Region	DSDP Hole	Depth in Hole (m)	Chlorinity (^{o}/oo)	Salinity (^{o}/oo)	Depth Below Sea Level (m)	Position Latitude	Longitude	Location	Reference
Western North Atlantic	390	3	19.6	35.2	2665	30°08.5'N	76°06.7'W	North rim of Blake Nose (limestone)	Benson, W. E., R. E. Sheridan, et al., 1978
	390	136	19.5	36.3					
	391A	152	20.1	34.6	4974	28°13.7'N	75°36.8'W	Blake–Bahama Basin	Benson, W. E., R. E. Sheridan, et al., 1978
	391A	655	19.8	33.6					
	392	0	19.7	36.3	2601	29°54.6'N	76°10.7'W	South rim of Blake Nose (limestone)	Benson, W. E., R. E. Sheridan, et al., 1978
	392	82	19.9	36.3					
Eastern North Atlantic	118	102	19.0	31.9	4901	45°02.9'N	09°00.5'W	Bay of Biscaye	Manheim et al., 1972
	118	449	19.8	33.1					
	119	51	19.6	35.6	4447	45°02.3'N	07°58.8'W	Cantabria Seamount	Manheim et al., 1972
	119	368	19.8	34.9					
	135	3	19.6	35.5	4152	32°20.8'N	10°25.5'W	Topographic high southeast of Horseshoe Abyssal Plain	Waterman et al., 1972
	135	435	19.9	34.1					
	136	136	19.68	35.8	4169	34°10.1'N	16°18.2'W	Abyssal hills north of Madeira and southwest of Gibraltar	Waterman et al., 1972
	136	282	19.12	34.1					
	137	58	19.4	37.4	5361	25°55.5'N	27°03.6'W	Abyssal hills 1000 km west of Cap Blanc, West Africa	Waterman et al., 1972
	137	381	20.5	37.4					
	138	58	19.7	35.2	5288	25°55.4'N	25°33.8'W	Foot of continental rise 870 km west of Cap Blanc	Waterman et al., 1972
	138	116	19.3	35.2					
	139	116	23.8	39.6	3047	23°31.1'N	18°42.3'W	Middle continental rise 250 km west of Cap Blanc	Waterman et al., 1972
	139	659	44.0	78.1					
	140	96	20.7	36.9	4483	21°45.0'N	21°47.5'W	Foot of continental rise 450 km west of Cap Blanc	Waterman et al., 1972
	140	647	32.2	52.8					
	141	13	19.7	35.5	4148	19°25.2'N	23°59.9'W	200 km north of Cape Verde Islands (on a basalt high)	Waterman et al., 1972
	141	197	19.4	33.0					
	336	6	19.4	34.4	811	63°21.1'N	07°47.3'W	Northern flank of the Iceland–Faeroe Ridge	Geiskes et al., 1978
	336	479	21.0	35.8					
	337	6	19.7	35.2	2631	64°52.3'N	05°20.5'W	Norway Basin	Geiskes et al., 1978
	337	93	19.8	35.5					
	338	5	19.2	35.2	1297	67°47.1'N	05°23.3'E	On Vøring Plateau	Geiskes et al., 1978
	338	401	20.5	35.2					
	341	13	19.2	33.3	1439	67°20.1'N	06°06.6'E	On Vøring Plateau	Geiskes et al., 1978
	341	434	16.1	32.4					
	342	6	19.6	35.2	1303	67°57.0'N	04°56.0'E	On Vøring Plateau	Geiskes et al., 1978
	342	124	20.0	34.9					
	343	2	19.2	34.5	3131	68°42.9'N	05°45.7'E	At base of Vøring Plateau	Geiskes et al., 1978
	343	272	20.5	36.6					
	366	2	19.2	35.2	2853	05°40.7'N	19°51.1'W	Sierra Leone Rise	Couture et al., 1977
	366	510	19.1	36.0					
	366A	5	19.5	35.2	2853	05°40.7'N	19°51.1'W	Sierra Leone Rise	Couture et al., 1977
	366A	239	20.1	36.3					
	367	8	19.3	34.9	4748	12°29.2'N	20°02.8'W	Cape Verde Basin	Couture et al., 1977
	367	779	18.9	32.4					

TABLE 3.   Chlorinity and Salinity of Interstitial Pore Water at Deep Sea Drilling Project (DSDP) Atlantic Sites (continued)

Region	DSDP Hole	Depth in Hole (m)	Chlorinity (^{o}/oo)	Salinity (^{o}/oo)	Depth Below Sea Level (m)	Position Latitude	Longitude	Location	Reference
Eastern North Atlantic	368 368	5 726	20.2 19.8	35.2 31.7	3366	17°30.4'N	21°21.2'W	Cape Verde Rise	Couture et al., 1977
	369A 369A	88 452	20.8 28.0	36.2 46.3	1752	26°35.5'N	14°59.0'W	Continental slope off Cape Bojador, Spanish Sahara	Couture et al., 1977
	370 370	6 1105	17.4 36.7	35.2 55.1	4214	32°50.2'N	10°46.6'W	Deep ocean basin off Morocco	Couture et al., 1977
Western South Atlantic	355 355	60 412	-- --	35.2 35.2	4886	15°42.6'S	30°36.0'W	Brazil Basin	Supko, P. R., K. Perch-Nielson, et al., 1977
Western South Atlantic	356 356	14 649	-- --	34.9 36.0	3175	28°17.2'S	41°05.3'W	São Paulo Plateau	Supko, P. R., K. Perch-Nielson, et al., 1977
	357 357	8 613	-- --	34.6 36.5	2086	30°00.3'S	35°33.6'	Rio Grande Rise	Supko, P. R., K. Perch-Nielson, et al., 1977
	358 358	54 757	-- --	35.2 31.1	4990	37°39.3'S	35°57.8'W	Argentine Basin	Supko, P. R., K. Perch-Nielson, et al., 1977
Eastern South Atlantic	360 360	88 420	-- --	34.6 34.1	2949	35°50.8'S	18°05.8'E	Upper continental rise west of Cape Agulhas, South Africa	Sotelo and Gieskes, 1978
	361 361	35 1267	19.2 19.9	35.2 34.6	4549	35°04.0'S	15°26.9'E	Lower continental rise west of Cape Agulhas, South Africa	Sotelo and Gieskes, 1978
	362 362	82 652	19.0 20.4	33.3 34.4	1325	19°45.5'S	10°32.0'E	Abutment Plateau portion of the Frio Ridge segment of Walvis Ridge	Sotelo and Gieskes, 1978
	363 363	38 505	19.7 20.2	35.5 36.6	2248	19°38.8'S	09°02.8'E	Isolated basement high on escarpment of Frio Ridge portion of Walvis Ridge	Sotelo and Gieskes, 1978
	364 364	13 728	19.4 35.5	35.2 59.4	2448	11°34.3'S	11°58.3'E	Seaward edge of salt plateau on continental slope southwest of Luanda, Angola	Sotelo and Gieskes, 1978
	365 365	229 685	19.9 75.6	35.2 122.1	3018	11°39.1'S	11°53.7'E	Continental slope off Angola	Sotelo and Gieskes, 1978

matic fluctuations and eustatic sea level changes worked in conjunction with tectonics to produce conditions favorable for the deposition of evaporites (Meyerhoff, 1970).

Conclusions

Review of the distribution of Atlantic evaporites reveals general features of the relation between evaporites and passive margins of a major opening ocean basin, as follows:

1. Evaporites do not occur continuously along passive continental margins but are grouped into provinces distinguished by age and location of the deposits.
2. The formation of an evaporite province at a passive margin is controlled by the tectonic setting of rifting and early opening that acts to restrict oceanic circulation, in conjunction with climatic factors that favor evaporation in excess of precipitation and inflow, and eustatic sea level changes that enforce barriers to circulation.
3. Tectonic settings favorable for evaporite deposition occur along those rifted portions of passive margins that trend nearly perpendicular to fracture zones of the adjacent ocean basin; tectonic settings along sheared portions of passive margins that trend nearly parallel to fracture zones are unfavorable.
4. Evaporite deposits of rifted continental margins may extend seaward hundreds of kilometers in basins that open seaward beneath the continental margin and adjacent ocean basin over both continental and oceanic crust.
5. The original thickness of an evaporite deposit beneath a passive continental margin may range up to about 5 km.
6. Evaporite deposits within a given province may transgress time either perpendicular to a passive margin and/or parallel to the margin as a consequence of the geometry of rifting and opening about poles of plate rotation, local structural irregularities, and facies migration of fractionation sequences of evaporite minerals.
7. Evaporites of widely different ages may be superposed in the stratigraphic column beneath passive continental margins where a province has been a recurrent region of evaporite deposition.
8. Evaporite deposition in basins both at passive margins and within cratons appears to coincide during certain periods. Stratigraphic correlation studies between these widely separated evaporite basins will help to decipher how the global factors of climatic fluctuation and eustatic sea level work in conjunction with tectonic setting to create conditions favorable for evaporite deposition at passive continental margins.
9. The association of evaporites, including kilometers-thick sections of rock salt, with passive continental margins has practical implications for the occurrence of hydrocarbons (oil and gas), and for use as a source of various salts, as a source of energy through salinity gradient energy conversion (Wicks and Isaacs, 1978), and as a possible container both for offshore storage of hydrocarbons and for offshore disposal of waste materials, especially radioactive waste.

References

Amery, G. B., Structure of Sigsbee Scarp, Gulf of Mexico, Amer. Assoc. Petrol. Geol. Bull., 58, 2480–2482, 1969.

Antoine, J. W., R. Martin, T. Pyle, and W. R. Bryant, Continental margins of the Gulf of Mexico, in Burk, C. A. and C. L. Drake, Geology of continental margins, New York, Springer-Verlag, pp. 683–694, 1974.

Asmus, H. E., and F. C. Ponte, The Brazilian marginal basins, in Nairn, A. E. M. and F. G. Stehli, editors, The ocean basins and margins, v. 1: New York, Plenum Press, pp. 87–133, 1973.

Audley-Charles, M. G., Triassic palaeogeography of the British Isles, Geol. Soc. London Quart. Jour., 126, 49–90, 1970.

Auzende, J. M., J. Bonnin, J. O. Olivet, G. Pautot and A. Mauffret, Upper Miocene salt layer in the Western Mediterranean basin, Nature Phys. Sci., 230, 82–84, 1971.

Aymé, J. M., The Senegal salt basin, in Salt basins around Africa, London, Institute of Petroleum, pp. 83–90, 1965.

Ball, M. M., W. Bock, C. G. A. Harrison, F. Nagle, Jr., and G. J. Williams, Diapirs of the Old Bahama Channel, EOS, Amer. Geophys. Union Trans., 55, 284, 1974.

Ball, M. M., B. P. Dash, C. G. A. Harrison, and K. O. Ahmed, Refraction seismic measurements in the northeastern Bahamas (abs.), EOS, Trans. Amer. Geophys. Union, 52, 252, 1971.

Bartlett, G. A., and L. Smith, Mesozoic and Cenozoic history of the Grand Banks of Newfoundland, Canadian Jour. Earth Sci., 8, 65–84, 1971.

Baumgartner, T. R., and T. H. Van Andel, Diapirs of the continental margin of Angola Africa, Geol. Soc. America Bull., 82, 793–802, 1971.

Beck, R. H., and P. Lehner, Oceans, new frontier in exploration, Amer. Assoc. Petrol. Geol. Bull., 58, 376–395, 1974.

Bell, W. A., Possibilities for occurrences of petroleum reservoirs in Nova Scotia, Nova Scotia Dept. Mines, 227 p., 1958.

Belmonte, Y., P. Hirtz, and R. Wenger, The salt basins of the Gabon and the Congo (Brazzaville). A tentative paleogeographic interpretation, in Salt basins around Africa, London, Institute of Petroleum, pp. 55–74, 1965.

Belt, E. S., Stratigraphy and paleogeography of Mabou Group and related Middle Carboniferous facies, Nova Scotia, Canada, Geol. Soc. America Bull., 76, 777-802, 1965.

Benavides, V., Saline deposits of South America, in Mattox, R. B., editor, Saline deposits, Geol. Soc. Amer. Spec. Paper 88, pp. 249-290, 1968.

Benson, W. E., R. E. Sheridan, et al., Initial reports of the Deep Sea Drilling Project, v. 44, Washington, U.S. Government Printing Office, 1005 p., 1978.

Benz, A., Relations between oil fields and sedimentary troughs in Northwest German Basin, in Weeks, L. D., editor, Habitat of oil, Tulsa, Oklahoma, Amer. Assoc. Petroleum Geologists, pp. 1054-1066, 1958.

Bidgood, D. E. T., The distribution and diapiric nature of some Nova Scotia evaporites - a geophysical evaluation, 3rd Symp. on Salt, Cleveland, Ohio, 1969.

Biju-Duval, B., J. Letouzey, L. Montadert, P. Courrier, J. F. Mugniot, and J. Sancho, Geology of the Mediterranean Sea basins, in Burk, C. A. and C. L. Drake, editors, The geology of continental margins, New York, Springer-Verlag, p. 695-721, 1974.

Birkeland, T. K., Perch-Nielsen, D. Bridgewater, and A. K. Higgins, An outline of the geology of the Atlantic coast of Greenland, in Nairn, A. E. M. and F. G. Stehli, The ocean basins and margins, v. 2, North Atlantic, New York, Plenum Press, pp. 125-159, 1974.

Blackett, P. M. S., Comparison of ancient climates with the ancient latitudes deduced from rock magnetic measurements, Proceedings of the Royal Society, A, pp. 1-30, 1961.

Bolli, H. M., W. B. F. Ryan, et al., Initial reports of the Deep Sea Drilling Project, v. 40, Washington, U.S. Government Printing Office, 1079 p., 1978.

Bonnard, E., A. Debourle, H. Hlauschek, P. Michel, V. Perebaskine, J. Schoeffler, R. Seronie-Vivien, and M. Vigneaux, The Aquitanian basin, southwest France, in Weeks, L. G., editor, Habitat of oil, Tulsa, Amer. Assoc. Petroleum Geologists, pp. 1091-1122, 1958.

Borchert, H., and R. O. Muir, Salt deposits, London, D. Van Norstrand Co., Ltd., 338 p., 1964.

Brink, A. H., Petroleum geology of Gabon basin, Amer. Assoc. Petrol. Geol. Bull., 58, 216-235, 1974.

Brinkmann, R., and H. Lögsters, Diapirs in Western Pyrenees and foreland Spain, in Braunstein, J. and G. D. O'Brien, Diapirism and diapirs, a symposium, Tulsa, Amer. Assoc. Petroleum Geologists, Memoir 8, pp. 275-292, 1968.

Brognon, G. P., and G. R. Verrier, Oil and geology in Cuanza basin of Angola, Amer. Assoc. Petrol. Geol. Bull., 50, 108-158, 1966.

Burk, C. A., M. Ewing, J. L. Worzel, A. O. Beall, Jr., W. A. Berggren, D. Bukry, A. G. Fischer, and E. A. Pessagno, Jr., Deep-sea drilling into the Challenger Knoll, central Gulf of Mexico, Amer. Assoc. Petrol. Geol. Bull., 53, 1338-1347, 1969.

Burollet, P., and A. Dumestre, Le diapir du Djebel Rhoeuis, Algeria, 19th International Geological Congress, S. 2, p. 6, 1952.

Butler, L. W., Shallow structure of the continental margin, southern Brazil and Uruguay, Geol. Soc. America Bull., 81, 1079-1096, 1970.

Campos, C. M. W., F. C. Ponte, and K. Miura, Geology of the Brazilian continental margin, in Burk, C. A. and C. L. Drake, editors, The geology of continental margins, New York, Springer-Verlag, pp. 447-461.

Choubert, G., and L. Hottinger, La série stratigraphique de Tarfaya et le probléme de la naissance de l'ocean Atlantique, Maroc. Service Geol. Notes et Mem., 31, 29-40, 1971.

Clark, S. O., Jr., editor, Handbook of physical constants, Geol. Soc. America, Memoir 97, 587 p., 1966.

Couture, R., R. S. Miller, and J. M. Gieskes, Interstitial water and mineralogical studies, Leg 41, in Lancelot, Y., E. Seibold, et al., Initial reports of the Deep Sea Drilling Project, v. 41, Washington, U.S. Government Printing Office, pp. 907-914, 1977.

Davies, G. R., Paleozoic evaporites of the Canadian Arctic archipelago, in Coogan, A. H., editor, Fourth Symposium on Salt, v. 1, Cleveland, Ohio, Northern Ohio Geol. Soc., Inc., pp. 119-128, 1974.

Davies, G. R., and W. W. Nassichuk, Subaqueous evaporites of the Carboniferous Otto Fiord formation, Canadian Arctic archipelago, A summary, Geology, 3, 273-278, 1975.

Dellwig, L. F., Origin of the Salina salt of Michigan, Jour. Sed. Petrology, 25, 83-93, 1955.

Demaison, G. J., The Triassic Salt in the Algerian Sahara, in Salt basins around Africa, London, Institute of Petroleum, pp. 91-100, 1965.

Dietz, R. S., Continent and ocean basin evolution by spreading of the sea floor, Nature, 190, 854-857, 1961.

Dietz, R. S., and J. C. Holden, Reconstruction of Pangaea: Breakup and dispersion of continents, Permian to present, Jour. Geophys. Res., 75, 4939-4956, 1970.

Dupouy-Camet, J., Triassic diapiric salt structures, southwestern Aquitaine basin, France, Amer. Assoc. Petrol. Geol. Bull., 37, 2348-2388, 1953.

Emery, K. O., J. Phillips, C. Bowin, and J. Mascle, Continental margin off western Africa: Angola to Sierra Leone, Amer. Assoc. Petrol. Geol. Bull., 59, 2209-2265, 1975.

Emery, K. O., E. Uchupi, J. D. Phillips, C. O. Bowen, E. T. Bunce, and S. T. Knott, Continental rise off eastern North America, Amer. Assoc. Petroleum Geologists, 54, 44-108, 1970.

Evans, R., Sedimentation of the Mississippian evaporites around Atlantic margin: An alternative model, Canadian Jour. Earth Sci., 7, 1349-1352, 1970.

Evans, R., Origin and significance of evaporites in basins around Atlantic margin, Amer. Assoc. Petrol. Geol. Bull., 62, 223-234, 1978.

Fonseca, J. I., Geological outline of the Lower Cretaceous Bahia supergroup, in Van Hinte, J. E., editor, Proc. 2nd West African Micropaleontological Colloquiom, Ibadan, pp. 49-71, 1966.

Franks, S., and A. E. M. Nairn, The equatorial marginal basins of West Africa, in Nairn, A. E. M. and F. G. Stehli, The ocean basins and margins, v. 1, South Atlantic, New York, Plenum Press, pp. 301-350, 1973.

Geiskes, J. M., J. R. Lawrence, and G. Galleisky, Interstitial water studies, Leg 38, Supplement to volumes 38, 39, 40, and 41 of the Initial Reports of the Deep Sea Drilling Project, Washington, U.S. Government Printing Office, pp. 121-133, 1978.

Gill, W. D., The Mediterranean basin, in Salt basins around Africa, London, Institute of Petroleum, pp. 101-111, 1965.

Gorell, H. A., and G. R. Alderman, Elk Point Group saline basins of Alberta, Saskatchewan, and Manitoba, Canada, in Mattox, R. B., editor, Saline deposits, Geol. Soc. Amer. Spec. Paper 88, pp. 291-318, 1968.

Gould, D. B., and de Mille, G., Piercement structures in Canadian Arctic islands, in Braunstein, J. and G. O. O'Brien, Diapirism and diapirs, a symposium, Tulsa, Amer. Assoc. Petroleum Geologists, Memoir 8, pp. 183-214.

Grow, J. A., R. E. Mattick, and J. S. Schlee, Multichannel seismic depth sections and interval velocities over continental shelf and upper continental slope between Cape Hatteras and Cape Cod, in Watkins, J., L. Montadert, and P. W. Dickerson, editors, Geological and geophysical investigations of continental margins, Amer. Assoc. Petroleum Geologists, Memoir 29, pp. 65-83, 1979.

Gussow, W. C., Carboniferous stratigraphy and structural geology of New Brunswick, Canada, Amer. Assoc. Petrol. Geol. Bull., 37, 1713-1816, 1953.

Harland, W. B., An outline structural history of Spitzbergen, in Geology of the Arctic, Toronto, Canada, Univ. Toronto Press, pp. 68-132, 1961.

Harland, W. B., Contribution of Spitzbergen to understanding of tectonic evolution of North Atlantic region, in Kay, M., editor, North Atlantic - geology and continental drift, Amer. Assoc. Petroleum Geologists, Memoir 12, pp. 817-851, 1969.

Hess, H. H., History of the ocean basins, in Petrologic studies - Buddington Memorial Volume, New York, Geol. Soc. of America, pp. 599-620, 1962.

Hite, R. J., Salt deposits of the Paradox Basin, southeast Utah and southwest Colorado, in Mattox, R. B., editor, Saline deposits, Geol. Soc. Amer. Spec. Paper 88, pp. 319-330, 1968.

Hodgman, C. D., R. C. Weast, R. S. Shankland, and S. M. Selby, editors, Handbook of chemistry and physics, Cleveland, Ohio, The Chemical Rubber Pub. Co., 3604 p., 1963.

Holmes, A., Principles of physical geology, New York, The Ronald Press, 1288 p., 1965.

Howie, R. D., and M. S. Barss, Upper Paleozoic rocks of the Atlantic provinces, gulf of St. Lawrence, and adjacent continental shelf, Geol. Surv. Canada, Offshore Geology of Eastern Canada, v. 2, Paper 74-30, pp. 3-50, 1974.

Howie, R. O., and L. M. Cummings, Basement features of the Canadian Appalachians, Geol. Surv. Canada, Bull. 89, 18 p., 1963.

Irving, E., Paleomagnetism, New York, Wiley, 399 p., 1964.

Jansa, L. F., F. P. Bujak, and G. L. Williams, Upper Triassic salt deposits of the western North Atlantic, Canadian Jour. Earth Sci., 17, 547-559, 1980.

Jansa, L. F., and J. A. Wade, Geology of the continental margin off Nova Scotia and Newfoundland, Geol. Surv. Canada, Offshore Geology of Eastern Canada, Paper 74-30, pp. 51-105, 1974.

Jux, V., The palynologic age of diapiric and bedded salt in the Gulf coastal province, Louisiana Geol. Survey, Geol. Bull. 38, 46 p., 1961.

Keen, M. J., A possible diapir in the Laurentian Channel, Canadian Jour. Earth Sci., 7, 1561-1564.

Kent, P. E., The geological framework of petroleum exploration in Europe and North Africa and the implications of continental drift hypotheses, in Hepple, P., editor, The exploration for petroleum in Europe and North Africa, London, Institute of Petroleum, pp. 3-17, 1969.

King, L. C., The morphology of the Earth, Edinburgh and London, Oliver and Boyd, 726 p., 1967.

King, L. H., and B. MacLean, A diapiric structure near Sable Island - Scotian shelf, Maritime Sediments, 6, 1-4, 1970.

Kirkland, D. C., and J. E. Gerhard, Jurassic salt, central Gulf of Mexico, and its temporal relation to circum-Gulf evaporites, Amer. Assoc. Petrol. Geol. Bull., 55, 680-686, 1971.

Kozary, M. T., J. C. Dunlap, and W. E. Humphrey, Incidence of saline deposits in geologic time, in Mattox, R. B., editor, Saline deposits, Geol. Soc. Amer. Spec. Paper 88, pp. 43-58, 1968.

Krömmelbein, D., and R. Wenger, Sur quelques analogies remarquables dan les microfaunes crétacées du Gabon et du Brésil Oriental (Bahia et Sergipe), in Reyre, D., editor, Sedimentary basins of the African coasts, Part I, Atlantic coast, Paris, Association of African Geological Surveys, pp. 193-196, 1966.

Kumar, N., and L. A. P. Gamboa, Evolution of the São Paulo Plateau (southeastern Brazilian margin) and implications for the early history of the South Atlantic: Geol. Soc. Amer. Bull. Part I, v. 90, pp. 281-293, 1979.

LaCoste, J., Etudes géologiques dans le Rif Meridional, Protectorat Repub. Francaise au Maroc,

Serv. Mines, Notes et Mem. 31 and 32, 2 vols., 660 p., 1934.

Landes, K. K., Preliminary map no. 40, the Salina and Bass Island rocks in the Michigan basin, U.S. Geol. Survey, Oil and Gas Inv., 1945.

Landes, K. K., G. M. Ehlers, and G. M. Stanley, Geology of the MacKinac Straits region, Michigan Geol. Survey, Pub. 44, Geol. Ser. 37, p. 192, 1945.

LePichon, X., Sea-floor spreading and continental drift, Jour. Geophys. Research, 73, 3661-3697, 1968.

Leyden, R., G. Bryan, and M. Ewing, Geophysical reconnaissance on African shelf: 2. Margin sediments from Gulf of Guinea to Walvis Ridge, Amer. Assoc. Petrol. Geol. Bull., 56, 682-693, 1972.

Leyden, R., J. E. Damuth, L. K. Ongley, J. Kostecki, and W. Van Stevenick, Salt diapirs on the São Paulo Plateau, Amer. Assoc. Petrol. Geol. Bull., 62, 657-666, 1978.

Lohman, H. H., Seismic recognition of salt diapirs, Amer. Assoc. Petrol. Geol. Bull., 63, 2097-2102, 1979.

Maher, J. C., and E. R. Applin, Geologic framework and petroleum potential of the Atlantic coastal plain and continental shelf, U.S. Geol. Survey Prof. Paper 659, 98 p., 1971.

Manheim, F. T., and J. L. Bischoff, Geochemistry of pore waters from Shell Oil Company drill holes in the continental slope of the northern Gulf of Mexico, Chemical Geol., 4, 63, 1969.

Manheim, F. T., and R. E. Hall, Deep evaporitic strata off New York and New Jersey - evidence from interstitial water chemistry of drill holes, Jour. Res., U.S. Geol. Survey, v. 4, pp. 497-502, 1976.

Manheim, F. T., and M. K. Horn, Composition of deeper subsurface waters along the Atlantic margin, Southeastern Geology, 9, 215-236, 1968.

Manheim, F. T., and F. L. Sayles, Brines and interstitial brackish water in drill cores from the deep Gulf of Mexico, Science, 170, 57-61, 1970.

Manheim, F. T., and F. L. Sayles, Composition and origin of interstitial waters of marine sediments, based on deep sea drill cores, in Goldberg, E. D., The sea, v. 5, New York, Wiley, 527-558, 1974.

Manheim, F. T., F. L. Sayles, and L. S. Waterman, Interstitial water studies on small core samples, Deep Sea Drilling Project: Leg 10, in Worzel, J. L., W. Bryant, et al., Initial reports of the Deep Sea Drilling Project, v. 10, Washington, U.S. Government Printing Office, pp. 615-623, 1973.

Manheim, F. T., F. L. Sayles, and L. S. Waterman, Interstitial water studies on small core samples, Deep Sea Drilling Project, Leg 12, in Laughton, A. S., W. A. Berggren, et al., Initial reports of the Deep Sea Drilling Project, v. 12, Washington, U.S. Government Printing Office, pp. 1193-1200, 1972.

Mattick, R. E., R. Q. Foote, N. L. Weaver, and M. S. Grim, Structural framework of United States Atlantic outer continental shelf north of Cape Hatteras, Amer. Assoc. Petrol. Geol. Bull., 58, Part II of II, pp. 1179-1190, 1974.

Mauffret, A., J. P. Fail, L. Montadert, J. Sancho, and E. Winnock, Northwestern Mediterranean sedimentary basin from seismic reflection profile, Amer. Assoc. Petrol. Geol. Bull., 57, 2245-2262, 1973.

Maync, W., The Permian of Greenland, in The geology of the Arctic, v. 1, Toronto, Ontario, Toronto University Press, pp. 214-223, 1961.

McIver, N. L., Cenozoic and Mesozoic stratigraphy of the Nova Scotia shelf, Canadian Jour. Earth Sci., 9, 54-70, 1972.

McKelvey, V. E., and F. F. H. Wang, Preliminary map, saline minerals, sulfur, phosphorite, manganese-oxide nodules, and metal-bearing mud, U.S. Geol. Survey, Misc. Geologic Investigations, Map I-632 (Sheet 4 of 4), 1969.

Mear, C. E., Upper Permian sediments in southeastern Permian Basin, Texas, in Mattox, R. B., editor, Saline deposits, Geol. Soc. Amer. Spec. Paper 88, pp. 349-358.

Meyerhoff, A. A., Continental drift II: High-latitude evaporite deposits and geologic history of Arctic and North Atlantic Oceans, Jour. Geol., 78, 406-444, 1970.

Meyerhoff, A. A., and C. W. Hatten, Diapiric structures in central Cuba, in Braunstein, J. and G. D. O'Brien, editors, Diapirism and diapirs, a symposium, Tulsa, Amer. Assoc. Petroleum Geologists, Memoir 8, pp. 315-357, 1968.

Meyerhoff, A. A., and C. W. Hatten, Bahamas salient of North America, in Burk, C. A. and C. L. Drake, The geology of continental margins, New York, Springer-Verlag, pp. 429-446, 1974.

Montadert, L., L. E. Winnock, J. R. Deltiel, and G. Grau, Continental margins of Galicia-Portugal and Bay of Biscay, in Burk, C. A. and C. L. Drake, editors, The geology of continental margins, New York, Springer-Verlag, pp. 323-342, 1974.

Morgan, W. J., Rises, trenches, great faults, and crustal blocks, Jour. Geophys. Research, 73, 1959-1982, 1968.

Murray, G. E., Salt structures in Gulf of Mexico basin - a review, in Braunstein, J. and G. D. O'Brien, editors, Diapirism and diapirs, a symposium, Tulsa, Amer. Assoc. Petroleum Geologists, Memoir 8, pp. 99-121, 1968.

Naylor, D., and S. N. Mounteney, Geology of the northwest European continental shelf, v. 1, London, Graham Trotman Dudley Publishers, Ltd., v. 1, 162 p., 1975.

Nettleton, L. L., Fluid mechanics of salt domes, Amer. Assoc. Petrol. Geol. Bull., 18, 1175-1204, 1934.

Odé, H., Review of mechanical properties of salt relating to salt dome genesis, in Mattox, R. B., editor, Saline deposits, New York, Geological Society of America, Special Paper No. 88, pp. 543-595, 1968.

Offshore Exploration Staffs, Amoco Canadian Petroleum Company, Ltd., and Imperial Oil, Ltd., Regional geology of the Grand Banks, Amer. Assoc. Petroleum Geologists, 58, Pt. II of II, pp. 1109-1123, 1974.

Olson, W. S., and R. J. Leyden, North Atlantic rifting in relation to Permian-Triassic salt deposition, in Logan, A. and L. V. Mills, editors, The Permian and Triassic systems and thier mutual boundary, Canadian Soc. Petroleum Geologists, Memoir 2, pp. 720-732, 1973.

Olson, W. S., Structural history and oil potential of offshore area from Cape Hatteras to Bahamas, Amer. Assoc. Petrol. Geol. Bull., 58, Part II of II, pp. 1191-1200, 1974.

Pamenter, B., Sands aplenty in Onondaga well, Oilweek, 22 November, pp. 22-24, 1971.

Parker, T. J., and A. N. McDowell, Model studies of salt-dome tectonics, Amer. Assoc. Petrol. Geol. Bull., 39, 2384-2470, 1955.

Pautot, G., J. M. Auzende, and X. LePichon, Continuous deep salt layer along North Atlantic margins related to early phase of rifting, Nature, 227, 351-354, 1966.

Pearson, W. J., Salt deposits of Canada, in Bersticker, A. C., editor, Symposium on Salt, Northern Ohio Geol. Soc., pp. 197-239, 1963.

Pegrum, R. M., G. Rees, and D. Naylor, Geology of the northwest European continental shelf, v. 2, The North Sea, London, Graham Trotman Dudley, Ltd., 225 p., 1975.

Pierce, W. G., and E. I. Rich, Summary of rock salt deposits in the United States as possible storage sites for radioactive waste materials, U.S. Geol. Survey Bull. 1148, 91 p., 1962.

Poag, C. W., Stratigraphy of the Atlantic continental shelf and slope of the United States, Ann. Rev. Earth Planet. Sci., 6, pp. 251-280, 1978.

Ponte, F. C., and H. E. Asmus, The Brazilian marginal basins: Current state of knowledge, in de Almeida, F. F. M., editor, International Symposium on Continental Margins of the Atlantic Type, São Paulo, Anais. da Academia Brasiliera de Ciências, 48, suplemento, pp. 215-239.

Querol, R., Regional geology of the Spanish Sahara, in Reyre, D., editor, Sedimentary basins of the African coasts, Part I, Atlantic coast, Paris, Association of African Geological Surveys, pp. 27-38, 1966.

Reyre, D., Particularités géologiques des Bassins de l'Oust Africain (essai de recapitulation), in Reyre, D., editor, Sedimentary basins of the African coasts, Part I, Atlantic coast, Paris, Association of African Geological Surveys, pp. 253-304, 1966a.

Reyre, D., Récapitulation Générale: Les formations Permiennes, Triasiques, Jurassiques et Crétacées dans les bassins cotiers Ouest-Africains, in Reyre, D., editor, Sedimentary basins of the African coasts, Part I, Atlantic coast, Paris, Association of African Geological Surveys, 304 p., 3 tables, 1966b.

Rickard, L. V., Stratigraphy of the Upper Silur-ian Salina Group, New York, Pennsylvania, Ohio, Ontario: New York State Museum and Science Service, Map and Chart Series No. 2, 57 p., 1969.

Rios, J. M., Saline deposits of Spain, in Mattox, R. B., editor, Geol. Soc. Amer. Spec. Paper 88, pp. 59-74, 1968.

Rona, P. A., Possible salt domes in the deep Atlantic off Northwest Africa, Nature, 224, 141-143, 1969.

Rona, P. A., Comparison of continental margins of eastern North America at Cape Hatteras and northwestern Africa at Cap Blanc, Amer. Assoc. Petrol. Beol. Bull., 54, 129-157, 1970.

Rona, P. A., Salt deposits of the Atlantic, Anais da Academia Brasileira de Ciências, 48, Suplemento, pp. 265-274, 1976.

Runcorn, S., Palaeo-magnetic evidence for continental drift and its geophysical cause, in Continental Drift, International Geophysics Series, no. 3, pp. 1-40, 1962.

Ryan, W. B. F., K. J. Hsu, et al., Initial reports of the Deep Sea Drilling Project, v. XIII, Parts 1 and 2, Washington, U.S. Government Printing Office, 1447 p., 1973.

Salvan, H. M., Les niveaux salifères marocains, leurs characterisistiques et leurs problèms, in Geology of saline deposits, UNESCO, Earth Sci. Ser., no. 7, pp. 147-159, 1972.

Sannemann, D., Salt-stock families in northwestern Germany, in Braunstein, J. and G. D. O'Brien, editors, Diapirism and diapirs, a symposium, Tulsa, Amer. Assoc. Petroleum Geologists, Memoir 8, pp. 261-270, 1968.

Sayles, F. L., F. T. Manheim, and L. S. Waterman, Interstitial water studies on small core samples, Leg 11, in Hollister, C. D., J. I. E. Wing, et al., Initial reports of the Deep Sea Drilling Project, v. XV, Washington, U.S. Government Printing Office, pp. 997-1008, 1972.

Schmalz, R. F., Deep water evaporite deposition: A genetic model, Amer. Assoc. Petrol. Geol. Bull. 53, 798-823, 1969.

Schneider, E. D., and G. L. Johnson, Deep-ocean diapir occurrences, Amer. Assoc. Petroleum Geologists, 54, 2151-2169, 1970.

Scientific Party, Basins and margins of the eastern South Atlantic, Leg 40 of the Deep Sea Drilling Project, Geotimes, 20, 22-24, 1975.

Sclater, J. G., S. Hellmayer, and C. Tapscott, The paleobathymetry of the Atlantic Ocean from the Jurassic to the present, Jour. Geol., 85, 509-522, 1977.

Sheridan, R. E., Atlantic continental margin of North America, in Burk, C. A. and C. L. Drake, editors, The geology of continental margins, New York, Springer-Verlag, pp. 391-407, 1974.

Sheridan, R. E., and W. L. Osburn, Marine geological and geophysical studies of the Florida-Blake Plateau-Bahamas area, in Yorath, C. J., E. R. Parker, and D. J. Glass, editors, Canada's continental margins, Calgary, Soc. Petrol. Geol., Memoir 4, pp. 9-32, 1975.

Société Cherifienne des Pétroles, Le bassin du

sud-ouest Marocain, in Reyre, D., editor, <u>Sedimentary basins of the African coasts, Part I, Atlantic coast</u>, Paris, Association of African Geological Surveys, pp. 5-12, 1966.

Sotelo, V., and J. M. Gieskes, Interstitial water studies, Leg 40: Ship board studies, in Bolli, H. M., W. B. F. Ryan, <u>et al</u>., Initial reports of the Deep Sea Drilling Project, v. 40, Washington, U.S. Government Printing Office, pp. 549-554, 1978.

Stewart, F. H., The Permian lower evaporites of Fordon in Yorkshire, <u>Proc. Yorks Geol. Soc.</u>, <u>34</u>, 1, 1963a.

Supko, P. R., and K. Perch-Nielsen, <u>et al</u>., Initial reports of the Deep Sea Drilling Project, v. 39, Washington, U.S. Government Printing Office, 1139 p., 1977.

Swain, F. M., Ostracoda from wells in North Carolina. 2. Mesozoic ostracoda, <u>U.S. Geol. Survey Prof. Paper 234-B</u>, pp. 59-152, 1952.

Tator, B. A., and L. E. Hatfield, Bahamas present complex geology, <u>Oil Gas Jour.</u>, <u>73</u>, no. 43, pp. 172-176; no. 44, p. 120-122, 1975.

Templeton, R. S. M., The geology of the continental margin between Dakar and Cape Palmas, in Delany, F. M., editor, <u>The geology of the East Atlantic continental margin. 2. Africa</u>, Rep. No. 70/16, Inst. Geol. Sci., pp. 44-60, 1971.

Toit, A. L. du, <u>Our wandering continents</u>, Edinburgh, Oliver, and Boyd, 366 p., 1937.

Tortochaux, F., Occurrence and structure of evaporites in North Africa, in Mattox, R. B., editor, Saline deposits, <u>Geol. Soc. Amer. Spec. Paper 88</u>, pp. 107-138, 1968.

Uchupi, E., K. O. Emery, C. O. Bowin, and J. D. Phillips, Continental margin off western Africa: Senegal to Portugal, <u>Amer. Assoc. Petroleum Geologists, 60</u>, 809-878.

Udden, J. A., Laminated anhydrite in Texas, <u>Geol. Soc. Amer. Bull.</u>, <u>35</u>, 347, 1924.

Van Houten, F. B., Triassic-Liassic deposits of Morocco and eastern North America, <u>Amer. Assoc. Petrol. Geol. Bull.</u>, <u>61</u>, 79-99, 1977.

Viana, C. F., Revisão estratigráfica de Bacia do Recôncavo-Tucano, <u>Bol. Tec. Petrobrás</u>, Rio de Janiero, v. 14, no. 3/4, pp. 157-192, 1971.

Viana, C. F., Stratigraphic distribution of ostracoda in the Bahia supergroup (Grazil), in Van Hinte, J. E., editor, <u>Proc. 2nd West African Micropaleontological Colloquium</u>, Ibadan, pp. 240-256, 1966.

Von Herzen, R. P., H. Hoskins, and T. H. Van Andel, Geophysical studies in the Angola diapir field, <u>Geol. Soc. Amer. Bull.</u>, <u>83</u>, 1901-1910, 1966.

Walton, H. S., and A. A. Berti, Upper Triassic and Lower Jurassic palynology of the Grand Banks area, Amer. Assoc. Stratigraphic Palynologists and Commission Internat. Microflore du Palèozoique, Abstracts with Programs, pp. 27-28, 1976.

Waterman, L. S., F. L. Sayles, and F. T. Manheim, Interstitial water studies on small core samples, Leg 14, in Hayes, D. E., A. C. Pimm, <u>et al</u>., Initial reports of the Deep Sea Drilling Project, v. XIV, Washington, U.S. Government Printing Office, pp. 753-762, 1972.

Watson, J. A., and G. L. Johnson, Seismic studies in the region adjacent to the Grand Banks of Newfoundland, <u>Canadian Jour. Earth Sci.</u>, <u>7</u>, 306-316, 1970.

Watson, J. M. and C. A. Swanson, North Sea - A major petroleum province, <u>Amer. Assoc. Petrol. Geol. Bull.</u>, <u>59</u>, 1098-1112, 1975.

Webb, G. W., Salt structures east of Nova Scotia, Earth Science Symposium on Offshore Eastern Canada, Geol. Survey Canada, Paper 71-23, pp. 197-218, 1973.

Wegener, A., <u>The origin of continents and ocean basins</u>, London, 212 p., 1924.

Wicks, G. L., and J. D. Isaacs, Salt domes: Is there more energy available from their salt than from their oil? <u>Science, 199</u>, 1436-1437, 1978.

Winnock, E., Exposé sussinct de l'evolution paléogelogique de l'Aquitaine, <u>Bull. Soc. Geol. France</u>, 7e Serie, v. XV, pp. 5-12, 1973.

Withington, C. F., Gypsum and anhydrite in the United States, exclusive of Alaska and Hawaii, U.S. Geol. Survey, <u>Mineral Inv. Resource Map MR-33</u>, scale 1:3,168,000, 1962.

Ziegler, P. A., Geologic evolution of North Sea and its tectonic framework, <u>Amer. Assoc. Petrol. Geol. Bull.</u>, <u>59</u>, 1073-1097, 1975.

CRUSTAL STRUCTURE AND DEVELOPMENT OF SHEARED PASSIVE CONTINENTAL MARGINS

R.A. Scrutton

Grant Institute of Geology, University of Edinburgh, West Mains Road, Edinburgh, Scotland

Abstract. Sheared passive continental margins
are ideally found in conjugate pairs on opposite
sides of an Atlantic-type ocean. Several variations from the ideal occur in nature, one of
which, the offset along the Gulf of Aqaba at the
northern end of the Red Sea, is investigated to
deduce its crustal structure. Its structure is
consistent with that seen at more mature sheared
margins which are in a later stage of development.
Little or no crustal thinning occurs under the
Gulf of Aqaba and, likewise, little occurs at
sheared margins. This structure is tabulated
along with other features of sheared margins to
show how different they are from rifted margins.
A qualitative model for the development of sheared
margins is now available and widely known,
although quantitative work is lacking. The
model is outlined, tracking the development from
continental shear stage to truly passive sheared
margin.

Introduction

The history of studies on sheared passive margins began with Wilson (1965) recognising that
major transform faults in the oceans may originate from pre-existing weaknesses in the adjacent
continental crust via passive margin offsets.
Indeed, sheared margins have been called offset
margins by some workers: at present, they are
commonly called transform margins. During the
International Geodynamics Project advances have
been made in our knowledge of the structures at
these margins and their origin (Keen and Keen,
1971; Le Pichon and Hayes, 1971; Falvey, 1972;
Francheteau and Le Pichon, 1972; Talwani and
Eldholm, 1973; Scrutton et al., 1974; Scrutton,
1976a; Mascle, 1976; Rabinowitz et al., 1976;
Wilson and Williams, 1979; Scrutton, 1979). A
two-fold classification of sheared margins into
ones with several, short offsets and ones with a
single, long offset was made by Bott and Scrutton
(1976); this may reflect the degree of inhomogeneity of the continental crust through which
the rift zone penetrated, but many more new data
are required to confirm this. Despite the
obvious need for studies designed specifically
to elucidate sheared margin structure and evolu-

tion, data appear to have been collected over
these margins only as parts of more regional
investigations. Consequently, a rather piecemeal
approach to the topics addressed in this paper
has been taken in the past.

The simplest way in which a sheared margin can
develop during continental breakup is the way
outlined by Le Pichon and Hayes (1971). They
proposed a simple rectilinear offset of the rift
zone where a continental shear zone develops into
a fracture zone and transform fault which offsets
the new sea-floor spreading centre by an amount
equal to the margin offsets. However, not all of
the major, well-documented sheared margins around
the two oceans most widely studied (the Atlantic
and Indian Oceans) are simple conjugate pairs
facing each other across the ocean at opposite
ends of the same fracture zone. Figure 1 shows
the locations of well-documented sheared margins
and Table 1 contains notes on their origins.
About a third depart in some way from the simple
conjugate mode: for example, where rift and
spreading centre jumps have caused asymmetries
they are a means of isolating continental fragments (Scrutton, 1976b). Anomalies of different
kinds occur at the Senja and Aqaba Fracture
Zones, as indicated in Table 1.

In this paper it is intended to illustrate the
crustal structure at sheared margins, to identify
and discuss common features of the structure and
of margin development, then to see if these will
throw more light on models for the origin of such
margins. It is not intended to discuss any one
margin in great detail.

Crustal Structure

This is summarised in the form of structure
sections across a variety of sheared margins
(Fig. 2). All the sections are redrawn from
the data sources to a common scale to facilitate
comparison between them. In addition to passive
margin sections there is a section across the
Queen Charlotte Fracture Zone, which is an
active sheared margin between the oceanic Pacific
and continental American plates, and one across
the active Gulf of Aqaba, which was drawn up for
the purposes of this paper. Also incorporated in

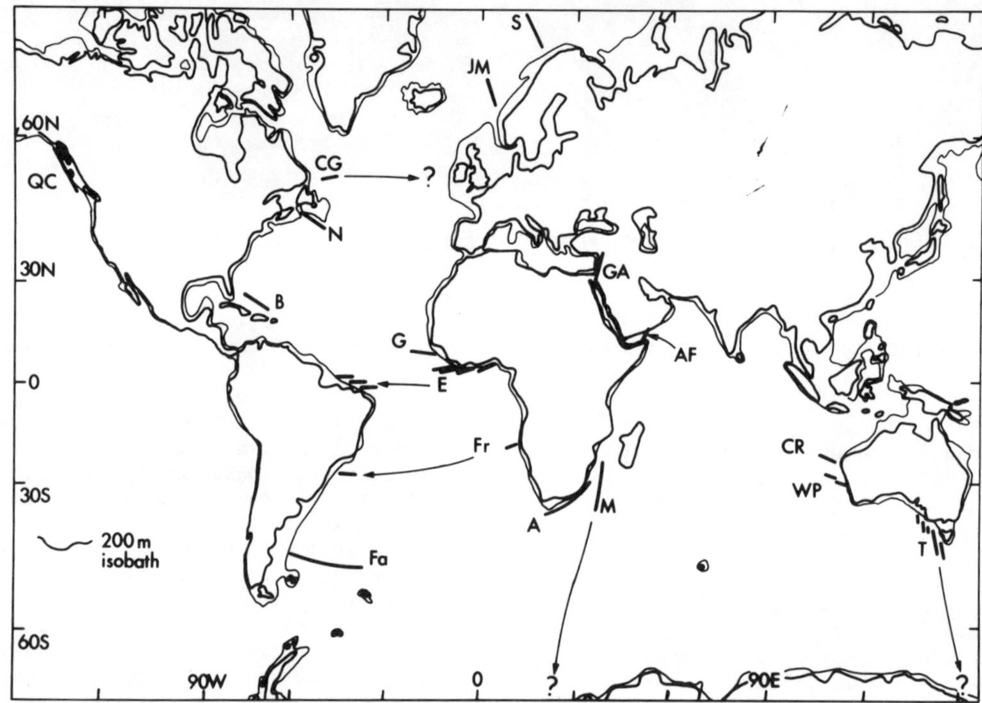

Fig. 1. The location of principal sheared passive margins. Associated fracture zone names
are: S-Senja, JM-Jan Mayen, CG-Charlie Gibbs, N-Newfoundland, B-Bahamas, G-Guinee,
E-Equatorial (Cape Palmas, St. Paul, Romanche, Chain), Fr-Frio, Fa-Falkland, A-Agulhas,
M-Mozambique, GA-Gulf of Aqaba, AF-Alula Fartak, CR-Cape Range, WP-Wallaby Perth, T-Tasman
and others, QC-Queen Charlotte.

the figure on the same scale is a section across
a "typical" rifted margin for comparison with the
sheared margins. It should be remembered that as
well as the relatively "obese" type of rifted margin, like the one illustrated here, there is also
the starved type, and within the latter group are
crustal structures that are not dissimilar in
cross section from those at sheared margins.
Section 15 of Scrutton (1976c) west of Cape Town,
for example, shows limited attenuation of continental crust at the continent's rifted edge.
Notwithstanding this possible similarity, it
can be quickly seen from Figure 2 that there are
fundamental differences in crustal structure
between rifted and sheared passive margins.
These will be summarised later.

Gulf of Aqaba - a sheared margin in the "rifting"
phase

Because the Gulf of Aqaba is clearly the site
of a sheared passive margin in the making, a
simple structure section across the Gulf (Fig. 2)
was made to give some indication of crustal
character at the stage just prior to the margin
coming into existence. Another example of a
continental shear zone about to become a sheared
margin is at the head of the Gulf of California,
but, although the data there are much more com-

prehensive, the Californian example cuts through
very complex basement structures and is itself a
more complex feature. Thus, the simpler Dead Sea
shear, which passes through only Precambrian
basement rocks at its southern end is preferred
for this study.

The cross section passes through the Aragonese
Deep where free-air gravity anomalies fall to
-200 m Gal. Allan (1970) described the gravity
field in the Gulf adequately for our purposes,
whilst Knopoff and Belshe (1966) have published
a crude map of regional gravity variations.

Seismic control on layer velocities (and hence,
densities) and crustal thickness was obtained
from Ginzburg et al.'s (1979) crustal refraction
experiments. A line of stations along the west
coast of the Gulf gives a depth to the Moho of
25 km. Detailed information on bathymetry and
sediment thicknesses in the Gulf was provided by
Ben-Avraham et al. (1979) who have shown that at
least 2-3 km of Pliocene and younger deposits are
present. From this modelling information, the
crustal structure shown in Figure 2 appears most
likely.

The most striking feature of the crustal section is the apparent absence of isostatic compensation for the Gulf and its low-density
sedimentary fill. If anything, the gravity
data require a crustal thickening here. This

TABLE 1. The geometry of sheared passive continental margins

Associated fracture zone	Equal conjugate margins?	Comments
Senja	Yes, but see "Comments"	Shearing between Greenland and Spitzbergen 55-13 myBP became rifting 13-0 myBP.
Jan Mayen	No	Spreading jump south of fracture zone isolated Jan Mayen Ridge and left sheared margin against Voring Plateau.
Charlie Gibbs	? No	Location of continent-ocean boundary in Rockall Trough unknown; a sheared margin may exist there.
Newfoundland	No	Original offset from Newfoundland to Tethys truncated at eastern end by spreading ridge jump north of the fracture zone to between Newfoundland and Portugal.
Bahama-Guinea	Yes	
Equatorial	Yes	Four or more offsets off S. America and Africa.
Frio	Yes	Very short offset.
Falkland-Agulhas	Yes	Detailed mapping indicates that the fracture zone may have been "leaky" in earliest stages of opening.
Mozambique	?	Structure off Antarctica yet to be determined.
Gulf of Aqaba	No	Fracture zone terminates at northern end in zone of compression on east side.
Alula Fartak	Yes	
Exmouth Wallaby-Perth	? ?	Possible conjugates now lost in zone of compression in S.E. Asia.
Tasman and others	? Yes	Several small offsets W. of Tasmania appear to have conjugates off Antartica.

is somewhat surprising, since the Gulf of Aqaba owes its existence to there being three elongate NNE-SSW rhomb-shaped, pull-apart depressions in the top of the crust, akin to the Dead Sea depression (Ben Avraham et al., 1979). Under these, crustal thinning or basaltic intrusion would be expected, as beneath the Salton Sea in California. However, as the depressions are elongated along the Gulf, any deep expression of them may be restricted to the E-W limits of the Gulf and not appear as crustal thinning beneath the flanks. As a result of this, there is no <u>wide</u> zone of crustal thinning associated with the shear zone, and there are apparently no large basic intrusions into the adjacent crust. Both of these features might be expected at rifted passive margins during the rifting phase.

Of the sedimentary deposits, those in the Gulf itself might be referred to as the rifting-phase deposits which will later be covered, probably unconformably, by drifting-phase deposits. Ben-Avraham et al. (1979) show that the Gulf deposits are extensively folded and faulted. Such deformation should be recognisable in the deeper sediments beneath the slopes of mature sheared margins, thus providing the rifting/drifting distinction used on rifted margins. Pelagic and turbiditic facies are present which, together with the aforementioned structures, could easily produce good traps for oil and gas. In the Sinai Peninsula west of the Gulf of Aqaba is a Mesozoic and Tertiary sedimentary basin, the southern end of which appears on the crustal model. Separating the basin from the Gulf is an upwarp in the Precambrian basement. A similar but much smaller basement upwarp occurs on the east side of the Gulf. Sedimentation has thus been restricted on the immediate flanks of the shear zone by basement highs that will later appear beneath the shelf break of the mature sheared margin.

More extensive studies of the Gulf of Aqaba and northern Red Sea structure with emphasis on its implications for sheared passive margin formation would be rewarding. The preliminary investigation described here has encountered a few features that will have some expression in the later evolutionary stages of a sheared margin. Indeed, Scrutton (1979) has already shown the value of looking at actualistic models for early sheared margin development.

<u>Structure and some other characteristics of sheared margins</u>

Two noteworthy papers summarising the crustal structure at sheared margins were presented at a conference in 1975 (Mascle, 1976; Scrutton, 1976a). All of what was concluded then still applies and is reassembled with some new observations in Table 2. In this table the structural features are compared with their equivalents on rifted margins. The sheared margins illustrated in Figure 2 are those at which the deep structure has

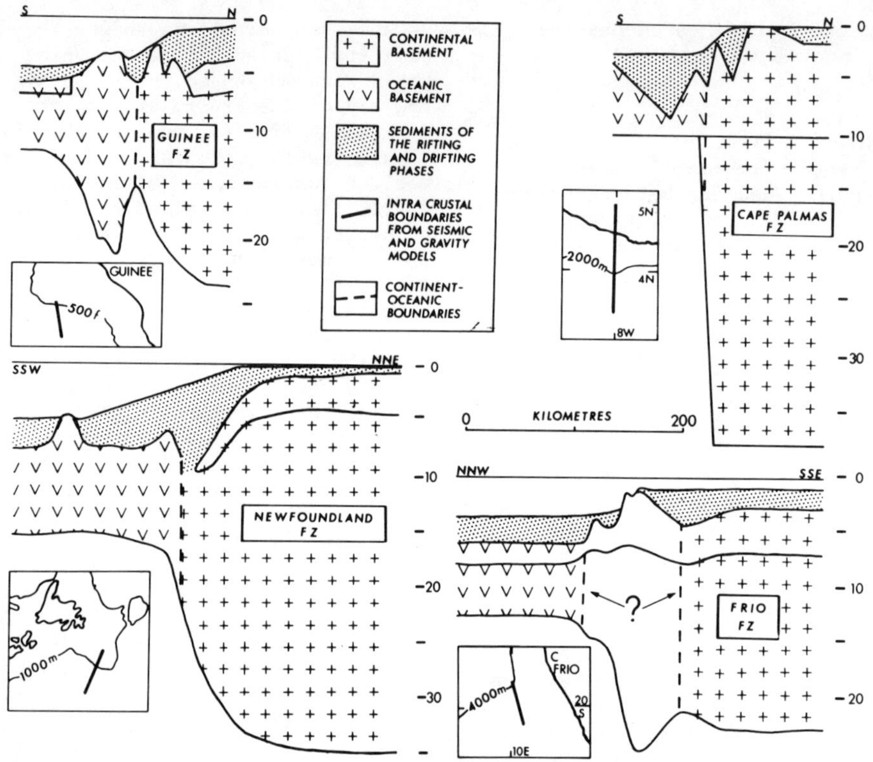

Fig. 2. Crustal structure sections across sheared margins, located in the insets. The sections are redrawn from the following publications: Newfoundland – Keen (this volume), Guinee – Jones (in press), Cape Palmas – Behrendt et al. (1974), Frio – Goslin and Sibuet (1975), Agulhas, west and east – Scrutton (1976c), Cape Range – Roots et al. (1979), Queen Charlotte – Dehlinger et al. (1971).
The section across the Gulf of Aqaba was calculated for the paper and is therefore shown with the observed and calculated gravity anomaly profiles.
For comparison with the sheared margin sections, the rifted margin section off Nova Scotia (Keen, this volume) is shown on the same scale. Vertical scales are in kilometres.

been modelled in anything more than a cursory fashion, and it is from these that the deep structural characteristics are obtained. The shallow structural characteristics are obtained from numerous studies in addition to these, especially those mentioned by Mascle (1976).

As well as the purely structural features observed at sheared passive margins there are other features common on these margins that give strong pointers to the mode of margin development. Firstly, as Wilson (1965) suggested, some fracture zones can be traced into the continent beyond the landward end of a sheared margin as some sort of geological "event". Either there is a pre-existing suture in the continental crust at which the rift zone was offset and a fracture zone formed, e.g. the Appalachian-Avalonian boundary northeast of Newfoundland that appears to be ancestral to the Charlie Gibbs Fracture Zone (Haworth and Lefort, 1979); or a line of alkaline, basic or ultrabasic major intrusions has been injected into the continental crust just before or at continental break-up, perhaps along a pre-existing suture that is not visible at the sur-

face, e.g. numerous intrusive centres around the South Atlantic described by Marsh (1973). Clearly, if such a pre-existing suture strikes obliquely to the opening rift zone (and most of them do), an offset can only follow that suture for a short distance, otherwise the offset would not follow a small circle of opening as the overriding plate tectonic forces require it to do. It is not surprising, therefore, that to date it is only the shorter offsets, those less than about 100 km long, that have been seen to be related to pre-existing sutures; the large offsets, like the Falkland and Agulhas margins, are not clearly related. The location of the latter must be controlled solely by plate tectonic forces.

Secondly, numerous authors have remarked upon the lack of igneous activity along sheared margins during their early, active phase. As both continental shear zones and active transform faults are seen to be free of igneous activity (compared to rift zones), this observation is not unexpected. Apparently, shear zones are not conduits for magma, probably simply because there

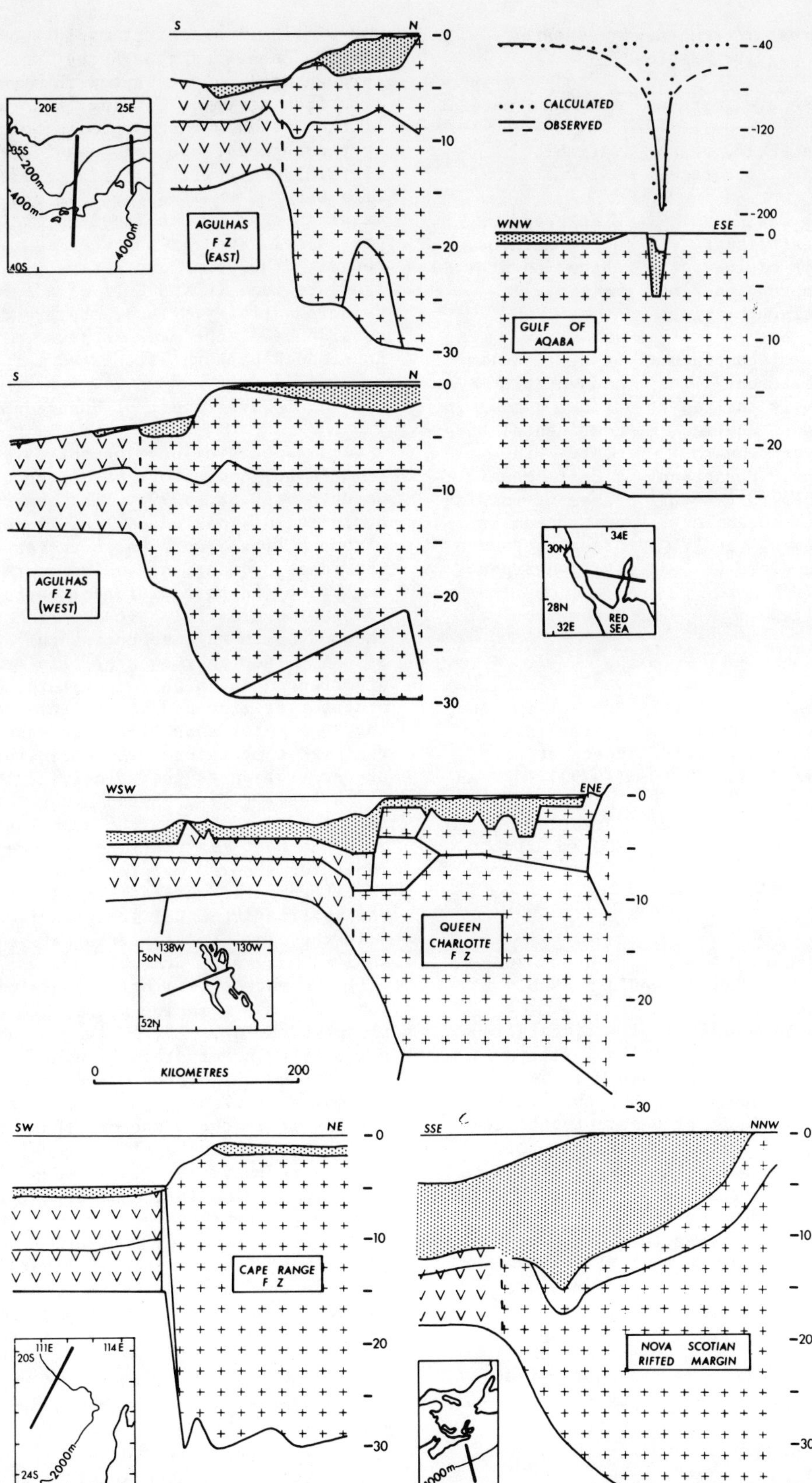

S
N
AGULHAS F Z (EAST)
20E
25E
35S
200m
400m
1000m
40S
S
N
AGULHAS F Z (WEST)
CALCULATED
OBSERVED
-40
-120
-200
WNW
ESE
GULF OF AQABA
34E
30N
28N
32E
RED SEA
WSW
ENE
QUEEN CHARLOTTE F Z
138W
130W
56N
52N
0
KILOMETRES
200
SW
NE
CAPE RANGE F Z
111E
114E
20S
24S
2000m
SSE
NNW
NOVA SCOTIAN RIFTED MARGIN
1000m

TABLE 2. Crustal structure at sheared
passive margins

Feature	At sheared margins	At rifted margins
Morphology	Wide shelf, steep slope, poor rise.	Variable
Sedimentary basins	Deep basins only beneath inner shelf or in fracture zone, not on continental rise.	Widespread on obese margins. Absent on starved margins.
Basement features	Edge of continental basement heavily faulted to produce shallow ridges beneath slope. Horst and graben in fracture zone. Oceanic basement rarely downwarped.	Listric faults and rotated blocks common in continental basement. Ridge in oldest oceanic basement in places; oceanic basement commonly downwarped.
Thinning of continental crust	Limited	Extensive
Thickening of oceanic crust	Absent, except in fracture zones exploited by vulcanism.	Present in some places (see Roots et al. (1979)).
Width of transition zone	Narrow, generally less than 100 km.	Variable, 100 to 300 km.

is no magma beneath them. Should a shear zone pass over a hot spot, then it would probably experience igneous activity.

Thirdly, two recent analyses of sedimentation at sheared margins reveal a distinct pattern to the subsidence history (Wilson and Williams, 1979; Dingle, this volume). Rather than there being an exponential decay of basement subsidence as at rifted margins, maximum subsidence rates are attained sometime after the margin begins to form. Wilson and Williams' data suggest that the maximum rate occurs when the ocean basin has opened to a width three times the length of the sheared margin offset and mechanical constraints between the diverging plates are relaxed, as described by Le Pichon and Hayes (1971). Dingle's data, from different basins, suggest that the maximum rate occurs when the ocean basin is no more than one offset length wide and it results from loss of support for the new margin from its conjugate. In either case a dynamic rather than thermal control is apparent. There may, however, be an underlying thermal effect, since following (and before, in Wilson

and Williams' examples) rapid subsidence the margins behave as though they were subsiding through cooling, perhaps with a small contribution from sediment loading. It is possible that at these stages the margin is being dragged down by the adjacent cooling oceanic crust. There is certainly very little evidence of active dislocation between the two types of crust at these margins once the sea-floor spreading centre has passed by. Even at the sheared margins associated with the young and active Alula Fartak fracture zone in the Gulf of Aden there is no seismic activity. Thus, the trend of thermal subsidence of the oceanic crust may underlie the subsidence pattern at sheared margins.

Development of Sheared Margins

Some aspects of development have just been alluded to, and others have been discussed by Scrutton (1976a, 1979). The general pattern of evolution of sheared margins is now widely known, although little work has been carried out to give the ideas a quantitative framework. A complicating factor is that different parts of the same sheared margin will have experienced different thermodynamic regimes during the developmental stages. The inner ends of offsets will have experienced the high temperatures of the continental rift zone and very little interplate shearing, the outer ends will have experienced both the high temperatures and extensive shearing, and the central parts the relatively low temperatures of an inter rift area and moderate shearing. It is intended to tackle the quantitative problem over the next few years to produce a geodynamic model for sheared margins that is comparable with that for rifted margins. Here, in conclusion, the qualitative model is summarised.

At the continental rifting stage the embryonic sheared margin is probably not very active. It will lie between the offset ends of two rift zones and may coincide with a pre-existing suture along which there may be igneous activity. Thermal uplift and crustal thinning by erosion will effect the ends of the offset and crustal stretching may take place adjacent to the rift zones. Perhaps it is these factors that ultimately allow a wide continental shelf to develop at sheared margins (Falvey, 1972). As soon as sea-floor spreading begins intracontinental shearing begins between the continental parts of the two plates now slipping past one another. Anastomosing faults most likely isolate slivers of continental crust in the shear zone, deep, tectonically-controlled sedimentary basins may develop and shear heating will occur in the fault zone. Temperatures may rise high enough to cause local melting or thermal expansion of the lithosphere. In the deeper parts of the crust a shear-parallel metamorphic fabric may be developed. This is the stage illustrated by the Gulf of Aqaba.

Simultaneously, oceanic crust is being emplaced against new sheared margins at the ends of the

offset. As the sea-floor spreading centres migrate along the margins they push half the
oceanic crust before them, which shears and
subsides against continental crust. Heat is
generated by shearing and the oceanic crust is
cooling, the effect being a rise in temperature
in the continental crust but a net fall in temperature in the oceanic crust. Limited uplift may
effect the edges of the continents at this stage
provided the subsiding oceanic crust is decoupled.
This is the stage illustrated by the Queen Charlotte Fracture Zone. The passing intrusion and
extrusion axis provides a transient heating event
that will briefly emphasise the effects begun by
the preceding oceanic crust. Behind the migrating sea-floor spreading centre the truly passive
sheared margin develops.

Once a new sheared margin becomes truly passive
there is no longer strike-slip faulting between
lithospheric plates to keep the edge of continental crust sharply truncated. Nevertheless,
there is apparently little change to the sharp
transition from continental to oceanic crust as
the margin matures, as can be seen in Figure 2.
The horizontal stresses that undoubtedly arise
from differing gravitational forces across the
margin (Bott, 1971) are not strong enough to produce deformation of the Moho surface, the latter
having been "annealled" during the shearing process. A possible mechanism for subsidence of the
continental crust is thus lost with the "anneall-
ing". Subsidence may take place, however,
through the oceanic crust left against the margin
cooling and subsiding, probably dragging the continental crust down with it, rapidly at first, as
discussed above. The release of mechanical constraints, mentioned above, may also allow subsidence to occur more rapidly. Opposing any
initial subsidence, though probably not effective enough to prevent it, is the transfer of
heat from the cooling oceanic lithosphere to the
continental lithosphere. Eventually, the continental lithosphere will cool as well, but as it
has already probably been dragged down out of
isostatic equilibrium this will only serve to
restore isostatic balance.

The cooling contracting oceanic lithosphere
will be thrown into tension, allowing some normal
faulting and horst and graben structures to develop in the fracture zone. It is here and on the
wide continental shelf that sedimentation will be
heaviest, since continental basement remains
shallow beneath the slope. The shallow basement
results from a buoyant sliver of continental
crust in the fracture zone or a marginal fracture
ridge (Le Pichon and Hayes, 1971), and the shelf
basin is inherited from a situation like the one
illustrated by the Gulf of Aqaba section (Fig. 2)
or like that outlined for the Agulhas Fracture
Zone (Scrutton and Dingle, 1976). If sedimentation is especially heavy or the sheared margin is
very mature, e.g. Newfoundland Fracture Zone
(Fig. 2), then the sub-slope basement high may
be overwhelmed and a smoother margin profile

built up. Finally, the sediment loading effect
must be considered. This is unlikely to be as
great and widespread as on rifted margins because
thick sedimentary deposits are restricted to
shelf basins and the fracture zone. There is
the possibility that loading in these two places
will increase the prominence of the shallow basement beneath the slope, however.

References

Allan, T.D., Magnetic and gravity fields over the
Red Sea, in N.L. Falcon, I.G. Gass, R.W. Girdler
and A.S. Laughton (Eds.), A discussion on the
structure and evolution of the Red Sea and the
nature of the Red Sea, Gulf of Aden and Ethiopia
rift junction, Phil. Trans. R. Soc. Lond., A267,
153-180, 1970.

Behrendt, J.C., J. Schlee, J.M. Robb, and M.K.
Silverston, Structure of the continental margin
of Liberia, Bull. geol. Soc. Am., 85, 1143-1158,
1974.

Ben-Avraham, Z., Z. Garfunkel, and G. Almagar,
Sediments and structure of the Gulf of Elat
(Aqaba) - northern Red Sea, Sed. Geol., 23,
239-268, 1979.

Bott, M.H.P., Evolution of young continental margins and formation of shelf basins, Tectonophysics, 11, 319-327, 1971.

Bott, M.H.P., and R.A. Scrutton, ICG Working
Group 8 Mid-Term Report, Geodynamics Internat.,
9, 20-28, 1976.

Dehlinger, P., R.W. Couch, D.A. McManus, and
M. Gemperle, Northeast Pacific structure, In:
A.E. Maxwell (Ed.) The Sea, 4, pt. 2, Wiley,
New York, 133-190, 1971.

Falvey, D.A., On the origin of marginal plateaux,
Unpub. Ph.D. Thesis, Univ. New South Wales,
242 pp., 1972.

Francheteau, J. and X. Le Pichon, Marginal fracture zones as structural framework of continental margins in the South Atlantic Ocean,
Bull. Am. Assoc. Petrol. Geol., 56, 991-1007,
1972.

Ginzburg, A., J. Makris, K. Fuchs, C. Prodehl,
K. Werner, and A. Uri, A seismic study of the
crust and upper mantle of the Jordan - Dead Sea
Rift and their transition toward the Mediterranean Sea, J. geophys. Res., 84, 1569-1582,
1979.

Goslin, J., and J-C Sibuet, Geophysical study of
the easternmost Walvis Ridge, South Atlantic:
Deep structure, Bull. geol. Soc. Am., 86, 1713-
1724, 1975.

Haworth, R.T., and J.P. Lefort, Geophysical evidence for the extent of the Avalon zone in
Atlantic Canada, Can. J. Earth Sci., 16, 552-
567, 1979.

Jones, E.J.W., The structure and evolution of the
West African continental margin off Guinee
Bissau, Guinee and Sierra Leone, In: R.A.
Scrutton and M. Talwani (Eds.), The Ocean Floor,
Wiley, Chichester, 000-000, in press.

Keen, M.J., and C.E. Keen, Subsidence and fracturing on the continental margin of Eastern Canada, In: P.J. Hood (Ed.), Offshore Eastern Canada, GSC Paper 71-23, 23-42, 1971.

Knopoff, L., and J.C. Belshe, Gravity observations of the Dead Sea Rift, In: T.N. Irvine (Ed.), The World Rift System, GSC Paper 66-14, 5-21, 1966.

Le Pichon, X., and D.E. Hayes, Marginal offsets, fracture zones and the early opening of the South Atlantic, J. geophys. Res., 76, 6283-6293, 1971.

Mascle, J., Atlantic-type continental margins: distinction of two basic structural types, In: F.F.M. de Almeida (Ed.), Continental Margins of Atlantic Type, Anais Acad. Brasil Ciencias, 48 (Supplement), 191-198, 1976.

Marsh, J.S., The origin of alkaline igneous rock lineaments in Africa and South America, Earth Planet. Sci. Lett., 18, 317-323, 1973.

Rabinowitz, P.D., S.C. Cande, and J.L. La Brecque, The Falkland Escarpment and Agulhas Fracture Zone: the boundary between oceanic and continental basement at conjugate continental margins, In: F.F.M. de Almeida (Ed.), Continental Margins of Atlantic Type, Anais Acad. Brasil Ciencias, 48 (Supplement), 241-252, 1976.

Roots, W.D., J.J. Veevers, and D.F. Clowes, Lithospheric model with thick oceanic crust at the continental boundary: a mechanism for shallow spreading ridges in young oceans, Earth Planet. Sci. Lett., 43, 417-433, 1979.

Scrutton, R.A., Continental break-up and deep crustal structure at the margins of Southern Africa, In: F.F.M. de Almeida (Ed.), Continent-al Margins of Atlantic Type, Anais Acad. Brasil Ciencias, 48 (Supplement), 275-286, 1976a.

Scrutton, R.A., Microcontinents and their significance, In: C.L. Drake (Ed.), Geodynamics: Progress and Prospects, A.G.U., Washington, 117-189, 1976b.

Scrutton, R.A., Crustal structure at the continental margin south of South Africa, Geophys. J.R. astr. Soc., 44, 601-623, 1976c.

Scrutton, R.A., On sheared passive continental margins, Tectonophysics, 59, 293-305, 1979.

Scrutton, R.A., A. Du Plessis, A.M. Barnaby, and E.S.W. Simpson, Contrasting structures and origins of the western and southeastern continental margins of Southern Africa, In: K.S.W. Campbell (Ed.), Proc. Third Internat. Gondwana Symp., Canberra, 651-661, 1974,

Scrutton, R.A., and R.V. Dingle, Observations on the process of sedimentary basin formation at the margins of Southern Africa, In: M.H.P. Bott (Ed.), Sedimentary basins of the continental margins and cratons, Tectonophysics, 36, 143-156, 1976.

Talwani, M., and O. Eldholm, Boundary between continental and oceanic crust at the margin of rifted continents, Nature, 241, 325-330, 1973.

Wilson, R.C.L., and C.A. Williams, Oceanic transform structures and the development of Atlantic continental margin sedimentary basins - a review, J. geol. Soc. Lond., 136, 311-320, 1979.

Wilson, J.T., A new class of faults and their bearing on continental drift, Nature, 207, 343-347, 1965.

THE STATE OF STRESS AT PASSIVE CONTINENTAL MARGINS

D. L. Turcotte

Department of Geological Sciences, Cornell University, Ithaca, New York 14853

Abstract. Passive continental margins are much
less active tectonically than active continental
margins where subduction is occurring. Nevertheless, significant amounts of vertical displacements occur on the boundary faults after the formation of a passive margin at a rift valley. As
a result high stress levels are expected. Based
on our current understanding of the rheology of
rocks, significant stress relaxation is not expected to occur at temperatures lower than about
600°C on geological time scales. The stress state
at continental margins is expected to be the result of both regionally and locally generated
stresses. Examples of regional stresses are
stresses associated with the driving mechanism
of plate tectonics and membrane stresses due to
the ellipticity of the earth. Examples of local
stresses are stresses due to changes in crustal
thickness, thermal stresses, and stresses due to
differential vertical displacements at the margin. Intraplate volcanism and the initiation of
subduction can be attributed to the stresses
concentrated at passive continental margins.

Introduction

The state of stress in the lithosphere is a subject of some controversy. These uncertainties
also affect any consideration of the state of
stress at passive continental margins. A primary
uncertainty is the rheological behavior of rock
within the lithosphere. Does the lithosphere or
some fraction of the lithosphere behave elastically
on geological time scales? If there is significant stress relaxation then the time at which
a stress is applied is important. It is also
necessary to understand the mechanism of failure
of the lithosphere. In near-surface rocks faulting appears to play an important role but the
deep structure of faults is poorly understood.

The state of stress at passive continental margins is a superposition of regionally generated
stresses and locally generated stresses. Examples
of regional stresses include:
(1) Stresses associated with the driving mechanism
of plate tectonics.
(2) Membrane stresses associated with the ellipticity of the earth.

Examples of locally generated stresses include:
(1) Stresses due to variations in crustal thickness.
(2) Stresses caused by differential vertical
subsidence.
(3) Thermal stresses due to cooling or heating.

Much important information would be available if
the state of stress could be measured directly.
Unfortunately, direct measurements of stress are,
at best, ambiguous. Stress measurements must be
made at relatively shallow depths, and it is not
clear that near-surface fractures and faults have
not strongly affected the stress state. Earthquake focal mechanism studies give important information on the sign of the stress (deviatoric
tension or compression) but not on the magnitude.

The concentration of stress at passive continental margins may have important implications
with regard to geological phenomena. There is
some evidence for the concentration of volcanic
activity at passive continental margins. Passive
continental margins may also be the location of
new subduction zones.

Rheology of the Lithosphere

Before any attempt can be made to calculate the
state of stress at passive continental margins it
is necessary to have an understanding of the rheology of the lithosphere. Studies of magnetic
anomalies and geological features such as fracture zones indicate that the interiors of the surface plates experience little deformation on geological time scales. In order to accommodate the
motion of the surface plates, the rock beneath the
lithosphere must flow on geological time scales.
The dominant creep process responsible for this
flow is thought to be thermally activated dislocation creep.

Although the rigidity of the lithosphere pre-
cludes large strains, the strain required to relieve tectonic stresses of the order of 100 bars
is only about 10^{-4}. It has been suggested by
Murrell [1976] and Kirby [1977] that the relaxation of elastic stress in the lithosphere will be
the result of thermally activated, solid-state
creep processes. Up to stress levels of about 2
kb the dominant deformation mechanism is expected

to be the movement of dislocations. It has been shown by Caldwell and Turcotte [1979] that the relaxation time for elastic stresses due to dislocation creep is

$$\tau_r = (3/2EB\sigma_o^2)\exp(A/RT) \qquad (1)$$

where E is Young's modulus, σ_o the initial stress level, R the gas constant, T the absolute temperature, and A and B are material properties. Taking A = 122 kcal/mole and B = 70 $s^{-1}bar^{-3}$ [Kohlstedt et al., 1976] the dependence of the relaxation time on temperature is given in Figure 1 for several values of the initial stress. It is seen that stress is not relaxed at temperatures less than about 600°C over geological times. The conclusion is that elastic stresses are not relaxed in the shallower parts of the lithosphere but the thickness of the elastic part of the lithosphere decreases with time.

There is ample observational evidence that near-surface crustal rocks fail by brittle fracture. There is also considerable observational evidence that faults in crustal rocks constitute zones of weakness that can be reactivated. Since passive continental margins are associated with continental rifts the normal faults associated with rifting are expected to act as zones of weakness on the continental side of passive continental margins.

Although there is ample evidence of brittle faulting in near-surface rocks, the mechanism of failure in deeper rocks is less well understood. It is expected from theory and laboratory experiments that rocks will fail by plastic deformation at high confining pressures. If crustal thinning occurs at passive continental margins it is likely to be due to plastic deformation of the deeper crustal rocks.

Regional Stresses

There is considerable evidence that the upper part of the lithosphere behaves as a stress guide on geological time scales. Stresses applied at

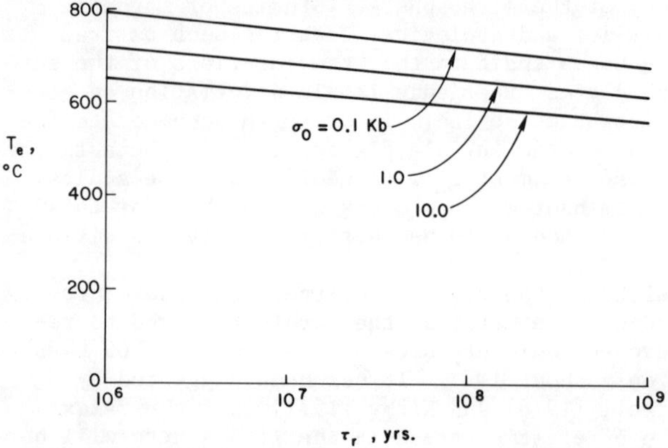

Figure 1. Dependence of the stress relaxation temperature on the time the load has been applied for several values of the initial stress.

large distances are transmitted through the elastic lithosphere. Examples include the driving stresses of plate tectonics and membrane stresses. These regional stresses constitute part of the stress field at passive continental margins.

Driving Stresses

The forces that drive the motions of the surface plates can be divided into three classes: (i) trench pull, (ii) ridge push, and (iii) basal tractions. Several authors [Forsyth and Uyeda, 1975; Chapple and Tullis, 1977] have argued that the dominant forces are the trench forces.

At passive continental margins the stresses associated with the driving mechanism are likely to change as the margin evolves. Early in the evolution of a passive margin the margin is closely associated with a regional tensional stress environment. Therefore, early in the evolution of a passive continental margin a tensional stress environment might be expected.

For developed passive continental margins such as the Atlantic the driving forces would be expected to result in compressional forces at the margins. The elevation of the ridges results in a ridge push on the adjacent plates. This can be thought of as gravitational sliding or as the horizontal pressure gradient associated with mantle convection cells. The topographic elevation of the ridge is the hydrostatic equivalent of a horizontal pressure gradient.

Membrane Stresses

Another type of regional stress is the membrane stress associated with the ellipticity of the earth. Since the earth is elliptical the motion of the surface plates results in the deformation of the plates; this deformation generates membrane stresses. These stresses will be distributed throughout each surface plate, and the magnitude and distribution will depend upon the size and previous movement of the plate. Membrane stresses are expected to be of the order of several hundred bars [Turcotte, 1974a].

Locally Generated Stresses

There are several mechanisms which directly generate stress at passive continental margins. These stresses must be added to the regionally generated stresses.

Stresses Due to Changes in Crustal Thickness

In order to maintain changes in crustal thickness, a distribution of stress is required. One result of these stresses is to generate a horizontal tension in the continental crust and a horizontal compression in the oceanic crust. The tensional stress would tend to thin the continental crust and the compressional stress would

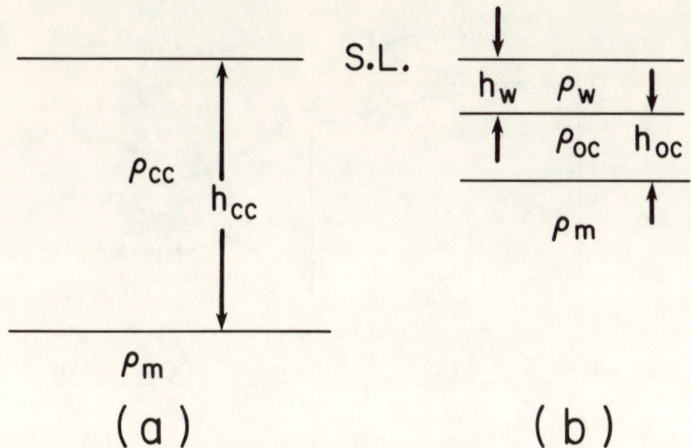

Figure 2.   Idealized structure of the continental crust (a) and oceanic crust (b).

thicken the oceanic crust in order to equilibrate the thicknesses.

In order to illustrate this process let us consider the simple crustal models illustrated in Figure 2. The continental crust has a thickness h_{cc} and a density ρ_{cc}; its upper surface is at sea level. The oceanic crust is covered with water of depth h_w and density ρ_w. The oceanic crust has a thickness h_{oc} and density ρ_{oc}. The mantle density is ρ_m.

Application of isostasy requires that

$$h_{cc} = [(\rho_m - \rho_w)h_w + (\rho_m - \rho_{oc})h_{oc}]/(\rho_m - \rho_{cc}) \qquad (2)$$

From the condition of isostasy the pressures in the mantle beneath the continental and oceanic crusts are equal. What we will show is that the pressure forces on the two crusts are not equal; the difference is the force that is driving the crusts to equal thickness.

The pressure as a function of depth z in the continental crust is given by

$$P_c = \rho_{cc} g\, z \qquad (3)$$

The pressures as a function of depth in the water above the oceanic crust, in the oceanic crust, and in the mantle beneath the oceanic crust are given by

$$
\begin{aligned}
0 < z < h_w & & & \rho_w g\, z_w \qquad (4)\\
h_w < z < h_w + h_{oc} & & P_o = & \; \rho_w g\, h_w + \rho_{oc} g\,(z - h_w)\\
h_w + h_{oc} < z < h_{cc} & & & \rho_w g\, h_w + \rho_{oc} g\, h_{oc}\\
& & & \quad + \rho_m g(z - h_w - h_{oc})
\end{aligned}
$$

The net difference in the hydrostatic pressure force between the continental and the oceanic crusts F is obtained by integrating the pressures over the thickness of the continental crust with the result

$$
\begin{aligned}
F = g[h_w h_{cc}(\rho_m - \rho_w) &+ h_{oc}\, h_{cc}(\rho_m - \rho_{cc})\\
- h_w h_{oc}(\rho_m - \rho_{oc}) &- 1/2\, h_w^2(\rho_m - \rho_w) \qquad (5)\\
- 1/2\, h_{oc}^2(\rho_m - \rho_{oc}) &- 1/2\, h_{cc}^2(\rho_m - \rho_{cc})]
\end{aligned}
$$

As we have previously discussed we expect that both the continental and oceanic lithospheres behave elastically. In order to prevent the thinning of the continental crust and thickening of the oceanic crust, elastic stresses must balance the difference in hydrostatic forces.

As a specific example we take $h_w = 5$ km, $\rho_w = 1$ gm/cm^3, $h_{oc} = 7$ km, $\rho_{oc} = 3.0$ gm/cm^3, $\rho_{cc} = 2.9$ gm/cm^3, and $\rho_m = 3.3$ gm/cm^3. From equation (2) the thickness of the continental crust is $h_{cc} = 34$ km. From equation (5) the difference in the total hydrostatic force is $F = 1.85 \times 10^{15}$ dyne/cm. The force in the continental crust is greater than the force in the oceanic crust and mantle. If the elastic stresses which balance this force are distributed over a depth of 50 km then the required stress is 370 b. In the continental crust this would be a tension as expected.

Detailed calculations of the stress distribution due to changes in crustal thickness at passive continental margins have been given by Bott [1971, 1973] and by Bott and Dean [1972]. These authors suggest that these stresses produce a flow in the lower crustal rocks that produces a thinning of the continental crust and thickening of the oceanic crust adjacent to the continental margin.

Thermal Stresses

Differential changes in temperature lead to stress in an elastic media. The magnitude of thermal stresses σ_{th} is related to a temperature change ΔT by

$$\sigma_{th} \simeq E\, \alpha_\ell \Delta T \qquad (6)$$

where E is Young's modulus and α_ℓ is the linear coefficient of thermal expansion. Taking $E = 6 \times 10^{11}$ dyne/cm^2 and $\alpha_\ell = 10^{-5}\,°C^{-1}$ we find that $\sigma_{th} = 600$ bars for $\Delta T = 100°C$. In general, thermal stresses are very large.

Since a passive continental margin is a region that has cooled and therefore thermally contracted, it might be expected that the thermal stresses would be tensional. However, this depends upon how the thermal stresses are relieved; detailed calculations [Turcotte, 1974b] show that thermal contraction can lead to near-surface compressional stresses.

Differential Vertical Subsidence

The importance of the cooling and subsidence of the oceanic lithosphere at a passive continental margin was recognized by Sleep [1971] and Sleep and Snell [1976]. As the oceanic litho-

sphere at the continental margin grows older it
cools and thickens. The cooling causes the density of the oceanic lithosphere to increase, resulting in subsidence. Since the continental
lithosphere is not cooling and subsiding there
is differential subsidence between the oceanic
and continental lithosphere at passive continental margins. This differential vertical subsidence must result in either lithospheric flexure
or displacements on faults at passive continental
margins. Watts and Ryan [1976] have considered
this problem with regard to the Gulf of Lyon and
the East Coast of the United States. It should
be noted, however, that there is horizontal heat
transport between the oceanic and continental
lithospheres at a passive continental margin and
as a result the differential vertical subsidence
would be expected to be spread over a distance of
about 100 km.

The loading of sediments at passive continental
margins is also expected to result in either lithospheric flexure or displacements on marginal
faults (or both). The flexure of the lithosphere
at a passive continental margin under a sediment
load has been considered by Gunn [1944], Walcott
[1972], and Cochran [1973]. These authors considered the flexure and subsidence of the lithosphere due to sediment loading but not due to
differential vertical subsidence.

A series of models for flexure and fault displacement at a passive continental margin, including both differential vertical subsidence and
sediment loading, have been given by Turcotte et
al. [1977]. These authors concluded that the low
observed gravity anomalies at passive continental
margins provided evidence that the marginal faults
remained active during the period of rapid subsidence of the oceanic lithosphere. Their preferred model is illustrated in Figure 3. In this
model a single active fault was assumed which
accommodated differential vertical movement. It
was assumed that the continental side of the margin was uplifted during the rifting event due to
heating, that the uplifted rocks were eroded, and
that subsequent subsidence of the continent due
to cooling of the continental lithosphere resulted in the formation of sedimentary basins on
the continental lithosphere. It was also assumed
that sediments were deposited to sea level on the
continental side of the continental margin. The
result is a sedimentary basin which thickens
toward the margin in accordance with the flexure
of the continental lithosphere during uplift and
sediment loading. The structures of the sedimentary basins on the East Coast of the United States
from North Carolina to New Jersey were shown to
be in good agreement with this model.

Shown in Figure 3 are the subsidence of basement, the free-air gravity anomaly and the near-
surface bending stress predicted by the theory.
The upper surface of the sediments is prescribed
to have an exponential shape. At the margin the
sediments are assumed to reach sea level. Seaward they are assumed to thin to the oceanic

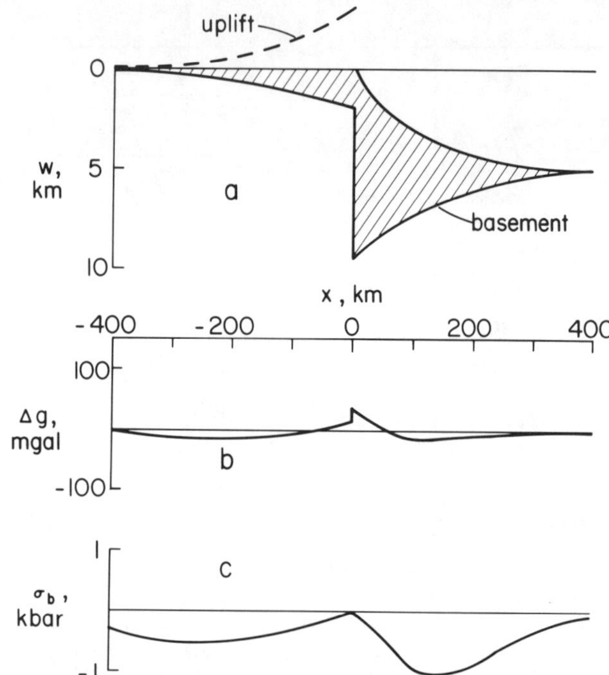

Figure 3. Dependence of the deflection (a),
free-air gravity anomaly (b), and the near-surface
bending stress on the distance from the continental margin.

basement. In the example given the sediments are
assumed to have a density of $2.2 \ gm/cm^3$ and the
upper boundary of the sediments is assumed to
have a scale length (1/e) of 100 km. The sediment loading flexes the oceanic lithosphere downward so that the maximum depth of the sedimentary
basin is 9.3 km. The maximum bending stress in
the continental lithosphere is 0.5 kbar and in the
oceanic lithosphere is 1 kbar. The maximum depth
of the sedimentary basin is 9.3 km. The maximum
bending stress in the continental lithosphere is
0.5 kbar and in the oceanic lithosphere is 1 kbar.
The maximum bending stresses in the continental
lithosphere occur about 250 km from the contin-
tal margin.

It should be emphasized that a number of approx-
imations are implicit in the model illustrated in
Figure 3. Probably the most unrealistic of these
is the single marginal fault. Fault displacements
should be distributed over a relatively large horizontal scale particularly on the continental
side.

Volcanism and the State of Stress at Passive Continental Margins

As we have previously discussed the passive continental margin is expected to be an area of
stress concentration. In order to relieve the
differential vertical subsidence lithospheric

fractures are expected. One possible consequence
of lithospheric fractures is volcanism [Turcotte
and Oxburgh, 1978]. Tensional failures of the
lithosphere are expected to allow magmas to reach
the surface.

The passive continental margins of the Atlantic
Ocean are characterized by localized areas of recent volcanism and short chains of volcanic centers adjacent to the continental margins, as illustrated in Figure 4. The best defined chain extends some 700 km off the east coast of South
America, terminating in the recently active volcanoes on the islands of Trinidade and Martin Vaz.
Somewhat farther north are the volcanic islands
Fernando Noronha and Rocas.

Very active volcanic centers along the continental margin of the west coast of Africa are the
Cape Verde Islands and the Canary Islands. The
association of the Canary Islands with a lithospheric fracture has been discussed by Anguita
and Herman [1975]. The island of Madeira is also
the site of recent volcanism. These volcanic centers do not take the form of volcanic chains.
Somewhat farther south on the continental margins
is the chain of islands Annobon, São Tomé, Prin-
cipe, and Fernando Poo, which terminate in the
active continental volcano Mount Cameroun. The
closeness of these centers of volcanism to the
continental margins suggests that there is a correlation. The large regional stresses may have
fractured the lithosphere, resulting in volcanism.

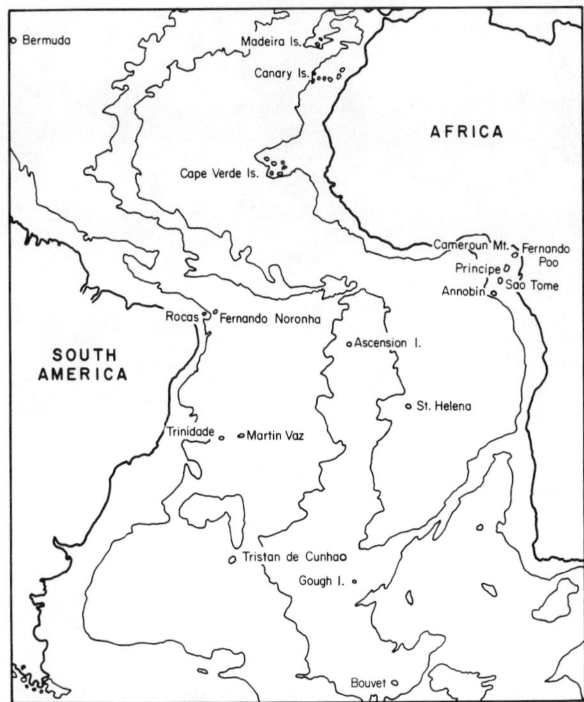

Figure 4. Volcanic islands of the central and
southern Atlantic Ocean.

Initiation of Subduction at Passive Continental Margins

It is clear from our understanding of plate tectonics that new trench systems must form. It is
expected that as the Atlantic Ocean continues to
open trench systems will develop which will lead
to its closure. There is considerable evidence
that this sequence of events occurred as the
proto-Atlantic Ocean developed. Since many trench
systems are immediately adjacent to continental
margins, it is reasonable to assume that new
trench systems develop at this location.

There are several reasons that trench systems
would be expected to form at continental margins.
The first is that a passive continental margin is
a zone of weakness as we have previously discussed.
There is considerable evidence from the continents
that major fault systems develop along previous
zones of weakness, i.e., older inactive fault
zones. A second possible cause of subduction at
a passive continental margin is the load of sediments. This depresses the ocean lithosphere and
may result in some transformation of basalt to
the denser phase eclogite [Fyfe and Leonardos,
1977].

The mechanics associated with the initiation of
subduction is not clear. Oceanic lithosphere,
older than a few tens of millions of years, is
more dense than the underlying mantle rocks and
is inherently unstable [Turcotte et al., 1977].
However, the lithosphere must bend through a substantial angle and hot mantle rock must override
it in order for subduction to begin.
McKenzie [1977] suggests that subduction begins
when the oceanic lithosphere is underthrust under
the continental lithosphere on a major thrust
fault in a compressional environment. However,
Turcotte et al. [1977] have suggested that subduction initiates in a tensional environment as
the oceanic lithosphere founders.

Acknowledgments. The preparation of this review article has been supported in part by the
Division of Earth Sciences, National Science Foundation under grant EAR 76-82457. Department of
Geological Sciences, Cornell University contribution 654.

References

Anguita, F., and F. Herman, A propagating fracture
  model versus a hot spot origin for the Canary
  Islands, Earth Plant. Sci. Let., 27, 11-19, 1975.
Bott, M.H.P., Evolution of young continental margins and formation of shelf basins, Tectonophys.,
  11, 319-327, 1971.
Bott, M.H.P., Shelf Subsidence in Relation to the
  Evolution of Young Continental Basins, Implications of Continental Drift to the Earth Sciences,
  eds. D. H. Tarling and S. K. Runcorn, Vol. 2,
  pp. 675-685, Academic Press, London, 1973.
Bott, M.H.P., and D.S. Dean, Stress at young continental margins, Nature, 235, 23-25, 1972.

Caldwell, J.G., and D.L. Turcotte, Dependence of the thickness of the elastic oceanic lithosphere on age, J.Geophys. Res., 84, 7572-7576, 1979.

Chapple, W.M., and T.E. Tullis, Evaluation of the forces that drive the plates, J. Geophys. Res., 82, 1967-1984, 1977.

Cochran, J.R., Gravity and magnetic investigations in the Guiana Basin, Western Equatorial Atlantic, Geol. Soc. Am. Bull., 84, 3249-3268,1973.

Forsyth, D., and S. Uyeda, On the relative importance of the driving forces of plate motion, Geophys. J. Roy. Astron. Soc., 43, 163-200, 1975.

Fyfe, W.S., and O.H. Leonardos, Speculations on the causes of crustal rifting and subduction, with applications to the Atlantic margin of Brazil, Tectonophys., 42, 29-36, 1977.

Gunn, R., A quantitative study of the lithosphere and gravity anomalies along the Atlantic Coast, J. Franklin Insti., 237, 139-154, 1944.

Kirby, S.H., State of stress in the lithosphere: Inferences from the flow laws of olivine, Pure Ap. Geophys., 115, 245-258, 1977.

Kohlstedt, D.L., C. Goetze, and W.B. Durham, Experimental deformation of single crystal olivine with application to flow in the mantle, The Physics and Chemistry of Minerals and Rocks, R.G.J. Strens, ed., pp. 35-49, John Wiley, London, 1976.

McKenzie, D.P., The initiation of trenches: A finite amplitude instability, Island Arcs, Deep Sea Trenches, and Back-Arc Basins, M. Talwani and W.C. Pitman, eds., Maurice Ewing Series 1, pp. 57-61, American Geophysical Union, Washington, D.C., 1977.

Murrell, S.A.F., Rheology of the lithosphere - experimental indications, Tectonophys., 36, 5-24, 1976.

Sleep, N.H., Thermal effects of the formation of Atlantic continental margins by continental break up, Geophys. J. Roy. Astro. Soc., 24, 325-350, 1971.

Sleep, N.H., and N.S. Snell, Thermal contraction and flexure of mid-continent and Atlantic marginal basins, Geophys. J. Roy. Astron. Soc., 45, 125-154, 1976.

Turcotte, D.L., Membrane tectonics, Geophys. J. Roy. Astro. Soc., 36, 33-42, 1974.

Turcotte, D.L., Are transform faults thermal contraction cracks?, J. Geophys. Res., 79, 2573-2577, 1974.

Turcotte, D.L., J.L. Ahern, and J.M. Bird, The state of stress at continental margins, Tectonophys., 42, 1-28, 1977.

Turcotte, D.L., W.F. Haxby, and J.R. Ockendon, Lithospheric instabilities, Island Arcs, Deep Sea Trenches, and Back-Arc Basins, M. Talwani and W.C. Pitman, eds., Maurice Ewing Series 1, pp. 63-69, American Geophysical Union, Washington, D.C., 1977.

Turcotte, D.L., and E.R. Oxburgh, Intra-plate volcanism, Phil. Trans. Roy. Soc. London, A288, 561-579, 1978.

Walcott, R.I., Gravity, flexure, and the growth of sedimentary basins at a continental edge, Geol. Soc. Am. Bull., 83, 1845-1848, 1972.

Watts, A.B., and W.B.F. Ryan, Flexure of the lithosphere and continental margin basins, Tectonophys.,36, 25-44, 1976.

STRESS BASED TECTONIC MECHANISMS AT PASSIVE CONTINENTAL MARGINS

Martin H.P. Bott

Department of Geological Sciences, University of Durham, Durham DH1 3LE, U.K.

Abstract. The flexural subsidence typical of the *drifting stage* of a passive continental margin is mainly attributable to loading and cooling of the lithosphere rather than to stress-based processes. However, large stresses must develop as the lithosphere bends in response to loading and differential cooling but there is not much indication of any tectonic response. Deviatoric tension must also occur in the adjacent continental crust because of crustal thickness variation across the margin and this may cause some regional or local subsidence by oceanward creep of hot lower crustal material. During the preceding *rifting stage*, a strong deviatoric tensile stress regime is implied by normal faulting, graben formation and dyke intrusion which is observed to have occurred at some margins at and around the time of continental splitting. It is suggested that the tension may be predominantly caused by the plate boundary suction force acting on the over-riding continental plate at subduction zones at those periods in earth history when a large continental mass is bordered by subduction zones on opposite sides. Most of the thinning of the continental crust at the margin probably occurs at this stage, producing substantial subsidence as exemplified by the North Biscay margin. This may result from intense stretching of the continental crust. Faulting and magma invasion may help to extend the brittle upper part of the crust. Alternatively, some of the crustal thinning may occur just after the margin has formed as a result of outslip of lower crustal material into the newly forming oceanic litho-sphere, or as a result of phase transformations.

Introduction

Passive continental margins typically undergo two main stages of tectonic development. A *rifting stage*, involving crustal stretching and normal faulting, may occur at and around the time of continental splitting, although evidence of this stage is not seen everywhere. The succeeding *drifting stage* is characterized by flexural subsidence of the margin as the newly formed ocean widens by sea-floor spreading. The rate of flexural subsidence decays approximately exponentially so that there is a youthful stage of rapid subsidence followed at about 60 My by a mature stage of slowing subsidence. The main geodynamical problem of passive margins is to understand the mechanism of subsidence which occurs during both stages of development. Three types of mechanism has been proposed: a gravity-based mechanism dependent on sagging of the lithosphere in response to loading, various thermal mechanisms dependent on heating of the lithosphere at time of continental splitting and subsequent cooling, and stress-based mechanisms dependent on response to the special stress environment at passive margins. Thermal mechanisms and associated sediment loading appear to dominate the drifting stage although stress based mechanisms may contribute. On the other hand, stress mechanisms are probably primarily responsible for the structures formed during the rifting stage, including the continental splitting mechanism itself. This paper reviews the stress-based mechanisms which may occur during both rifting and drifting stages of evolution.

Critical Evidence from the North Biscay Margin

One of the difficulties in assessing the tectonic mechanisms operating at passive margins has been the lack of information about the rifting phase. It is particularly important to relate the time-scales of continental splitting, marginal subsidence and development of the crustal transition. We also need to know whether the splitting episode is associated with uplift and erosion. Evidence of great import-ance in these respects has come from the recent DSDP drilling of the north Biscay margin (Montadert, Roberts *et al.*, 1977) and from associated geophysical investigations (De Charpal *et al.*, 1978). Sediments post-dating the formation of the margin are relatively thin so that the underlying formations can be investigated by drilling and geophysical methods without great difficulty, in contrast to the typical sediment loaded margin where they may be buried beneath 5 to 10 km of later

sediments. The continental splitting at the
north Biscay margin occurred in about Aptian
(Lower Cretaceous) time. The continent-ocean
crustal contact has been located with some
confidence near the foot of the slope.

Three groups of sediments occur beneath the
slope. Pre-rifting sediments of Upper Jurassic
(Kimeridgian to Portlandian) age were penetrated
in DSDP hole 401; these were deposited in shallow
water epicontinental seas. Seismic evidence and
dredging indicates that the pre-rift sequence
(Triassic-Jurassic) was deposited on a continental
basement including granites. The seismic
reflection records show that this sequence has
been extensively faulted in early Cretaceous time,
now forming a series of rotated and tilted fault
blocks beneath the slope. The faults predominantly dip towards the oceanward side of the margin
and are of listric normal type, flattening into a
plane of discontinuity at around 10 km below sea
level as indicated by the seismic reflection data.
The extension of the upper crust beneath the slope
by the faults is about 10%. A second syn-rifting
group of sediments were deposited in the half-
grabens in pre-Aptian time. Unfaulted post-rift
sediments of Aptian age drape the fault blocks.
These were deposited in about 2000 m of water,
indicating that the margin subsided by about 2 km
during the rifting stage rather than suffering
uplift and erosion. Subsequently the slope has
suffered further flexural subsidence amounting to
over 2000 m at the outer slope to attain the
present water depths.

Seismic refraction lines show that the
continental crust has been thinned greatly beneath
the slope, the Moho being about 13 km deep beneath
the outer slope and at closely similar depth to
the adjacent oceanic Moho. This is what would be
expected on isostatic grounds. It is probable
that the crustal thinning occurred contempor-
aneously with the rifting stage, outweighing the
thermal effects of the split which would tend to
give rise to uplift rather than subsidence, so
that the combined effect gives a 2 km subsidence
at the crustal contact decreasing to nearly zero
at the shelf edge.

This evidence indicates that the rifting phase
lasted up to about 30 My between the end of the
Portlandian (140 My) and the start of the Aptian
(110 My). This phase involved (1) extension of
the upper crust beneath the slope of about 10%,
(2) thinning of the continental crust beneath the
slope by an average of about 40%, and (3)
subsidence of 2 km adjacent to the continent-ocean
contact, decreasing landwards beneath the slope.
The observed extension appears to be inadequate
to account for the inferred crustal thinning.

Stress-based Mechanisms during Drifting Stage

Separate treatment is needed for stresses
associated with passive margins during the post-
split drifting stage when plate boundaries
become progressively more distant (this

section), and for those applicable during the
rifting stage (next section).

<u>Flexure of the Lithosphere</u>

The lithosphere becomes stressed whenever a
surface load or an internal load is applied to
it. Such loading occurs at a passive margin as
sediments are progressively deposited on shelf,
slope and rise. It also occurs as the oceanic
lithosphere and the continental lithosphere
immediately adjacent suffer progressive cooling
and consequent thermal contraction; where the
surface is above sea-level the load may be
decreased by erosion but where it is below sea-
level the increasing water depth will give rise
to an additional load. The lithosphere
responds to loading by flexure with a time-
constant of the order of 0.1-1.0 My dependent
on the effective viscosity of the underlying
region, the dimension of the load, and the
thickness of the "flexural" lithosphere.

Lithospheric flexure can be modelled with
acceptable accuracy by using the theory of
elastic or visco-elastic bending of thin beams
or sheets (e.g. Heiskanen and Vening Meinesz,
1958; Walcott, 1977). Using beam theory, the
vertical displacement of a uniform elastic two-
dimensional lithosphere of thickness T subjected
to a surface loading $P(x)$ is given by

$$D\frac{d^4w}{dx^4} + (\rho_m - \rho_w)wg = P(x)$$

where $D = ET^3/12(1 - \sigma^2)$ is the flexural
rigidity, E is Young's modulus, σ is Poisson's
ratio, and ρ_m and ρ_w are the densities of the
displaced materials below and above the
lithosphere respectively (e.g. mantle material
and sea-water). For loading by a line mass at
the origin, the solution takes the form

$$w = A \cos(x/a) \exp(-x/a)$$

where the flexural parameter a is given by
$a^4 = 4D/(\rho_s - \rho_w)g$. For loads of wavelength
greater than about $7a$ the response should be
close to local isostatic equilibrium, and for
those of less than $2a$ the load is effectively
borne by the strength of the lithosphere without
significant deformation. The downbending of the
elastic layer below the load produces
compression in the upper half and tension in the
lower half; the opposite situation applies to
the regions lateral to the load where the bending
is in the opposite sense. A positive gravity
anomaly would be expected above the load and
smaller negative anomalies on the flanks.

The flexural rigidity of the oceanic
lithosphere to long-term loading is estimated to
be about 2×10^{23} N m (Walcott, 1970). A value
of about 2×10^{23} N m also appears to be

appropriate to passive margins (Walcott, 1972;
Cochran, 1973). This corresponds to an
effective elastic lithospheric thickness of 30 km
and a flexural parameter of about 75 km. The
flexural lithosphere is thus significantly
thinner than the seismological lithosphere, the
reason being that it responds to long-term
loading at least partly by visco-elastic
deformation. The values above are appropriate
to loading over a period of 10 My or thereabouts.

Several studies of flexure at passive margins
have been made. Walcott (1972) showed that
flexure of the lithosphere can account for the
great accumulation of sediment at the Mississippi
and Niger Deltas, and that the observed gravity
anomalies agree with the predictions. He calculated that the observed bending stresses are of
the order of 1 kbar. Cochran (1973) found that
the deformation of the lithosphere by the
Amazon sediment cone and the associated gravity
anomalies are best fitted by a flexural
rigidity of 2×10^{23} N m for long-term loads,
with calculated maximum bending stress of about
4.5 kbar beneath the load, and subsidiary
lateral maxima of about 1.3 kbar on either side.
Turcotte *et al*. (1977) studied the combined
flexural effect of sediment loading and progressive cooling of the oceanic lithosphere
adjacent to the margin, (1) where the oceanic
and continental lithospheres are locked together
and (2) where they are free to move vertically
independently with a fault between them. They
argued that the observed gravity anomalies are
best explained by the presence of a major fault
between oceanic and continental lithospheres
which has remained active over most of the
lifespan of the margins. This conclusion appears
to conflict with observations which generally
indicate that the adjacent lithospheres remain
locked together after initial formation.

Much of the post-split downsagging at passive
margins can be accounted for by the flexural
response of the lithosphere to cooling of the
oceanic and marginal continental lithospheres and
to sediment loading. However, the detailed
working-out of the flexural hypothesis needs much
further attention now that the sediment piles are
becoming better defined by seismic reflection
work. The theoretical bending stresses
associated with margins appear to be locally
exceptionally large and it is possible that they
become reduced by some stress relaxation
process such as transient creep. It is
surprising that we do not have more evidence of
their existence.

Crustal creep hypothesis

A differential stress system occurs across a
passive margin as a result of the change in
thickness of the low density crust from
continental to oceanic sides and the associated
change in water depth (Bott and Dean, 1972).
This differential loading and upthrust gives
rise to a horizontal deviatoric tension in the
continental crust, relative to the oceanic side.
The greatest stress difference is about 300 bar
and occurs between depths of about 5 and 10 km
in the continental crust, falling off to zero
upwards towards the surface and downwards towards
the Moho. Kusznir and Bott (1977) have further
shown that if the lower continental crust behaves
as a stiff visco-elastic material with overlying
elastic upper crust, then the deviatoric tensions
become increased in the upper elastic layer of
the crust to 1 kbar or thereabouts. This tensile
deviatoric stress sytem characteristic of the
continental crust adjacent to passive margins
should act to thin the continental crust; whether
or not this occurs depends on the rheological
properties.

Bott (1971, 1973) suggested that this stress
system may cause creep of lower continental
crustal material from beneath the shelf and slope
towards the sub-oceanic upper mantle, thereby
thinning the continental crust and causing
progressive subsidence. A complementary uplift
of the continental rise would be expected, as low
density continental crustal material accumulates
below it. It now appears that this mechanism is
probably of less importance than thermal and load
produced subsidence in producing the general
downsagging during the drifting stage at a
typical passive margin. On the other hand, local
regions of strong differential subsidence
producing deep sedimentary basins adjacent to
passive margins, such as Hatton Basin on Rockall
Bank (Matthews and Smith, 1971; Bott, 1972) and
the deep basin beneath Porcupine Bight in the
north eastern Atlantic seem to be most satisfactorily explained by crustal outflow towards
the oceans, possibly in regions where the lower
crustal temperature is relatively high.

Stress based mechanisms during the rifting phase

The rifting phase is best characterized by
normal faulting (listric type?) affecting the
continental crust adjacent to the incipient
margin. This may give rise to graben formation
(Burke, 1976) or to other patterns of block
faulting (De Charpal *et al*., 1978). Dyke
intrusions may extend the continental crust
during this stage in regions where the lithosphere is notably hot such as off East Greenland
adjacent to Denmark Strait, but evidence is
lacking for similar invasion of the North Biscay
margin. The evidence from the North Biscay
margin shows that local thinning of the
continental crust adjacent to the newly forming
margin by more than 50% can occur during this
stage. It has been suggested (Hsu, 1965; Sleep
1971, 1973) that uplift of the continental
lithosphere by thermal expansion producing an
elevated dome structure can occur during the
rifting stage, but the evidence from the North
Biscay margin indicates that any such uplift is
more than annulled by crustal thinning.

Detailed evidence on the processes occurring
during the rifting phase is only available for a
few localities because the structures are
typically buried beneath thick sediments of later
age. The available evidence suggests that two
contrasting types of development may occur.
Firstly, there are the hot spot regions such as
central East Greenland typified by extensive
igneous invasion of the continental crust
including dyke swarms and by topographical uplift
of the continental region. Secondly, there are
the more normal incipient passive margins where
dyke intrusion and uplift are not evident, but
where normal faulting is commonly observed. It
is possible that continental splitting is
initiated in the hot spot regions. Both appear
to develop under crustal tension.

We need to discuss the origin of the tensile
stress system and the mechanism by which this
may cause crustal thinning.

Origin of the tensile stress

Plate boundary forces associated with the plate
driving mechanism are the most obvious source of
the persistent stressing of the lithosphere. If
we accept that plates are driven by forces acting
on their edges rather than by the drag of underlying convection currents acting on the base of
the lithosphere, then upwelling of low density
material to form new lithosphere at ocean ridges
causes compression, and the sinking lithosphere
at convergent boundaries may cause tension in
both the subducting and overriding plates. Of
particular relevance to the rifting stage of
passive margin development is the tension in the
continental crust which may be caused by the
suction forces exerted on the overriding plate
at a subduction zone (Elsasser, 1971; Forsyth
and Uyeda, 1975). At periods in earth history
when subduction occurs on opposite sides of a
large continental mass, such as Pangaea during
the early Mesozoic, this would give rise to
deviatoric tension throughout the continental
lithosphere. This situation does not apply to
any major continental regions at the present
time and the effects of the suction force are now
restricted to the vicinity of convergent plate
boundaries, such as the marginal basins of the
western Pacific.

Stress in the crust may also occur as a result
of crustal thickness variations (Bott, 1971;
Artyushkov, 1973). Regions where elevated
surface topography is isostatically compensated
by thick crust, or by underlying low density
upper mantle, are subjected to deviatoric
tension relative to the adjacent regions. It was
pointed out earlier that the continental side of
passive margins would be expected to be in tension
relative to the oceanic side as a result of the
stresses associated with the change of crustal
thickness. A tensile stress of similar origin
must occur locally within the crust in plateau
uplift regions such as East Africa and western
United States, where high surface elevation is
supported by a low density region in the underlying upper mantle (Bott and Kusznir, 1979).
This stress system may give rise to rifting and
graben formation. It might ultimately give rise
to continental splitting, although there is some
difficulty in understanding how the split could
propagate into the adjacent more normal
continental regions.

The tensile stresses produced by plate
boundary forces acting on the lithosphere, or by
isostatically compensated surface elevation, may
be enhanced by a phenomenon discussed by Kusznir
and Bott (1977). If the lower part of the
lithosphere behaves as visco-elastic material
and the upper part can be regarded as truly
elastic, then relaxation occurs in the visco-
elastic lower part of the lithosphere and the
stresses in the upper part are comparably
increased. The deviatoric stresses in the
elastic part are amplified by a factor
approximately equal to the ratio of the lithospheric thickness to that of the elastic layer.
Thus tensile stresses greater than 1 kbar can
readily develop from plate boundary forces or in
plateau uplift regions due to this phenomenon.
Furthermore, the stress amplification is greatest
in regions where the elastic part of the lithosphere is thinnest. This suggests that any
heating of the lithosphere which thins the
elastic layer will cause the deviatoric
stresses in the upper part to be increased, in
addition to decreasing the strength. Given an
initial tensile stress in the continental crust,
the effect of lithospheric heating will be both
to weaken the lithosphere and to increase the
tensile stress, making such a region vulnerable
to the development of normal faulting and dyke
intrusion, and possibly to continental
splitting.

Tensile membrane stresses may also be produced
locally in the lithosphere as a plate migrates
in latitude and accommodates to the changing
radius of curvature of the surface (Turcotte
1974). If a plate moves towards the equator,
the outer part is thrown into compression and
the inner part into tension. If it moves away
from the equator, then the opposite situation
applies. Oxburgh and Turcotte (1974) attributed
the initiation of the East African rift system
(an incipient continental split?) to tension
developed in the African plate as it has migrated northwards towards the equator over the
last 100 My.

Mechanism of continental splitting

Two contrasting viewpoints have been prevalent
concerning the mechanism of continental
splitting dependent on whether the tensile stress
system or the thermal event is the primary cause.
According to Turcotte and Oxburgh (1976), the
cracking caused by tensile stresses of membrane
type is primary and magma is developed as a

secondary feature by upwelling along the newly formed crack. Other workers (e.g. Morgan, 1972; Burke and Kidd, 1975) have suggested that the thermal event producing a hot upper mantle is primary event. Bott and Kusznir (1979) took this latter viewpoint in a modified form, suggesting that upwelling beneath the continental lithosphere produces a magma rich upper mantle and a heated lithosphere. Plateau uplift occurs in response to the reduced density of the hot mantle region below, and tensile stresses are developed as indicated above. Continental splitting can then be initiated in the region by dyke intrusion provided that the dykes can spread laterally far enough to join up with other plate boundaries.

A different viewpoint was taken by Bott (1981) who suggested that continental splitting is initiated by local weakening of the continental lithosphere by hot spot activity at a period in earth history when a continent is subjected to widespread regional tension produced by the plate boundary suction force acting on opposite sides. The split would start in the thinned lithosphere above the hot spot. Once a crack had formed, the tension would be concentrated at both ends of the crack which would then propagate laterally beyond the hot spot region. Immediately after continental splitting had occurred, the general tension would be counteracted by the newly developed ridge push force at the new spreading centre. This would explain the abrupt change in stress regime between rifting and drifting stages of passive margin development.

Mechanism of crustal thinning

Interpretations of the crustal transition at passive margins based on seismic refraction and gravity observations indicate that the main change in crustal thickness typically occurs over a horizontal distance of about 40-100 km. Until recently, it has not been clear how the change in crustal thickness relates to the position of the continent-ocean crustal contact and whether the main changes in crustal configuration occur during the rifting or drifting phase. If the situation at the North Biscay margin can be taken as typical, then there is indication now that the gradation in crustal structure is mainly due to thinning of the continental crust during the rifting phase and that the location of the continent-ocean contact is near the foot of the slope, as suggested by Kinsman (1975).

One possibility is that extreme thinning of the continental crust occurs by plastic necking in response to tension prior to continental splitting, as suggested for rift valleys by Artemjev and Artyushkov (1971). McKenzie (1978) has applied this idea to explain the North Sea subsidence by stretching of the continental lithosphere, this phenomenon causing subsidence because of the thinned low density crust and an uplift of lesser amplitude because of the thermal effect of rise in the asthenosphere beneath the thinned lithosphere. This idea can be taken over to explain the crustal thinning during the rifting phase of passive margins and the associated thermal effects. The main difficulty with this hypothesis is that the uppermost 10 km or so of the crust is brittle and probably cannot extend by 20% or more without bodily disruption, which is not indicated on present evidence except where extensive dyke intrusion affects the upper crust. At the North Biscay margin the extension of the upper crust beneath the slope by block faulting is estimated to be about 10% (De Charpal *et al.*, 1978), which is much less than is needed to account for the observed crustal thinning averaging about 40%.

Another possible mechanism for thinning the crust is that the extension of the upper brittle layer by normal faulting, or even elastic stretching, continues back for a large distance, of the order of 1000 km, into the continental interior. The stretching of the lower part of the lithosphere, on the other hand, is concentrated over about 50-100 km on either side of the incipient split. In this way, quite a small extension may be seen in the upper crust whereas the lower crust is substantially thinned beneath the incipient slope. This mechanism implies that there is a decollement between the elastic and ductile parts of the crust and lithosphere which extends some 1000 km back into the continental interior regions.

An alternative mechanism is that the crustal thinning occurs shortly after new oceanic crust starts to form by the crustal creep mechanism of Bott (1971). The continental crustal material outflows beneath the newly forming oceanic region possibly being partly incorporated in the newly forming oceanic crust and partly still residing in the uppermost mantle adjacent to the margin. Another possibility is that the crust thins as a result of phase transitions.

Understanding the process of continental crustal thinning during or near to the time of rifting is one of the outstanding problems of passive margin development. Although the evidence from North Biscay throws some light on this problem, a full understanding must await more accurate knowledge of the relationship of the phenomenon of the rifting stage to the early stages of sea-floor spreading, and more detailed evidence on the nature of the crustal and upper mantle beneath the transition zone.

Conclusions

During the *drifting stage* of passive margin development, characterized by flexural down-sagging of the margin, stress-based mechanisms of subsidence appear to be subordinate to thermal and gravity loading mechanisms. However, stress systems are caused by two phenomena associated

with the margin and its development. Firstly,
the bending of the lithosphere in response to
sediment loading and differential thermal
subsidence must cause quite large stress
differences, but there is a surprising lack of
evidence of any tectonic response to such
stresses. Secondly, the crustal thickness
change across the margin must produce a
deviatoric tensile stress system on the continental side relative to the oceanic side. This
may cause oceanward creep of lower continental
crustal material where and when the temperature
is sufficiently high to allow creep to occur.
Subsidence of some local sedimentary basins at
passive margins, such as Hatton basin on
Rockall Plateau, may originate in this way.

The earlier *rifting stage* of development
involves continental splitting, possibly graben
formation or other normal faulting which may
typically start prior to the split, and dyke
intrusion. If the North Biscay margin is
typical, substantial thinning of the continental
crust adjacent to the margin occurs at this
stage, so that the continental and oceanic crust
are of nearly the same thickness at their
contact. Stress based mechanisms are characteristic of this stage. The evidence of the
faulting and dyke intrusion indicates that the
incipient margin is subjected to a strong
deviatoric tensile stress system. It is
suggested that the tension may be caused
predominantly by the suction force acting on
the overriding continental plate at subduction
zones at periods in earth history when a large
continental region has subduction zones on
opposite sides. The intense thinning of the
continental crust during this stage may
possibly be caused by stretching of the crust.

References

Artemjev, M.E., and E.V. Artyushkov, Structure and
isostasy of the Baikal rift and the mechanism of
rifting, J. Geophys. Res., 76, 1197-1211, 1971.

Artyushkov, E.V., Stresses in the lithosphere
caused by crustal thickness inhomogeneities,
J. Geophys. Res., 78, 7675-7708, 1973.

Bott, M.H.P., Evolution of young continental
margins and formation of shelf basins,
Tectonophysics, 11, 319-327, 1971.

Bott, M.H.P., Subsidence of Rockall Plateau and of
the continental shelf, Geophys. J. Roy. Astron.
Soc., 27, 235-236, 1972.

Bott, M.H.P., Shelf subsidence in relation to the
evolution of young continental margins, in
Implications of continental drift to the earth
sciences, Vol.2, edited by D.H. Tarling and
S.K. Runcorn, 675-683, Academic Press, London
and New York, 1973.

Bott, M.H.P., Mechanism of continental splitting
(abstract), Geophys. J. Roy. Astron. Soc., 65,
247, 1981.

Bott, M.H.P., and D.S. Dean, Stress systems at
young continental margins, Nature (Phys. Sci.),
235, 23-25, 1972.

Bott, M.H.P., and N.J. Kusznir, Stress distributions associated with compensated plateau
uplift structures with application to the
continental splitting mechanism, Geophys. J.
Roy. Astron. Soc., 56, 451-459, 1979.

Burke, K., Development of graben associated with
the initial ruptures of the Atlantic Ocean,
Tectonophysics, 36, 93-112, 1976.

Burke, K, and W.S.F. Kidd, Earth, heat flow in,
in McGraw-Hill yearbook of science and technology, pp.165-170, McGraw-Hill, New York, 1975.

Cochran, J.R., Gravity and magnetic investigations
in the Guiana Basin, western equatorial Atlantic,
Bull. Geol. Soc. Amer., 84, 3249-3268, 1973.

De Charpal, O., P. Guennoc, L. Montadert, and
D.G. Roberts, Rifting, crustal attenuation and
subsidence in the Bay of Biscay, Nature, 275,
706-711, 1978.

Elsasser, W.M., Sea-floor spreading as thermal
convection, J. Geophys. Res., 76, 1101-1112,
1971.

Forsyth, D., and Uyeda, S., On the relative
importance of the driving forces of plate
motion, Geophys. J. Roy. Astron. Soc., 43,
163-200, 1975.

Heiskanen, W.A., and F.A. Vening Meinesz, The
Earth and its gravity field, 470 pp., McGraw-
Hill, New York, 1958.

Hsu, K.J., Isostasy, crustal thinning, mantle
changes, and the disappearance of ancient land
masses, Am. J. Sci., 263, 97-109, 1965.

Kinsman, D.J.J., Rift valley basins and sedimentary history of trailing continental margins, in Petroleum and global tectonics,
edited by A.G. Fischer and S. Judson, 83-126,
Princeton University Press, Princeton, New
Jersey, 1975.

Kusznir, N.J., and M.H.P. Bott, Stress concentration in the upper lithosphere caused by
underlying visco-elastic creep,
Tectonophysics, 43, 247-256, 1977.

Matthews, D.H., and S.G. Smith, The sinking of
Rockall Plateau, Geophys. J. Roy. Astron. Soc.,
23, 491-498, 1971.

McKenzie, D., Some remarks on the development of
sedimentary basins, Earth Planet. Sci. Lett.,
40, 25-32, 1978.

Montadert, L., D.G. Roberts et al., Initial
reports of the Deep Sea Drilling Project,
Vol. 48, Washington (U.S. Government Printing
Office), 1979.

Morgan, W.J., Plate motions and deep mantle
convection, Mem. Geol. Soc. Amer., 132, 7-22, 1972.

Oxburgh, E.R., and D.L. Turcotte, Membrane
tectonics and the East African rift, Earth.
Planet. Sci. Lett., 22, 133-140, 1974.

Sleep, N.H., Thermal effects of the formation of
Atlantic continental margins by continental
break up, Geophys. J. Roy. Astron. Soc., 24,
325-350, 1971.

Sleep, N.H., Crustal thinning on Atlantic

continental margins: evidence from older margins, in *Implications of continental drift to the earth sciences*, Vol.2, edited by D.H. Tarling and S.K. Runcorn, 685-692, Academic Press, London and New York, 1973.

Turcotte, D.L., Membrane tectonics, *Geophys. J. Roy. Astron. Soc.*, 36, 33-42, 1974.

Turcotte, D.L., J.L. Ahern, and J.M. Bird, The state of stress at continental margins, *Tectonophysics*, 42, 1-28, 1977.

Turcotte, D.L., and E.R. Oxburgh, Stress accumulation in the lithosphere, *Tectonophysics*, 35, 183-199, 1976.

Walcott, R.I., Flexural rigidity, thickness, and viscosity of the lithosphere, *J. Geophys. Res.*, 75, 3941-3954, 1970.

Walcott, R.I., Gravity, flexure, and the growth of sedimentary basins at a continental edge, *Bull. Geol. Soc. Amer.*, 83, 1845-1848, 1972.

THERMAL CONTRACTIONS BENEATH ATLANTIC MARGINS

Norman H. Sleep*

Dept. of Geological Sciences, Northwestern University, Evanston,
IL 60201. *Current address: Dept. of Geophysics, Stanford University, Stanford, CA 94305

Abstract. Thermal contraction of the lithosphere following continental break-up may cause
subsidence of Atlantic continental margins. The
subsidence rate should decrease exponentially with
time as at mid-ocean ridges. The observed subsidence rate at Atlantic margins generally obeys
this relationship, if sedimentological effects
and flexure of the lithosphere are included. The
subsidence time relationship and rheology of the
lithosphere have not been deduced in detail as
the sedimentological effects are difficult to
remove. Sediment temperatures in the past can
potentially be obtained from the subsidence rate.
The thermal contraction hypothesis may also be
applicable to classical miogeosynclines.

Introduction

Atlantic continental margins form when a new
mid-oceanic ridge begins spreading beneath a pre-
existing continent. The history of these margins
subsequent to break-up can be attributed to cooling of lithosphere which becomes heated during
break-up. The process can be summarized as
follows (Vogt and Ostenso, 1967; Schneider, 1969a,
1972; Sleep, 1971):

1) During continental break-up the lithospheric
mantle near the margin is replaced by hot mantle
from below. The isostatic freeboard of the continental crust is reduced by loading by intrusions and necking of the crust.

2) Erosion of the uplifted continental margin
further reduces freeboard. This erosion and
subsidence due to thermal contraction of hot
material beneath the margin reduces the elevation
to sea-level in a few tens of million years.

3) Further subsidence accommodates a wedge of
sediments which is several kilometers thick at
the shelf break and more than 100 km in width.
The subsidence rate decreases with time as the
lithosphere cools. The sea-ward thickening of
sediments is expected because heating of the
lithosphere and crustal thinning would be more
intense near the line of eventual break-up. The
rate of subsidence increases seaward to that of
oceanic crust.

This sequence of events explains the general geometry of continental shelf sedimentation beneath margins and also the regional unconformity
following break-up.

The rate of subsidence on the continental shelf
can be easily predicted if the base of the lithosphere is considered to be an isothermal boundary (Vogt and Ostenso, 1967; Sleep, 1971). The
temperature beneath the continental margin (neg-
lecting higher order terms) is then

$$T(z,t) = \frac{T_\lambda z}{\lambda} + T_1(x)\sin(\frac{\pi z}{\lambda})\exp(\frac{-\kappa\pi^2 t}{\lambda^2}) \qquad (1)$$

where T is the temperature, T_λ is the temperature
at the base of the lithosphere, $T_1(x)$ is a slowly
varying temperature function of horizontal position, z is depth, λ is lithosphere thickness,
κ is thermal diffusivity, and t is the time
after some reference time. Neither the lithosphere thickness nor the diffusivity is precisely
determined independently of thermal phenomena.
Therefore, the heat flow

$$q = k\frac{\partial T}{\partial z} = q_o(x)\exp(-t/t_o) \qquad (2)$$

and elevation change

$$u = \frac{\rho_m}{(\rho_m-\rho_s)}\alpha\int_o^\lambda (T - \frac{T_\lambda z}{\lambda})dz \equiv u_o(x)\exp(-t/t_o) \quad (3)$$

are used to determine the time constant $t_o =
\lambda^2/\kappa\pi^2$. The parameter k is thermal conductivity,
ρ_m and ρ_s are densities of the crust and mantle,
and α is the volume thermal expansion coefficient.

For Atlantic margins, the elevation change
through time is best inferred from the thicknesses of sediments accumulated during the subsidence. Assuming for simplicity that deposition always occurs at an unchanging sea-level
and that all the sediments are the same density

154

and never compact, the depth to a bed of an age is

$$\text{Depth} = u_o(x) \left\{ \exp\left(\frac{\text{age}}{t_o} \right) - 1 \right\} \qquad (4)$$

History of Hypothesis

The causes of Atlantic margin subsidence appear to have been seldom discussed prior to plate tectonics. Wegener (1929, p. 208), who considered Atlantic margins to be the trailing edges of continental blocks, believed that thermal contraction should cause subsidence of the sea-floor with increased age and that the youngest parts of the ocean should have the least sediments. He applied this concept only to the difference between the Atlantic and Pacific oceans and thus did not discover sea-floor spreading. Hess (1962) noted that thermal contraction might cause ridges to subside but did not consider Atlantic margin basins. Dietz (1963) proposed that sediment loads on the continental rise caused the shelf to subside. In his model Atlantic margins episodically collapsed to become fold belts but were not related to continental break-up. Hsu (1965) attributed Atlantic margin subsidence to mantle density changes perhaps due to thermal contraction and discussed thinning of the crust by erosion during uplift prior to subsidence. He did not relate these processes to continental break-up.

The applicability of thermal contraction toward explaining the geological history of Atlantic continental margins was first recognized by Vogt and Ostenso (1967). The thermal origin of the topography of mid-oceanic ridges was proposed in detail at about the same time (Langseth et al., 1966; Vogt and Ostenso, 1967; McKenzie and Sclater, 1969). Sleep (1969) used a topographic calculation to show that the 1200°C temperature of basalts inferred by petrologists satisfied the data better than the 550°C ridge temperature used by McKenzie (1967) and McKenzie and Sclater (1969). After a paper by Sclater and Francheteau (1970), the thermal origin of ridges became well accepted in the literature.

The thermal origin of Atlantic margin subsidence was not well received at first. Other than Sleep (1970) and Eric Schneider (1969a,b), an associate of Vogt and Ostenso, no discussion of the hypothesis was published following Vogt and Ostenso (1967) until 1971. The slow recognition of Vogt and Ostenso's (1967) work can be partly attributed to its inclusion in a quickly outdated paper on other aspects of what is now called sea-floor spreading. Schneider's work appeared in abstracts, a trade journal, and in 1971 in a French colloquium volume. Other than the 1970 abstract, Sleep was unable to get his work published until late in 1971.

Opposition and the failure of other marine geophysicists to rediscover the thermal contraction hypothesis can be attributed in part to the great weight American geophysicists placed on the quiet magnetic boundaries of the Atlantic Ocean. Drake et al. (1968) considered that a proto-ocean existed (and thus precluded continental break-up). Emery et al. (1970) considered break-up to be Permian (and thus too old for thermal contraction to cause Cretaceous subsidence). The thick Triassic (or Early Jurassic) sill upon which a major oceanography institution rests apparently was not considered a good indication of the time of continental break-up. Opposition to Triassic continental break-up was cited as a reason for rejection of Sleep's first attempt to publish on the thermal origin of margin subsidence. Deep ocean drilling data and magnetic anomaly interpretations now favor a Late Triassic or Early Jurassic time of formation of the U.S. Atlantic margin (Vogt and Einwich, 1980).

Sleep (1971) tested the thermal contraction hypothesis using data from Florida, and the Gulf and East Coasts of the United States. He proposed plausible sedimentological causes for the moderate deviations of depth-age curve from an exponential. Kinsman (1975), Falvey (1974), and Keen and Keen (1973) presented their work on the thermal contraction hypothesis orally at about this time. Kinsman was thus instrumental in introducing the hypothesis to the petroleum industry, while Falvey's variant of the hypothesis (he includes phases changes) was introduced to Australians and the staff of Lamont-Doherty Geological Observatory.

The presence of vocal proponents of the thermal contraction hypothesis at meetings led to a general cognizance of the hypothesis by the marine geological community. By 1974, thermal contraction was cited by several authors in a volume on continental margins (Burk and Drake, 1974). Drake (1976) included thermal contraction as a possible mechanism for Atlantic margin subsidence in the introduction for a volume from a 1975 symposium.

Current Topics

Current research to further appraise the hypothesis that thermal contraction causes Atlantic margin subsidence is largely concerned with separating the thermal subsidence from extraneous sedimentological effects. Sleep (1971) qualitatively considered compaction, loading due to sediments, eustatic changes, and changing water depth. These effects have been considered in detail by Steckler and Watts (1978) and Watts and Steckler (1979) for the East coast of the United States and by Keen (1979) for the east coast of Canada. The data are compatible with thermal subsidence and a eustatic change in sea-level of 150 m since Late Cretaceous time. The sedimentary history of the East coast is sufficiently complex that detailed inference from the thermal subsidence curve is unlikely.

Flexure

Loads due to thermal contraction or to sediments are redistributed by the resistance of the lithosphere to bending. Sleep (1971, 1973) qualitatively considered this effect for a viscoelastic lithosphere which undergoes long term creep when loaded. During early rapid stages of subsidence the regions of maximum subsidence are buoyed up by their surroundings and the regions of minimum subsidence are dragged down by their surroundings. Later when the subsidence rate due to thermal contraction has decreased lithosphere creep relaxes the stress built-up during the early stages of subsidence. This causes uplift on the flanks of the basin and increased subsidence at the center. Sleep and Snell (1976) showed a viscosity of 10^{25} poise for a 100 km thick lithosphere would produce the observed outcrop patterns.

A difficulty in deducing the viscosities of the lithosphere from subsidence data is that eustatic changes in sea-level may also expose the beds on the flanks of the Atlantic marginal basins. Other than for the relaxation discussed above, the flexure of a thin (25 km) elastic lithosphere is not distinguishable from the flexure of a thicker (100 km) viscoelastic lithosphere on a time scale of 10^6 to 10^8 years. The upper layer of the thin elastic lithosphere may also conceivably undergo creep. A viscosity of about 2×10^{26} poise for a thin (25 km) viscoelastic lithosphere would produce flexure similar to a thick (100 km) viscoelastic lithosphere with a viscosity of 10^{25} poise. At present the mechanical (as opposed to the thermal) thickness of the lithosphere cannot be uniquely determined from basin subsidence studies.

Sediment temperatures

The maturation of petroleum source beds is strongly affected by their temperature history. The thermal contraction hypothesis permits constraints on the thermal history without requiring data from deep drilling.

The sedimentary column is sufficiently thin that it can be ignored when computing heat loss from the lithosphere (Turcotte and Ahern, 1977). The temperature in a bed at a given time is thus

$$T = T(\text{surface}) + ksq_{out} \tag{5}$$

where k is thermal conductivity, s is the actual depth at that time, and q_{out} is heat flow.

In any column of lithosphere the heat balance is (Parsons and McKenzie, 1978)

$$q_{out} - \gamma H - q_\lambda = \frac{\partial T_a}{\partial t} \, \rho c \lambda \tag{6}$$

where H is the radioactive heat generation, λ is lithosphere thickness, T_a is the average

lithospheric temperature and q_λ is the heat flow at the base of the lithosphere. For half space cooling $q_\lambda = 0$. Thus one obtains

$$q_{out} - \gamma H = \frac{\partial T_a}{\partial t} \, \rho c \lambda \tag{7}$$

where γ is crustal thickness.

The exponential subsidence law was obtained in equation (4) by assuming that the base of the lithosphere was isothermal. Because the excess heat is centered within the lithosphere

$$q_{out} - f\gamma H - \frac{T_\lambda K}{\lambda} = 1/2 \, \frac{\partial T_a}{\partial t} \, \rho c \lambda \tag{8}$$

where the constant f depends on the geometry of the heat sources in the lithosphere. It is 1/2 if the heat sources are uniformly distributed and 1 if they are concentrated near the surface. As the latter case is likely (8) can be written as

$$q_{out} - q_{out}(t=\infty) = 1/2 \, \frac{\partial T_a}{\partial t} \, \rho c \lambda \tag{9}$$

Following Parsons and McKenzie (1978), the amount of thermal contraction is

$$\frac{\partial u}{\partial t} = \alpha \lambda \, \frac{\partial T_a}{\partial t} \tag{10}$$

This expression can be combined with either (7) or (9) to obtain a relationship between the amount of thermal contraction and heat flow. Equation (7) is applicable to young margins and (9) is more applicable to older margins. For intermediate ages the two formulas should give similar results.

Thus, the paleotemperature at any depth and at any time in a succession can be evaluated. The approach requires several operations and assumptions that should cause no difficulties to petroleum geologists:

(1) The conductivity and thickness of sediments required in (5) must be corrected for compaction.

(2) The driving subsidence rate $\partial u/\partial t$ must be deduced from the sedimentary record correcting for compaction, eustatic changes, sediment loads, and changes in water depth (Watts and Ryan, 1976).

(3) The heat flow at $t=\infty$ in (9) can be determined as measured heat flow for old margins. For younger margins the radioactive heat generation may be deduced from the range of heat flow values. For very young crust this term is not important.

It is uncertain how useful the paleotemperatures obtained in this (or a similar) way will be to petroleum exploration. It is possible that an independent appraisal of the thermal contraction

hypothesis will be obtained from the paleotemperature data.

Older margins

The similarity of Atlantic margins to classical
geosynclines has been noted for some time. Dietz
(1963) discussed this similarity with respect to
sea-floor spreading. The thermal contraction
hypothesis provides a further means of relating
fossil geosynclines to Atlantic-type margins.
If the subsidence-time relation is an exponential similar to modern margins (as in the lower
Paleozoic of the White Mountains, California,
U.S.A., Stewart and Suczek, 1977), then the geosyncline can be considered an Atlantic-type
margin. In older poorly dated rocks, the thermal
contraction hypothesis can be used to give crude
absolute timing of the duration of sedimentation.

The oldest subsidence attributed to thermal
contraction is that following the deposition
the iron formation in the Upper Peninsula of
Michigan (Cambray, 1977). The 3.3 B.Y. old Moodies
group in South Africa is the oldest sequence re-
cognizably attributable to an Atlantic-type
margin (Eriksson, 1979). More sedimentological
studies are needed to recognize continental
shelf deposits in the Precambrian to which the
thermal contraction model can be applied.

Acknowledgments

Support was provided by the National Science
Foundation grants DES-74-22337, EAR-76-22499,
and EAR-76-02952.

References

Burk, C.A., and C.L. Drake, eds., The Geology of
Continental Margins, Springer Verlag, New
York, 1009 pp, 1974.

Cambray, F. Wm., Plate tectonics as a model for
the environment of deposition and deformation
of the early Proterozoic (Precambrian X) of
northern Michigan. Geol. Soc. of Amer.
Abstracts w/Programs 10, 376, 1978.

Dietz, R., Collapsing continental rises: an
actualistic concept of geosynclines mountain
building, J. Geol. 71, 314-333, 1963.

Drake, C.L., Atlantic continental margins: Observation and ideas, in Continental Margins of
Atlantic Type, in Anais de Academia Brasileira
de Ciencias, 48, suplemento, Sao Paulo, pp.
9-14, 1976

Drake, C.L., J.E Ewing, and H. Stockard, The
continental margin of the eastern United States,
Canad. J. Earth Sci. 5, 993-1010, 1968.

Emery, K., E. Uchupi, J. Phillips, C. Bowin, E.
Bunce, and S. Knot, Continental rise off
eastern North America, Am. Assoc. Petro. Geol.
Bull. 54, 44-108, 1970.

Eriksson, K.A., Marginal marine depositional processes from the Archean Moodies group, Bar-
berton mountain land, South Africa: Evidence
and significance, Precambrian Res. 8, 153-
182, 1979.

Falvey, D.A., The development of continental margins in plate tectonic theory, J. Austr. Petr.
Expl. Assoc. 14, 95-106, 1974.

Hess, H.H., History of ocean basins, in Engle
et al., eds., Petrologic studies: a volume
in honor of A.F. Buddington, Geol. Soc. Amer.,
599-620, 1962.

Hsu, K.J., Isostasy, crustal thinning, mantle
changes, and the disappearance of ancient land
masses, Amer. J. Sci. 263, 97-109, 1965.

Keen, C.E., Thermal history and subsidence of
rifted continental margins -- evidence from
wells on the Nova Scotian and Labrador Shelves,
Can. J. Earth Sci. 16, 505-522, 1979.

Keen, M. and C. Keen, Subsidence and fracturing
of the continental margin of eastern Canada,
Earth Science Symposium on Offshore Eastern
Canada, Geol. Surv. Can. Ppr. 71-23, 23-
42, 1973.

Kinsman, D.J.J., Rift valley basins and sedimentary history of trailing continental margins,
in Fischer, A.G. & S. Judson, eds., Petroleum
and Global Tectonics, Princeton Univ. Press,
pp. 83-162, 1975.

Langseth, M., X. LePichon and M. Ewing, Crustal
structure of mid-ocean ridges, 5, Heat flow
through the Atlantic Ocean Floor and Convection currents, J. Geophys. Res. 71, 5321-5355,
1966.

McKenzie, D., Some remarks on heat flow and gravity anomalies, J. Geophys. Res. 72, 6261-6273,
1967.

McKenzie, D.P. and J.G. Sclater, Heat flow in
the Eastern Pacific and sea-floor spreading,
Bull. Volcanol. 33, 101-118, 1969.

Parsons, B. and D. McKenzie, Mantle convection
and thermal structure of the plates, J.
Geophys. Res. 83, 4485-4496, 1978.

Schneider, E.D., The deep-sea--a habitat for
petroleum, Undersea Technology 10, 32-
57, 1969a.

Schneider, E.D., Models for rifted and compressional continental margins, Geol. Soc. Amer.
Abstracts w/Programs 1, 291-292, 1969b.

Schneider, E.D., Sedimentary evolution of rifted continental margins, Geol. Soc. Amer. Mem.
132, 109-118, 1972.

Sclater, J.G. and J. Francheteau, The implications of terrestrial heat flow observations on
current tectonic and geochemical models of the
crust and upper mantle of the earth, Geophys.
J. R. astr. Soc. 20, 509-542, 1970.

Sleep, N., Sensitivity of heat flow and gravity
to the mechanism of sea-floor spreading, J.
Geophys. Res. 74, 542-549, 1969.

Sleep, N.H., Thermal effects of the formation of
Atlantic continental margins by continental
break-up, Trans. Amer. Geophys. Un. 51, 429
(abstract T65), 1970.

Sleep, N.H., Thermal effects of the formation of
Atlantic continental margins by continental

break-up, Geophys. J. R. astr. Soc. 24, 325-
350, 1971.

Sleep, N.H., Crustal thinning on Atlantic continental margins: evidence from older margins,
in Implications of Continental Drift to the
Earth Sciences 2, Academic Press, London,
685-692, 1973.

Sleep, N.H. and N.S. Snell, Thermal contraction
and flexure of mid-continent and Atlantic
marginal basins, Geophys. J.R. astr. Soc. 45,
125-154, 1976.

Steckler, M.S., and A.B. Watts, Subsidence of
the Atlantic-type margin off New York, Earth
and Planet. Sci. Lett. 41, 1-13, 1978.

Stewart, J.H. and C.H. Suczek, Cambrian and
latest Precambrian paleogeography and tec-
tonics in the western United States, in
Paleozoic Paleogeography of the Western
United States, Pacific Coast Paleogeography
Symp. 1, Soc. Econ. Paleontologists and
Mineralogists, Los Angeles, 1-17, 1977.

Turcotte, D.L. and J.L. Ahren, On the thermal and

subsidence history of sedimentary basins, J.
Geophys. Res. 82, 3762-3766, 1977.

Vogt, P. and N. Ostenso, Steady state crustal
spreading, Nature 215, 810-817, 1967.

Vogt, P.R. and A.M. Einwich, Magnetic anomalies
and sea-floor spreading in the western North
Atlantic, and a revised calibration of the
Keathley (M) geomagnetic reversal chronology,
Deep Sea Drilling Project Int. Rept. 43,
857-876, 1980.

Watts, A.G. and W.B.F. Ryan, Flexure of the
lithosphere and continental margins basins,
Tectonop. 36, 25-44, 1976.

Watts, A.B. and M.S. Steckler, Subsidence and
eustasy at the continental margin of eastern
North America, Ewing Vol., Implications of Deep
Drilling Results in the Atlantic Ocean: Con-
tinental margins and Paleoenvironment, Amer.
Geophys. Union, 218-234, 1979.

Wegener, A., The Origin of Continents and Oceans,
(English trans., 1966), Dover Publ., New
York, 246 p., 1929.

A THERMAL MODEL FOR THE ELEVATION OF CONTINENTAL FRAGMENTS

L. T. Long and R. P. Lowell

School of Geophysical Sciences, Georgia Institute of Technology, Atlanta, Georgia 30332

Abstract. The process of sea-floor spreading has caused some fragments of continental material to be separated from the main continental mass. The heat produced in the fragments is greater than that in the surrounding oceanic or transitional lithosphere; consequently the rate of cooling with time is diminished, and the rate of subsidence of the oceanic lithosphere is retarded near the continental fragment. Such dislocated fragments will exhibit sustained topographic prominence over geologic time.

Analytical and numerical thermal models were developed to evaluate the magnitude of the thermal effect. The analytical model showed that an elevation difference of 700 meters could be explained by inhomogenieties in heat production. The numerical model included also the influence of the initial temperature of formation of new oceanic lithosphere and indicates nearly one kilometer of elevation difference. The numerical model compares favorably with the elevations of the pre-Cretaceous surface over the Ocala uplift in Florida and very favorably with the basement topography across the Orphan Knoll and Flemish Pass east of Newfoundland.

Introduction

The development of the plate tectonics hypothesis has provided a conceptual framework which accomodates not only the large-scale horizontal motions of the lithosphere but also contemporary movement as indicated by the associated earthquake activity at plate margins. In plate interiors, however, vertical movements are not so easily accomodated by the simple plate tectonics hypothesis. The vertical movements require the traditional tectonic hypothesis involving isostasy, erosion or sedimentation, and thermal expansion or contraction. Of particular interest are the tectonic processes at continental margins such as the East and Gulf Coast of North America. These margins, which are within a contemporary plate interior, represent large-scale changes in crustal structure which had their origin at a plate margin. This paper deals with the tectonics of these continental margins during and after integration into the interior of a plate. In particular, this paper will discuss the role of changes in crustal structure and thermal contraction in determining the elevation of continental margins.

The principle tectonic activity of an Atlantic type continental margin following the initial uplift and rifting is subsidence, and it has been suggested [Sleep, 1971] that the subsidence of such continental margins may be due in part to thermal contraction as the continental boundaries move away from the ridge axis. The same mechanism of subsidence via thermal contraction has been shown to explain the variation of ocean ridge topography with age [Sclater, et al., 1971] and may also partially explain the formation of mid-continent basins [Sleep and Snell, 1976].

The subsidence of continental margins is somewhat more complex than the subsidence of ocean floor. Local variations in heat production, variations in sedimentation rate, and isostatic effects may significantly affect vertical movements. The accumulation and release of elastic stresses associated with differential vertical movements may provide a mechanism for some continental edge faults or fragmentation of the continental margins into blocks. Furthermore the process of sea floor spreading has caused some fragments of continental lithosphere to be separated from the main continental block. Such fragmentation may result from relocations or bifurcation of the spreading center, from transform faulting or by some combination of these mechanisms. Current examples of fragmentation are Baja California, which is due to transform faulting [Larson et al., 1968], and the Afar triangle, which may be due either to relocation [McKenzie et al., 1970] or to bifurcation [Mohr, 1970] of spreading zones east and west of the Danakil horst. Examples of fragments of continental lithosphere left behind following rifting could be part of the Florida peninsula or the Flemish cap east of Newfoundland.

After separation is achieved, the dislocated subcontinental fragment would be surrounded by oceanic and/or transitional lithosphere; sedimentation subsequent to cessation of a short term spreading episode could conceal much evidence of fragmentation. In general, a short separation

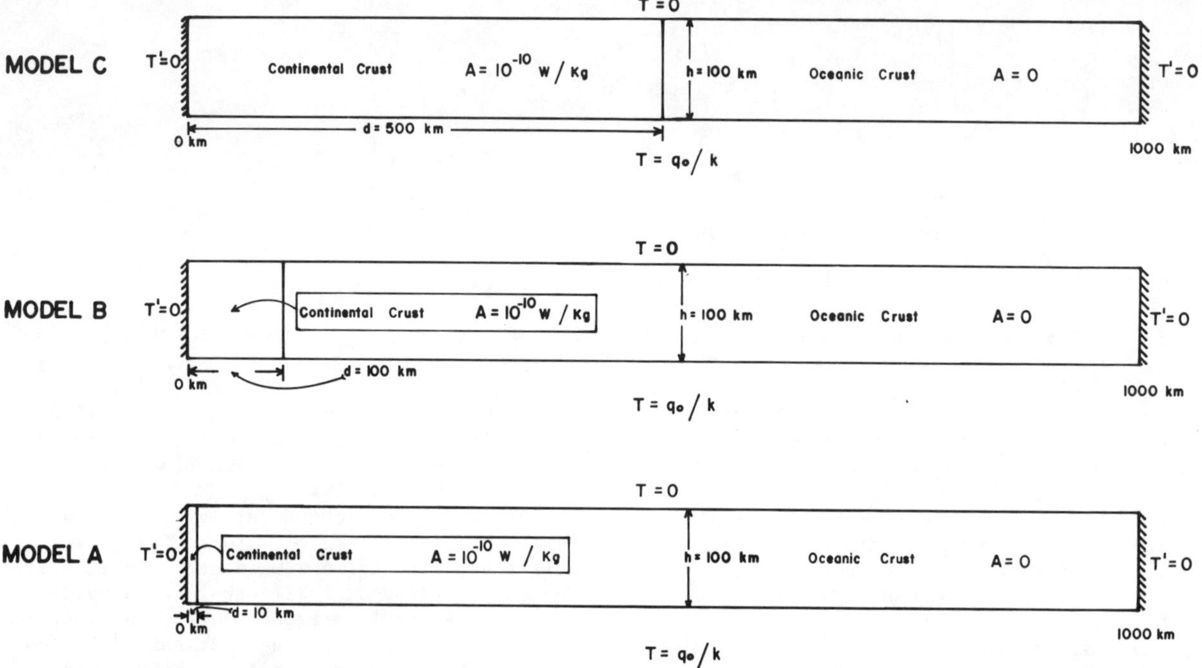

Fig. 1. Thermal models of continental lithospheric segments of varying width emplaced in idealized oceanic lithosphere.

distance or a slow spreading rate would allow essentially continuous sedimentation over the intervening newly created lithosphere. The initial cooling of the intervening newly created lithosphere would result in contraction and subsidence relative to the continental fragment. Also, because of the greater heat production in continental-type crustal rocks relative to the surrounding transitional or oceanic lithosphere, the dislocated block would retard the subsidence and exhibit sustained prominence over geologic time. Thus we propose that the prominence of some sections of continental margins formed by separation can be explained in part by the mechanism of thermal contraction and the influence of phenomena resulting from the greater heat production in continental rocks.

In order to examine the process involved, two theoretical models are developed. The first is an analytical solution to an abrupt interface between continental and oceanic lithosphere. The initial condition is that of a uniform thermal gradient so that the model allows us to examine the effects of heat generation in the continental lithosphere on the elevation of the adjacent oceanic floor. The second is a finite difference model of an idealized continental fragment separated from the continent by oceanic lithosphere. We assume in this model that the oceanic lithosphere is injected at mantle temperatures so that the model includes the influence of thermal contraction of oceanic lithosphere as well as heat production and thermal expansion in the

continental lithosphere. The stresses generated by the thermal contraction have not been included in this study. We have only assumed that the time constant for a release of stress is short compared to the thermal processes and the horizontal components of stress are completely decoupled. Much of this work has been developed in earlier papers: the analytical model in Lowell and Long [1977] and the finite difference model in Long and Lowell [1973].

The Analytical Model

Since continental lithosphere usually has greater heat production than oceanic lithosphere, one would expect significant inhomogeneties in heat production along a traverse perpendicular to a continental margin or a displaced continental fragment. We have constructed a simple two-dimensional model for conductive heat transport along such a traverse in order to evaluate the influence of heat generation on uplift in the continental blocks relative to adjacent oceanic lithosphere. The higher heat production in the continental lithosphere is expected to lead to differential subsidence of the continental margin.

The thermal model used for the heat transfer calculation is shown in Figure 1. Within the central block, which is of width 2d, heat is being produced at a constant rate, A, per unit mass relative to the adjacent blocks. The upper surface is assumed to be maintained at temperature $T = 0$ and no heat is lost through the ends of the system. A constant heat flux q_o is assumed to enter the lower surface and the initial temperature distribution is assumed to be $q_o z/k$ where k is the thermal conductivity and z is the depth. Hence, our model assumes an initially uniform temperature distribution and computes the perturbation introduced by insertion of radiogenic heat production in part of the model. The horizontal coordinate x is measured from the center of the central block, being positive to the right. Due to the symmetry of the system, we need only to solve for the region $0 \leq x \leq L$ and $0 \leq z \leq h$, the boundary condition at $x = 0$ being the absence of horizontal heat transfer. The pertinent heat diffusion equation for this system is then

$$\partial T/\partial t = a(\partial^2 T/\partial x^2 + \partial^2 T/\partial z^2) + A(x,z)/c \qquad (1)$$

where a is the thermal diffusivity, c is the specific heat and

$$A(x,z) = \begin{cases} 0, & d < x \leq L \\ A, & 0 \leq x \leq d \end{cases}$$

The boundary and initial conditions are

$$T(x,0,t) = 0$$

$$\frac{\partial T}{\partial x}(0,z,t) = \frac{\partial T}{\partial x}(L,z,t) = 0$$

$$\frac{\partial T}{\partial z}(x,h,t) = q_o/k \qquad (2)$$

$$T(x,z,0) = q_o z/k$$

Equation (1) with conditions (2) can be solved using Green's functions [Carslaw and Jaeger, 1959], the Green's function being determined by the method of eigenfunction expansion. The result is

$$T(x,z,t) = \frac{q_o z}{k} - \frac{2Adh^2}{Lac\pi^3} \sum_{n=0}^{\infty} \left\{ \frac{\sin[(n+1/2)\pi z/h]}{(n+1/2)^3} \right.$$

$$\left. \cdot \left\{ 1 - \exp[-a(n+1/2)^2 \pi^2 t/h^2] \right\} \right\}$$

$$\qquad (3)$$

$$+ \frac{4A}{ac\pi^4} \sum_{\substack{n=0 \\ m=1}}^{\infty} \left\{ \frac{\sin[(n+1/2)\pi z/h]\, \cos[m\pi x/L]\, \sin[m\pi d/L]}{m(n+1/2)\left\{ \left(\frac{n+1/2}{h}\right)^2 + \left(\frac{m}{L}\right)^2 \right\}} \right.$$

$$\left. \cdot \left\{ 1 - \exp[-a\pi^2 t((\frac{n+1/2}{h})^2 + (\frac{m}{L})^2)] \right\} \right\}$$

The deviation of the crustal layer relative to a reference elevation is found from

$$E = \alpha \int_0^h T^*(x,z,t)\, dz \qquad (4)$$

where α is an apparent thermal expansion coefficient and $T^* = T - q_o z/h$. Integration of (3) over z gives

$$E = \frac{-2Adh^3 \alpha}{Lac\pi^4} \sum_{n=0}^{\infty} \left\{ 1 - \exp[-a(n+1/2)^2 \pi^2 t/h^2] \right\}/(n+1/2)^4$$

$$+ \frac{4A\alpha h}{ac\pi^5} \sum_{\substack{n=0 \\ m=1}}^{\infty} \left\{ \frac{\cos[m\pi x/L]\, \sin[m\pi d/L]}{m(n+1/2)^2 \left\{ \left(\frac{n+1/2}{h}\right)^2 + \left(\frac{m}{L}\right)^2 \right\}} \right. \qquad (5)$$

$$\left. \cdot \left\{ 1 - \exp[-a\pi^2 t((\frac{n+1/2}{h})^2 + (\frac{m}{L})^2)] \right\} \right\}$$

Equation (5) for elevation has been evaluated for three cases; $d = h/10$, $d = h$, and $d = 5h$. Solutions for 50, 100, 200 and ∞ m.y. are given in Figure 2 for the three models. Reasonable values for the thermal parameters were used:

$$A = 10^{-10} w/kg, \; h = 10^5 m, \; \alpha = 2 \times 10^{-5}/°C$$

$$L = 10^6 m, \; a = 10^{-6} m^2/sec, \; c = 10^3 j/kg°C$$

The first model (Figure 1A) corresponds to the inclusion of a small (20 km wide) crustal fragment in oceanic crust. The model illustrates the influence of heat generation in a relatively small crustal fragment on the normal subsidence of surrounding oceanic crust. The second model (Figure 1B) corresponds to a significant segment of continental crust (200 km wide) surrounded on either side by oceanic crust (e.g. the Florida peninsula). The third model (Figure 1C) is representative of a simple abrupt continental margin.

Discussion of the Analytical Model

In all models (Figure 1A-C) the heat flow is expected to be greater over the continental lithosphere than in the adjacent basin. For model parameters in Figure 1B the temperature gradient should be about $3°/km$ greater over the central part of the continental lithosphere. The simple models suggests that the lateral inhomogeneity in heat production retards the subsidence adjacent to the continental lithosphere and prevents the oceanic lithosphere from reaching its normal depth equilibrium position with age. In addition, heat lost from the continental lithosphere ultimately caused contraction and subsidence of the continental lithosphere adjacent to the margin.

The resulting elevations at 50, 100, 200 and ∞ m.y. for models A, B and C are given in Figure 2.

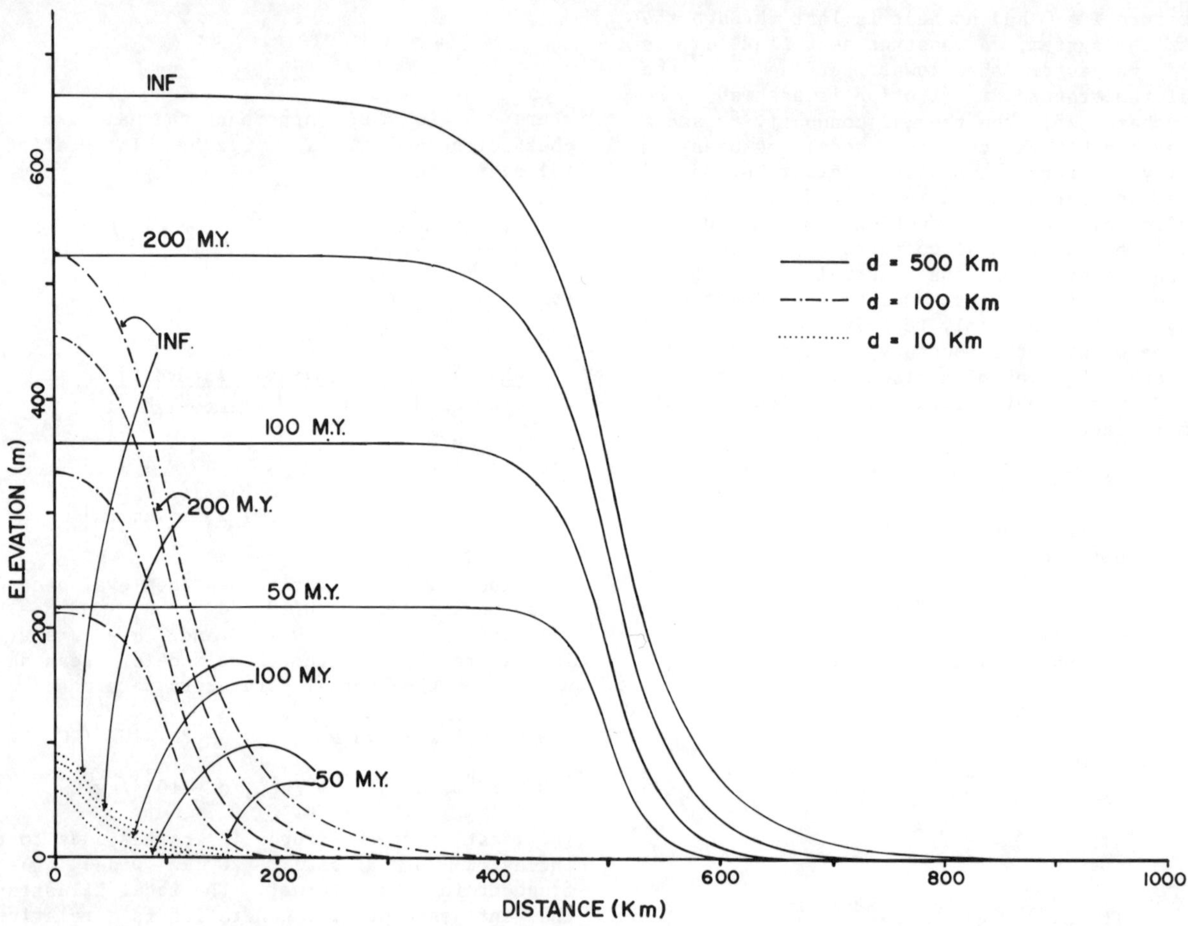

Fig. 2. Surface elevation of selected times for models A, B, C of Figure 1.

The maximum elevation difference (about 700 m) is derived from the difference in equilibrium temperatures for continental and oceanic lithosphere, given an initially uniform temperature gradient. We consider the profiles representing time at infinity to be the most significant since they best represent perturbation from an equilibrium condition induced by radiogenic heat production. In all models the surface heat flow is expected to be greater over the continental fragment than in the adjacent basin.

This simple model (Figure 2) suggests that lateral inhomogeneities in heat production retard the subsidence of oceanic lithosphere adjacent to continental fragments. Elevation differences of up to 700 meters can be explained by thermal contraction alone. A map of the pre-Cretaceous surface of the Florida platform [Kinney, 1967] shows the relative elevation of the Peninsular Arch above the adjacent basement. The thermal effects derived above certainly contribute to the differential subsidence of the continental margin; however, our computations do not include the effects of sediment loading or erosion prior to subsequent deposition. This effect may be substantial [Rona, 1974] and perhaps explains the order of magnitude difference between our computed 700 meters and the observed depths [Kinney, 1967]. We should point out, however, that the sediment loading may in fact be coupled to the thermal effect in that the differential sedimentation may partially result from the Penninsular Arch remaining above sea level during small scale sea water transgressions. It may also be that phase changes contribute to the subsidence [Sheridan, 1974]. This effect is also coupled to the thermal and sediment loading effects, however, the treatment of these interrelated effects represents a rather formidable mathematical problem.

Finite Difference Models

The analytical model assumed the initial temperature gradient to be uniform. Hence, these results cannot be used to evaluate elevation changes during or immediately following the rifting. In particular, the pre-rift elevation rise induced by heat from magmas rising from the mantle or the contraction of new oceanic lithosphere and the formation of boundary faults in part by thermal contraction is not evident in the simple analytical model. We have developed [Long and Lowell, 1973] a more complex finite difference model which includes some of these phenomena. The model, Figure 3, consists of idealized continental lithosphere separated by oceanic lithosphere. The dislocated block was separated from the main block by 25 km. The continental crust was assumed to be 50 km thick and to overlie an additional 50 km of mantle material.

The temperature within the model was found by solving the heat transport equation (equation 1) by the method of finite differences. The thermal diffusivity and the specific heat were assumed to be constant everywhere. The heat production was assumed to be $A = 10^{-9}$ w/kg and $A = 3 \times 10^{-10}$ w/kg in the upper and lower 25 km of the continental crust, respectively, whereas $A = 0$ in the remainder of the material. To the equation we applied the boundary conditions that the upper surface was maintained at 300°K and the lower surface was at 1500°K. There was assumed to be no lateral heat flow through the ends of the model. At time $t = 0$ we assumed an equilibrium geothermal gradient throughout the continental crust and underlying mantle material and assumed that a 25 km wide slab of oceanic and mantle material was injected at 1500°K between the continental blocks. For convenience, the initial temperature of the material to the right of the dislocated block was set at 1500°K.

The temperature distribution within the model was calculated for different times. Figure 4 shows the temperature distribution after 4 m.y.; at this time a substantial amount of cooling in the oceanic section is observed. Figure 5 depicts the temperature distributions after 100 m.y. The heating of the continental material relative to the oceanic material is now quite apparent. At $t = 100$ m.y., the thermal effects of sedimentary cover were simulated by replacing the uppermost layer of the model by a layer with lower conductivity. Figure 6 shows that the main thermal effect of the sedimentary cover is to increase temperatures at depth.

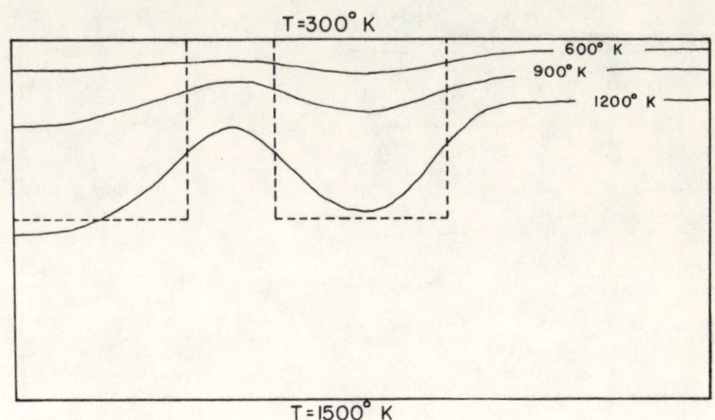

Fig. 4. Temperature structure at about 4 m.y.

We then computed elevation as a function of time, relative to the initial elevation, as a result of thermal expansion. We have not included isostatic adjustment for deposition or erosion of elevated areas. These effects would tend to exaggerate movements on a relatively short time scale, whereas our primary concerns were for initial displacements and the implications of the long-term perturbation in elevation due to the non-uniform heat source distribution. Figure 7 shows elevation profiles of the model at selected times. The initial response combines an expansion of the continental lithosphere with contraction in the newly implaced oceanic lithosphere to form a rise on the sides of the rift and a depression on the oceanic side. The stresses created by the contrast in thermal expansion and contraction could result in block edge faults with a near-surface throw of about 0.5 km in this case. The throw would decrease with depth as the temperature contrast decreases. The implied faulting suggests that rift zones associated with magma injection could develop without horizontal tension. With increasing time and cooling the block-edge effect attenuates and the total section contracts.

In the later time periods, the retarding influence of the heat generated in the crustal

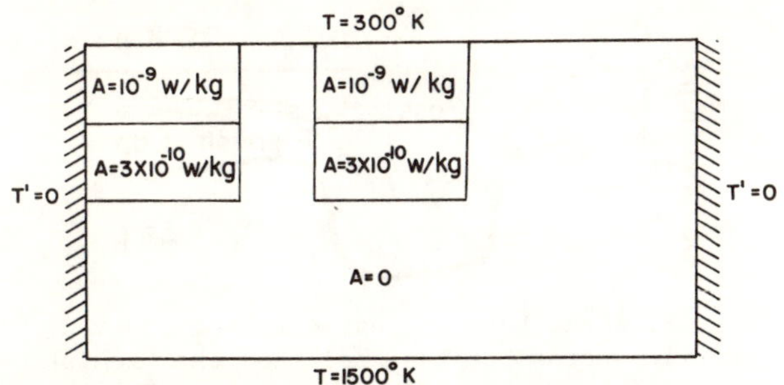

Fig. 3. Thermal model of dislocated continental block surrounded by idealized oceanic lithosphere.

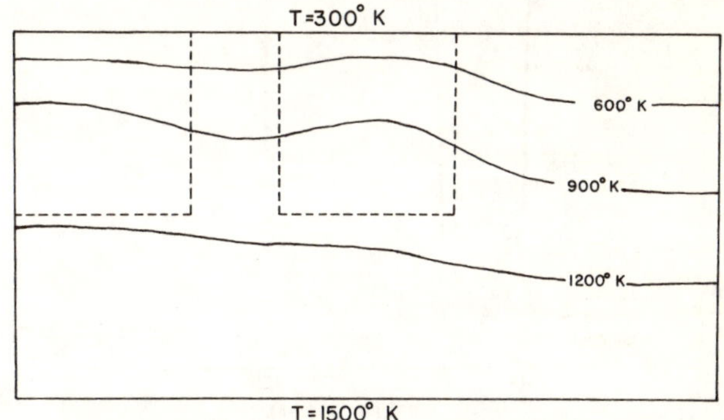

Fig. 5. Temperature structure at about 100 m.y.

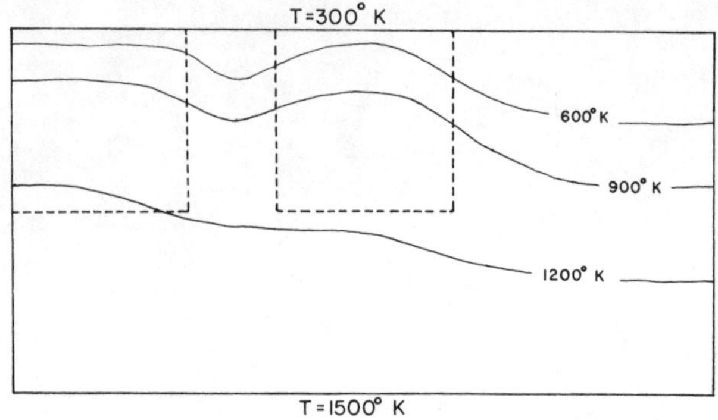

Fig. 6. Temperature structure at about 150 m.y.

blocks is manifested in the development of a prominent rise over the dislocated block. The elevation at the final time considered shows that areas adjacent to the dislocated continental fragment are significantly affected by lateral heat flow. The thermal effect of the sedimentary cover is manifested by the general uplift of the entire section after 100 m.y.

Discussion of the Finite Difference Model

The finite difference model was inspired by a study of the origin of prominent positive Bouguer gravity anomalies in southern Georgia [Long et al., 1972] and their corresponding basement rocks. The positive peaks indicate high-density basic crustal rocks below the cover of Cretaceous sediments. The magnitude and shape of the anomalies are comparable to volcanos associated with rifts. Recent analysis of magnetic data [Popenoe and Zietz, 1978] confirms the existence of rift related tectonics under Coastal Plain

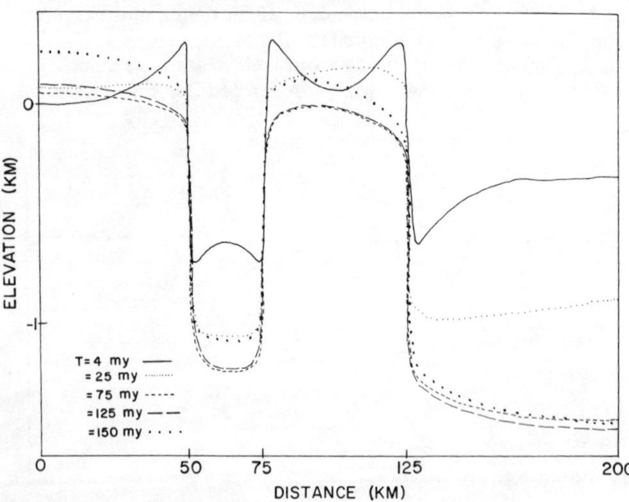

Fig. 7. Profiles of elevation change at selected times.

sediments, and also confirms the proposal [Long and Lowell, 1973] that the continental lithosphere under northern Florida is separated from the continental lithosphere of Georgia by a zone of transient spreading and (or) extensive intrusions which ceased before Cretaceous time. Following the transient spreading the region subsided in accord with the thermal contraction mechanism described by Sclater et al. [1971]. Due to the higher rate of heat production of the northern Florida continental lithosphere, however, the rate of subsidence has been retarded and the Ocala Uplift area exhibits prominence relative to adjacent areas on the Atlantic continental margin. Observed elevations of the pre-Cretaceous surface in the southeastern United States [Kinney, 1967] agree reasonably well with the calculated elevations furnished by our simple model (Figure 7). Also, the thermal perturbation effect suggested by the finite difference model helps explain the apparent time variations in the subsidence rate of the Florida peninsula which were pointed out by Sleep [1971]. A zone of expansion in south Georgia during the Triassic would help alleviate the overlap of Florida and Africa which Bullard et al. [1965] obtained in their pre-drift reconstruction of the continents. Zones of continental and oceanic-type crust have also been proposed by Krivoy and Pyle [1972] in an analysis of gravity data from Florida and the Gulf of Mexico. Thus, the actual geometry of the dislocated continental fragments may be consider-

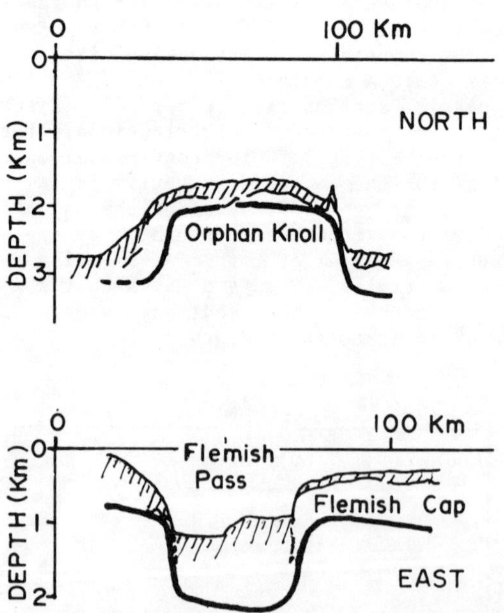

Fig. 8. Ocean bottom, sediment, and basement profiles across Orphan Knoll and the Flemish Pass (after Grant, 1972 lines PQ and TU respectively). Heavy line is finite difference theoretical basement profile from Fig. 7.

ably more complex than assumed by our simple thermal model, but the mechanism would be similar.

Perhaps, a better application of the models is to the Flemish Cap east of Newfoundland. Seismic reflection profiles [Grant, 1972] `(Figure 8) across the Flemish Pass onto the Flemish Cap show many of the characteristics shown in Figure 7 for ages beyond 25 m.y. Both the Cap and main continent show rounded edges. The Flemish Pass, which is about 25 km across shows a depression of at least 1.0 km (the profile did not detect basement or extend below 1.6 km). Also a profile across Orphan Knoll [Grant, 1972] just north of the Flemish cap shows similar characteristic but is separated by more than 100 km from the mainland. The edges on the Orphan Knoll are also rounded and it drops off almost exactly 1.0 km on the north edge where basement was detected, in close agreement with the estimate from the finite difference model. The curvature of the sediments (after removal of displacement of the edge fault) is similar to that predicted by the analytical model. The greater elevation difference could be related to sediment load. A steeper slope could imply that the model assumed a crustal thickness greater than appropriate.

Conclusions

A complete treatment of the thermal history of some portions of passive continental margins would include the interconnected effects of erosion and sedimentation (and the concomitant isostatic response) together with the thermal phenomena related to continental fragmentation and heat production differences between continental and oceanic lithosphere. The simple analytical and finite-difference models presented here indicate the possible importance of the latter phenomena; and these have not been included in continental margin studies to date (but see Zielinski [1979], which appeared after the preparation of this manuscript). At present, the full treatment of the thermal problem is beyond the scope of this paper.

References

Bullard, E.C., J.E. Everett, and A.G. Smith, The fit of the continents around the Atlantic, Royal Soc. London Philos. Trans., ser. A, 258, 41-51, 1965.

Carslaw, H.S., J.C. Jaeger, Conduction of Heat in Solids, 2nd ed., The Clarendon Press, Oxford, 510 p., 1959.

Grant, A.C., The continental margin off Labrador and eastern Newfoundland - morphology and geology, Can. J. Earth Sciences, 9, 1374-1430, 1972.

Kinney, D.M., ed., Basement map of North America, U. S. Geol. Survey, scale 1:5,000,000, 1967.

Krivoy, H.L., and T.E. Pyle, Anomalous crust beneath west Florida shelf, Am. Assoc. Petroleum Geologists Bull., 56, 107-113, 1972.

Larson, R.L., H.W. Menard, and S.M. Smith, Gulf of California: A result of ocean floor spreading and transform faulting, Science, 161, 781-784, 1968.

Long, L.T., S.R. Bridges, and L.M. Dorman, Simple Bouguer gravity map of Georgia, Georgia Geol. Survey, 1972.

Long, L.T., and R.P. Lowell, Thermal model for some continental margin sedimentary basins and uplift zones, Geology, 1, 87-88, 1973.

Lowell, R.P., and L.T. Long, Thermal model of the Florida Crust, in: The Geothermal Nature of the Floridan Plateau, ed. by D. L. Smith and G. M. Griffin, special publication no. 21 of State of Florida, Dept. of Natural Resources, Tallahassee, p. 149-161, 1977.

McKenzie, D.P., D. Davies, and P. Molnar, Plate tectonics of the Red Sea and East Africa, Nature, 226, 243-248, 1970.

Mohr, P.A., The Afar triple junction and sea-floor spreading, Jour. Geophys. Research, 75, 7340-7352, 1970.

Popenoe, P. and I. Zietz, The nature of the geophysical basement beneath the coastal plain of South Carolina and northeastern Georgia, U.S. Geol. Survey, Prof. Paper, 1028-I, 119-137, 1977.

Rona, P.A., Subsidence of the Atlantic continental margin, Tectonophysics, 22, 283-299, 1974.

Sclater, J.G., R.N. Anderson, and M.L. Bell, Elevation of ridges and evolution of the central-eastern Pacific, Jour. Geophys. Research, 76, 7888-7915, 1971.

Sheridan, R.E., Atlantic continental margin of North America, in: The Geology of Continental Margins, ed. by C.A. Burke and C.L. Drake, Springer-Verlag, N.Y., p. 391-408, 1974.

Sleep, N.H., Thermal effects of the formation of Atlantic continental margins by continental break up, Geophys. Jour. Royal Astron. Soc., 24, 325-350, 1971.

Sleep, N.H. and N.S. Snell, Thermal contraction and flexure of mid-continent and Atlantic marginal basins, Geophys. J. Roy. Astron. Soc., 45, 125-154, 1976.

Zielinski, G.W., On the thermal evolution of passive continental margins, thermal depth anomalies, and the Norwegian-Greenland Sea, Jour. Geophys. Res., 84, 7577-7588, 1979.

METASTABLE PHASE TRANSITIONS AND PROGRESSIVE DECLINE OF GRAVITATIONAL ENERGY:
ASPECTS OF ATLANTIC TYPE MARGIN DYNAMICS

Horst J. Neugebauer

Institut für Geophysik, Technische Universität Clausthal
D 3392 Clausthal-Zellerfeld, Fed. Rep. Germany

Tilman Spohn

Department of Earth and Space Sciences, University of California,
Los Angeles, CA 90024, USA

Abstract. The comparison of phase boundaries of
the basalt (gabbro)-garnet granulite - eclogite
transition with oceanic and continental geotherms
leads to the conclusion that gabbro and basalts
are metastable under crustal conditions. A transformation to a stable phase assemblage may occur
beneath sedimentary basins. This will have consequences for the dynamics of passive continental
margins. The reason for the metastability of the
crustal rocks is considered to lie on kinetic
grounds. Applying a nucleation and growth model
to the transition and using transition times and
run durations reported from laboratory studies the
values of two parameters of the model were determined, the activation energy for diffusion
(Q = 56 kcal/mole) and the diffusion constant
(D_0 = 0,1 cm^2/s). With the help of the latter model, the values of the cited parameters, and some
reasonable assumptions on the average crustal
grain size (0,01 - 1 mm) one can show that basaltic
rocks become thermodynamically metastable as they
are carried away from the ridge by the moving
plate.
  Beneath sedimentary basins adjacent to passive
continental margins the crust will become rewarmed
again due to thermal blanketing by the accumulated
sedimentary layer. Accordingly a transition of the
metastable rock to a stable phase assemblage may
occur in the lower crust. A one dimensional thermal model of such a sedimentary basin has been
evaluated and the results will be summarized. If
the sedimentation rate remains constant there will
be a stable phase layer developing with a rate of
thickening comparable to the sedimentation rate
but lagging behind at a certain time delay. This
time delay as a function of the model parameters
has been calculated. The stable phase layer is
separated from the metastable phase layer by a
zone where the transformation has not yet been
completed. Realistically low sedimentation rates
imply time delays of more than 100 Ma.

The body forces resulting from the phase change
may help to explain the further subsidence of the
cited basins. However, passive margin structure is
anomalous as it combines a thick continental crust
and a thin oceanic crust. The associated difference in the gravitational potentials is released
by body forces which cause a progressive decay of
the structural anomaly. The corresponding subsidence of the sedimentary basins decreases with
age. In addition, stable phase layers as well as
transformation zones might cause body forces
which, through concentration below the rise,affect
the tectonic development of the entire mature
margin.
  To quantify the influence of these two processes
on the development of the passive margin as a
function of age, a finite element model of a marginal cross section has been used.
  Well data and seismic refraction data have been
analyzed to compile sedimentation rates. It will
be demonstrated that the cited model calculations
serve well to explain both the temporal and spatial distribution of rates across passive continental margin structures.

1. Transition of Metastable Basalt to a
Stable Phase
A Geodynamic Driving Force

The basalt (gabbro) to garnet granulite to eclogite phase change has been of considerable interest to geodynamicists ever since it has been discovered that it might occur in the upper part of
the lithosphere. Lovering (1958) discussed uplift
and subsidence of the Earth's crust according to
volume changes resulting from the phase change.
Kennedy (1959) tried to explain the stability of
continental shields by assuming that the crust
mantle boundary was a phase change, separating a
gabbroic lower crust from an eclogite mantle.
Furthermore, he gave a model to explain orogeny

as being caused by the movement of the phase boundary after a temperature rise at depth. The development of geosynclines in the context of the
"Moho" as a phase change hypothesis has been discussed in great detail by McDonald and Ness
(1960), Wetherill (1967), van de Lindt (1967),
O'Connell and Wasserburg (1967, 1972) and Joyner
(1967). These authors studied the movement of the
phase change boundary after perturbation of pressure and temperature. The problem to solve mathe-
matically is known as Stefan's problem. The velocity of the phase change boundary is determined
by the rate at which the latent heat can be removed. Ito and Kennedy (1971) and Kennedy (1972)
have modified the original phase change interpretation of the Moho on the basis of their experimental results. They proposed that the shallow
seismic velocity discontinuity, known as the Con-
rad discontinuity, found in some continental
areas, may map a boundary, where temperature is
high enough to allow for recrystallization of
metastable gabbro to stable garnet granulite. On
the other hand, the Mohorovicic discontinuity is
interpreted as being caused by the equilibrium
transition from garnet granulite to eclogite.

Green and Ringwood (1966) and Ringwood (1975)
prefer to interpret the "Moho" in terms of a chemical discontinuity but state that in some areas of
high heat flow the gabbro to eclogite phase change
may explain the gradual increase in seismic velocities often found in these regions. A related
hypothesis proposed by Green and Ringwood (1966)
is that gabbro in the oceanic crust may transform
to eclogite at subduction zones, thus providing at
least part of the body force driving the mantle
flow associated with plate tectonics.

In the context of the development of sedimentary
basins adjacent to mature passive continental margins Neugebauer and Spohn (1978) and Spohn and
Neugebauer (1978) have proposed the transformation
of metastable basalt to stable garnet granulite
or eclogite as a probable cause of subsidence. The
metastable nature of the oceanic crust and the
development of a stable phase layer beneath a sedimentary basin has been discussed by Spohn (1979).
Furthermore the dynamics of the progressive decay
of a marginal transition zone associated with the
decline of gravitational energy was investigated
by means of nonlinear numerical experiments (Neugebauer, 1979).

In this paper we will present an advanced investigation on the transition of metastable basalt
and its implications as a geodynamic driving
force. Metastability of oceanic basalt and the
constraints for a transition to a stable phase
will be discussed in detail. The results will be
tested by means of numerical models in order to
estimate the possible consequences of the derived
loading system on the dynamics of lithospheric
structures associated with sedimentary basins.

Metastability of Basalt (Gabbro) in the Crust

It has become evident that knowledge of the temperature and pressure stability fields of the

transformation and its transition rate as a function of pressure and temperature are required.
There have been extensive experimental investigations of the stability fields of the phases (for
references see Spohn and Neugebauer, 1978). Fig. 1
shows the results of more modern work by Green and
Ringwood (1967) revised by Ringwood (1975) and Ito
and Kennedy (1971) together with theoretical phase
boundaries determined by Ahrens and Schubert
(1975). While the slopes of the boundaries seem
to be in good accordance (~ 15 bar/^{o} for the basalt to garnet granulite boundary, ~ 20 bar/^{o} for
the garnet granulite to eclogite boundary), the
width of the garnet granulite field differs significantly. This is mainly attributed to the different chemistries of the basalts under study
(Ringwood, 1975; Spohn, 1979). Presumably, the
aluminium content of the basalts is of major importance to the location of the stability fields
of the phase assemblages in the P.T.-plane. Spohn
(1979) found a positive correlation between the
degree of aluminium oversaturation and the width
of the garnet granulite field for 10 basalts including the ones studied by Green and Ringwood
(1967) and Ito and Kennedy (1971). The degree of
Al-oversaturation of an average oceanic tholeiite
from Engel et al. (1965) fits closer to the quartz
tholeiite investigated by Green and Ringwood
(1967) than the olivine tholeiite investigated by
Ito and Kennedy (1971). It may therefore be concluded that phase boundaries determined by Green
and Ringwood (1967) and revised by Ringwood
(1975) are more representative of the oceanic
crust than others.

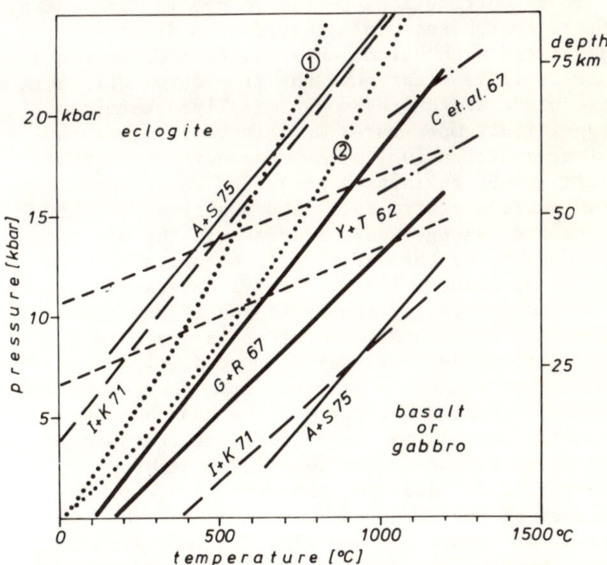

Fig. 1. Basalt – garnet granulite – eclogite
phase diagram according to Green and Ringwood
(1967) and revised by Ringwood (1975) (G +
R 67), Ito and Kennedy (1971) (I + K 71), and
Ahrens and Schubert (1975) (A + S 75).
(1) continental shield geotherm and (2)
oceanic geotherm after Clark and Ringwood
(1964).

Fig. 1 additionally shows the oceanic and continental shield geotherms as given by Clark and Ringwood (1964). Although these geotherms may not well represent mantle temperatures, as convective heat transport has not been considered, they may well serve as reasonable representatives of a temperature distribution in the crust in tectonically inactive regions. Relating the geotherms to the phase boundaries of the basalt (gabbro) to eclogite transformation one finds that basalt and gabbros in the Earth's crust should mainly be in a metastable state. The reason for this is likely to lie on kinetic grounds. It is therefore essential to investigate the reaction rate of the cited transformation. In general the lack of knowledge of transition rates of most crustal and mantle phase transformations and chemical reactions is the main limiting factor in our assessment of their tectonic implications.

Kinetics of the Transformation

A large majority of polymorphic solid-solid transitions are described by a process known as nucleation and growth (Rao and Rao, 1978). Such a transition will only occur in an assemblage if the formation of the new phase lowers its Gibbs free energy. But even if the new phase becomes thermodynamically stable it will not necessarily nucleate. Nucleation comes about as a result of thermal and compositional fluctuations. The formation of a volume of the new phase causes a decrease in Gibbs free energy but a new interface will be formed also. This causes an increase in Gibbs free energy which opposes its formation. Further contributions to the change in free energy across the phase change result from distortions associated with volume changes. From the above arguments it is clear that the transition will not take place at the thermodynamically determined transition temperature. At this temperature the volume contribution to the change in Gibbs free energy is by definition zero. The contribution from surface energy is always positive, so that Gibbs free energy would increase if the new phase would form. As this never will occur, a certain amount of undercooling is always necessary. Fig. 2 shows schematically the variation of the nucleation rate as a function of temperature. The nucleation rate increases from zero at the transition temperature to a maximum, from whereon it decreases again. The latter is the case because nucleation depends on the diffusional transport of the atoms to the nucleation sites. Heterogeneous nucleation differs from homogeneous nucleation because of preferred nucleation sites. The existence of impurities and imperfections such as grain boundaries, dislocations, strained areas etc. contributes to the energy balance and usually reduces the amount of undercooling.

From the experiments of Green and Ringwood (1967) and Ito and Kennedy (1971) it may be concluded that the amount of overlap between nucleation rate (NR) and growth rate (GR) is less than ± 50°. But as glasses were used as starting materials this

168   NEUGEBAUER AND SPOHN

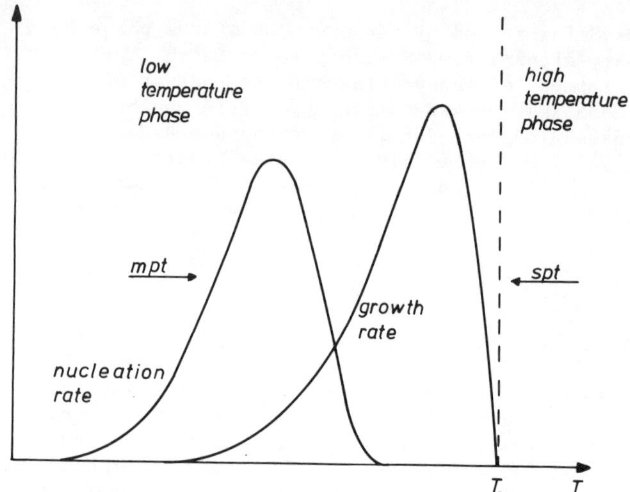

Fig. 2. Nucleation (NR) and growth rate (GR) of a high temperature to low temperature phase transition. T_0 is the thermodynamically defined transition temperature. The arrows indicate the direction of the temperature change during a "stable phase transition" (spt) and during a "metastable phase transition" (mpt).

value may possible serve only as a lower bracket. Once a nucleus has become stable, it may grow by diffusional transfer of atoms to the nucleus. At the thermodynamically defined transition temperature the growth rate as function of temperature (Fig. 2) passes through zero as jumps to and from the product phase site are equally probable. With sinking temperature the probability of back jumps sharply decreases, the growth rate increases and finally passes through a maximum which usually (Salmang and Scholze, 1968) lies at significantly higher temperatures then the maximum of the nucleation rate curve.

The latter fact has some interesting consequences, as the transition rate of a phase change and the resulting grain sizes may depend on the pressure and temperature path through which the assemblage passes. Consider an isobaric phase change from a high temperature phase to a low temperature phase as illustrated in Fig. 2. If the high temperature phase assemblage is brought to the transition temperature T_0 from its stable phase region and is further cooled slowly it first passes through an interval of increasing and maximum growth rate having no or only few nuclei to grow. With sinking temperature it enters an interval of increasing nucleation rate but its growth rate sinks. The results will be a low transition rate together with large grain sizes and/or metastable phases. If, on the other hand, the high temperature phase assemblage is brought in a metastable state into the low temperature phase region and slowly warmed up, it will pass through the cited intervals in a much more efficient way. It first passes through the interval of increasing and maximum nucleation rate, the nuclei so produced then experience increasing and maximum

growth rate. This results in comparatively higher
transition rate and lower grain size.

The basalt to garnet granulite to eclogite
transformation is not a simple solid-solid phase
transition but a complex series of transitions and
reactions with many rate curves superimposed, and
these may not be independent of each other. But the
simple arguments from the theory above seem to be
applicable as some observations reported by Cohen
et al. (1967) show. They report (p 501) that "a
difference in paragenesis resulting from different
paths to pressure and temperature is reproducible".
The runs, that were brought to the transition temperature and pressure (1300° C, 19,9 kb) from the
low temperature side resulted in crystallization
of 1μ to 2μ large garnet and pyroxenes
(30:70 %). On the contrary large metastable pyroxenes crystallized from runs that were brought to
the phase boundary from the high temperature side.
The starting material was a basaltic glass. Further observations on the difficulty of nucleating
garnet in "stable phase transitions" were reported
to some extent by Green and Ringwood (1967).

The basaltic crust is formed thermodynamically
stable at the mid-oceanic ridges. Being carried
away from the ridges, as part of the plate system,
it cools down and becomes apparently metastable
as mentioned earlier. If a transition to garnet
granulite, for instance, occurs, it would be a
"stable phase transition". The low efficiencies of
such transitions may serve as a first explanation
of the metastability of the oceanic crust. But we
will come back to this problem in more detail later.

The Diffusivity

The diffusion coefficient

$$D = D_o \exp \left(- \frac{Q+PV}{RT}\right) \qquad (1)$$

is a parameter of major importance in both nucleation and growth rate models (D_o: diffusion
constant, Q: activation energy, P: pressure, V:
activation volume, R: gas constant, T: temperature). It is therefore desirable to know its value as a function of temperature. The variation
with pressure is only slight in crustal environments as the activation volume usually is around
10 cm^3/mole (Manning, 1968; Kittel, 1976; Nicolas
and Porier, 1976). The contribution to the numera-
tor in the exponential function would then be
0,2 kcal/mole kbar. Ahrens and Schubert (1975),
in their study on the subject, have collected some
data on diffusion constants and activation energies from the literature. They selected data on
tracer diffusion experiments on cations that take
part in the basalt to eclogite transformation. The
data on diffusion constants scatter over more than
12 orders of magnitude (compare Fig. 4), while the
data on activation energies span from around
29 kcal/mole to 167 kcal/mole. Unfortunately, the
data on aluminium, possibly the rate limiting ca-

tion, are rather scarce. Furthermore, the question rises to what extent data on tracer diffusion experiments are applicable to solids undergoing a phase transition.

Spohn (1979), therefore, tried to determine
these data from the times necessary to attain
equilibrium in laboratory runs on the transformation. These values should then not be applied
to a specific cation, but serve as an apparent
or average value for the transformation itself.
It is nevertheless felt that these values should
be of comparable order of magnitude to those
determined in diffusion experiments.

The transformation times as a function of temperature may be plotted in an Arrhenius-diagram.
If a sufficient number of nuclei are present and
if the growth rate model is applicable and furthermore if we are not too close to the transition temperature, than the data should be fitted
by a line, the slope of which gives the activation energy. The intercept with the ordinate
should be proportional to L^2/D_o, with L the grain
size.

Unfortunately, we do not have many data of ex-
plicit transformation times. In fact, we have two
estimates by Ito and Kennedy (1971) at 800° C and
1200° C and one by Green and Ringwood (1967) at

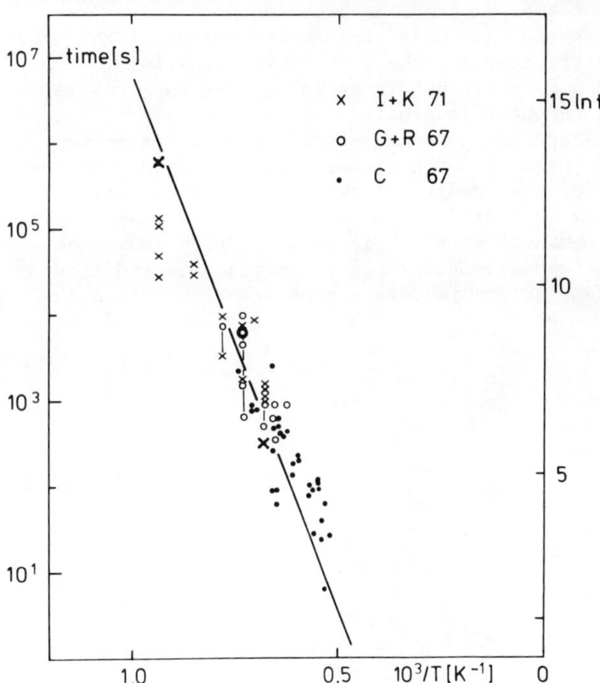

Fig. 3. Transformation times (heavy signs)
and run durations as a function of the reci-
procal temperature. The slope of the regression line gives the activation energy for
diffusion, while the abscissa of the intersection with the 10³/T-axis is proportional to
the diffusion const. Abbreviations as in
Fig. 1, C 67: Cohen et al. (1967).

1100° C. But we do have sufficient data on the run durations. These data may serve as upper limits to the transformation times. As can be envisioned from Fig. 3 the transformation times can be well fitted by a line whose correlation coefficient is $r = 0.97$. The data on run durations, if corrected for pressure assuming an activation volume of 10 cm^3/mole, closely fit to the line and thus support the estimate. After some reasoning on the grain size and an accommodation factor, we found:

$$Q = 56 \text{ kcal/mole} \pm 25 \text{ \%}$$

and
$$2{,}2 \cdot 10^{-4} \leq D_o \leq 13 \text{ cm}^2/\text{s}$$

with
$$D_o = 5 \cdot 10^{-2} \text{ cm}^2/\text{s}$$

being the most probable value. More than 80 % of the diffusion constants gathered by Ahrens and Schubert (1975) fall into the cited interval. Disregarding an extreme value of $D_o = 10^{19}$ cm^2/s, their data may be fitted by a normal distribution, the mean value of which ($D_o = 3{,}7 \cdot 10^{-2}$ cm^2/s) lies close to the one determined by Spohn (1979). Additionally, the variances are in good accordance.

Application of the Model to the Oceanic Crust

Having determined reasonable values of the parameters entering the diffusivity equation, one may turn back to the discussion on the metastability of the oceanic crust.

At present it seems to a large extent to be accepted that the oceanic crust is formed at the midoceanic ridges by chemical differentiation from the peridotitic mantle. The temperature of the mantle material is thought to be above the solidus temperature of dry peridotite and above the liquidus temperature of basalt (Forsyth, 1977).

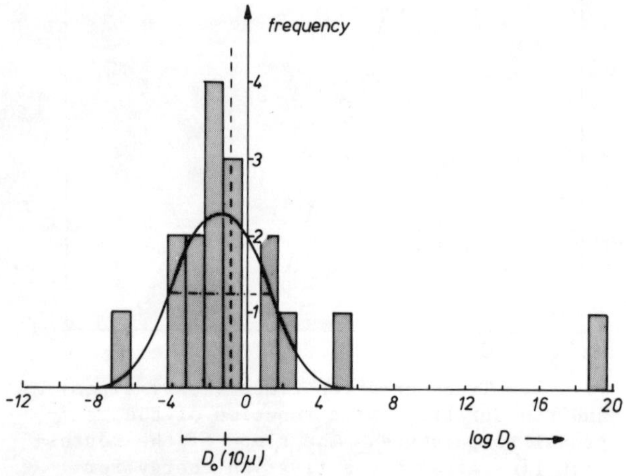

Fig. 4.   Frequency distribution of diffusion constants (D_o) as compiled by Ahrens and Schubert (1975).   The diffusion constant marked below has been determined from Fig. 3.

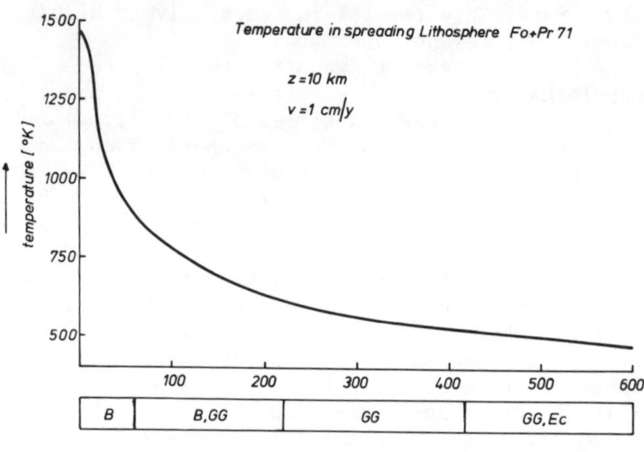

Fig. 5.   Temperature in spreading lithosphere according to Forsyth and Press (1971).   The depth is 10 km and a spreading rate of 1 cm/a has been assumed. The phases stable according to Green and Ringwood (1967) and Ito and Kennedy (1971)   as marked below.

The gravitationally unstable basalt melt intrudes the crust and forms sheeted dikes. Some melt pierces the crust and solidifies as pillow lava and some crystallises at lower crustal levels as gabbro. Beneath the crust, ultrabasic mantle material rests as depleted mantle.

The oceanic lithosphere cools on its way away from the ridges. The basalts, originally thermodynamically stable, approach the transition temperatures to other phase assemblages, while the pressure acting on them does not change significantly. Fig. 5 shows the temperature as a function of the distance from the ridge axis at a depths of 10 km according to Forsyth and Press (1971). The plate velocity is assumed to be 1 cm/a. Below this we marked the phase assemblages that are stable in the given depth and the respective temperature intervals. Both the phase diagrams of Green and Ringwood (1967) and Ito and Kennedy (1971) have been taken into consideration. Basalt is the stable assemblage only in the immediate surroundings of the ridge axis. According to Ito and Kennedy (1971), or, as discussed earlier, in the case of a basalt with relatively high degree of Al-oversaturation, the transition temperature to garnet granulite will be reached at 50 km distance from the axis according to Green and Ringwood. In the case of less oversaturated basalt, the transition temperature will be reached at 220 km distance. Within little more than 400 km distance, corresponding to 40 my from the ridge axis, the less oversaturated rock will reach the transition temperature to eclogite, while the more saturated rock will remain in the garnet granulite field. As the slopes of the oceanic geotherm and the phase boundaries (Fig. 1) do not differ significantly, the conclusions above are equally valid at greater depths.

To answer the question whether these transitions
are able to occur one may apply the nucleation and
growth model. As has been pointed out earlier, the
transitions would be "stable phase transitions"
and one would expect difficulties in nucleation of
the phases. In the following we will, on the contrary, assume that there are sufficient nuclei
present and will ask whether these are able to
grow to a given grain size. To do this we calculate the time necessary to transform a grain of
given size. If we are not too close to the thermodynamically defined transition temperature, the
transformation time may be calculated from

$$t = a \frac{L^2}{D} \qquad (2)$$

with a = 0.22 an accommodation factor (Spohn and
Neugebauer, 1978). Some examples obtained by eva-
luating (2) are illustrated in Fig. 6. It gives
the transition time as a function of the reci-
procal temperature. The curve parameters are activation energies, diffusion constants and grain sizes. The intersections with the isotherms give the
respective transition times which may be read in
years from the ordinate axis. The transformation
from basalt to garnet granulite seems to be most
probable for a highly oversaturated basalt.
Assuming a grain size of 1 mm and taking the diffusion constant to be 0,1 cm^2/s the transition at
isotherm 1 takes times between 10^3 to 10^8 years
according to the activation energy assumed. If the
grain size were only a micron, the transition times
would be much shorter. Taking the transformation to
occur at 560 K (isotherm 2), then the transition
time would amount to more than 10^8 y even for a
one micron grain size. The transformation to eclogite seems to be impossible in geological times
under crustal conditions. Furthermore, the crust
will go on cooling after having reached the transformation temperature. So the transformation time
will lengthen still further. Fig. 7 shows the increase in transformation time during the process.
The origin of the time scale is the time after
which the transition temperature has been reached.
The crust is supposed to cool with rates taken
from Fig. 5. After 5·10^5 y to 10^6 y the transition
time will be doubled and then will increase further exponentially. Transition times above 10^5
years will therefore lengthen to infinity as they
go on. Unless the grain size is not smaller than
1μ, the crust will therefore run into metastability on its way away from the ridge.

Application of the Model to Sedimentary Basins

A metastable rock may transform to a stable
phase assemblage via a "metastable phase transition". In the oceanic crust this may occur in
subduction zones or beneath sedimentary basins
adjacent to passive continental margins. If the
lower continental crust is composed of metastable
gabbro, a "metastable phase transition" may occur
beneath sedimentary basins or in regions of high

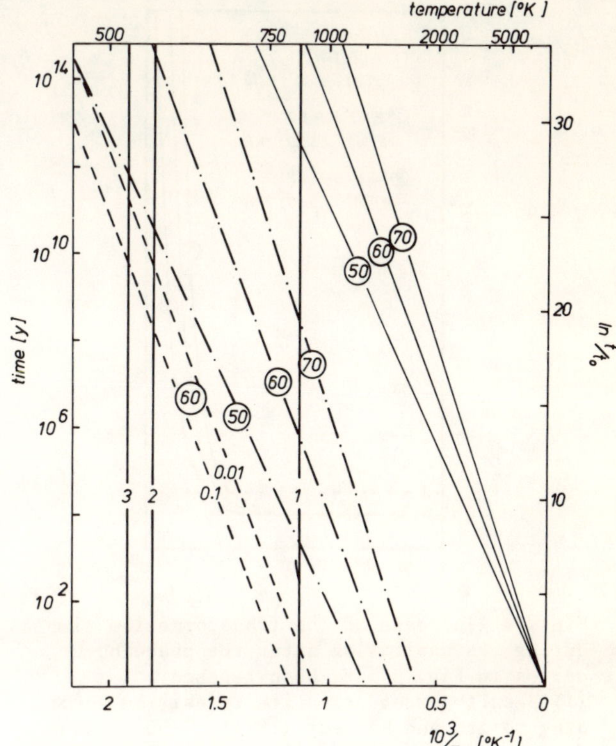

Fig. 6. Characteristic transformation time
as a function of temperature at a pressure of
3 kbar. The ordinate axis to the right gives
the transformation times normalized by

$$t_o = a \frac{L^2}{D_o}.$$ The curve parameters are activation energies in kcal/mole, the activation
volume is 10 cm^3/mole. From these curves one
may determine the transformation times after
fixing the activation energy, the diffusion
constant (D_o) and the grain size (L). Some
examples are given. dash-dotted lines: D_o =
0.1 cm^2/s, L = 0.1 cm, dashed line: D_o =
0.1 cm^2/s and 0.01 cm^2/s, L = 1. The intersections with the isotherms 1,2,3 give the
transformation times of the basalt-garnet
granulite transition according to Ito and
Kennedy (1971) (1) to Green and Ringwood
(1967) (2) and from garnet granulite to eclogite after Green and Ringwood (3).

heatflow. Haxby et al. (1976) have proposed such
a model to explain the subsidence of the Michigan
basin. But the composition of the lower continental crust is still a matter of debate (Ringwood,
1975). There may be, on the other hand, metastable
phase transition in more silicic rock systems as
well. This should be a subject of further study.
There has been considerable effort to relate the
development of sedimentary basins adjacent to passive continental margins to a modern geosynclinal
concept and to plate tectonics (e.g. Drake et al.,
1959; Mitchel and Reading, 1969; Dewey and Bird,

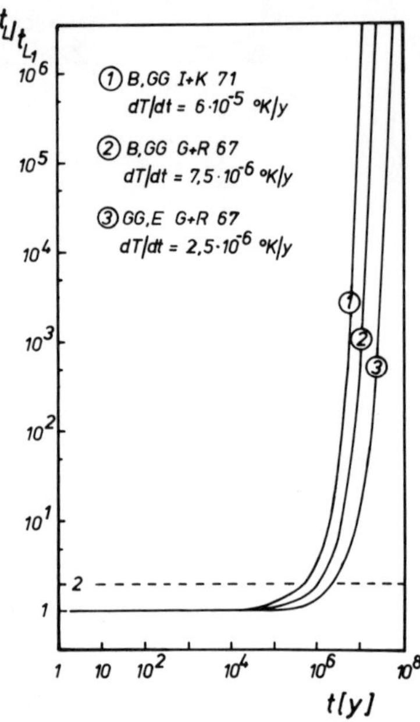

Fig. 7.   Increase of the transformation time during its expiration after the phase boundaries in Fig. 5 have been reached.
(1) Basalt-garnet granulite transition according to Ito and Kennedy (1971), ~50 km away from the ridge axis.   (2) Basalt-garnet granulite transition according to Green and Ringwood (1967), ~220 km away from the ridge axis.   (3) Garnet granulite - eclogite transition according to Green and Ringwood (1967), ~400 km away from ridge axis.

1970; Falvey, 1974). Thus the first depressions may form by tectonic movements during the break up of the continents and the formation of the crust of the opening ocean. Further cooling of the hot crust will result in further deepening of the performed depressions. Sediments will accumulate and act as an additional load. After some time (>80 Ma) cooling may cease and in turn a rewarming may occur due to thermal blanketing by the accumulated sediments. The sedimentary basins of the east coast of the U.S. or the west coast of Northern Africa may be in such a stage of development now. To calculate the development of a stable phase layer in the oceanic crust beneath the continental rise basin two mathematical models have been constructed (for details see Spohn and Neugebauer, 1978).

Both the models consider in one dimension the thermal evolution of the crust beneath a sedimentary layer. The first model gives the thickness of the stable phase layer as a function of time, the stable phase layer being the layer where the transformation has been completed. The second model solves the differential equation of diffusion at each point of the model and calculates a degree of transformation from the concentration of the relevant cations. The transition between stable and metastable phase is not sharp but gradational, the transition zone will be represented in the second model. The models have been evaluated numerically by means of the finite difference method and the following parameters have been varied: activation energy and activation volume for diffusion, diffusion constant, grain size, sedimentation rates, the initial heat flux out of the crust and the initial thickness of the crust. The initial conditions are such that sedimentation starts at the origin of time and the sediments accumulate on a crust with an initially stationary temperature distribution according to the oceanic geotherm by Clark and Ringwood (1964). The thermal conductivity varies with depth such that the temperature field remains stationary when no sediments accumulate on top and that the heat flux out of the crust would be one or two hfu. The boundary con-

TABLE 1.   Parameters of the standard model. These values were used during the discussion unless otherwise stated.

Parameter	Value	Source
ρ_e (Density of stable phases)	from 3.0 g/cm^3 (basalt) to 3.5 g/cm^3 (eclogite)	Ringwood (1975)
ρ_b (Density of metastable phase)	3.0 g/cm^3	Ringwood (1975)
ρ_s (Density of sediments)	2.3 g/cm^3	Drake et al. (1959)
K_s (heat conductivity of sediments)	0.0035 cal/cm	O'Connell and Wasserburg (1967)
L (grain size)	0.01 cm	this study
D_o (Diffusion constant)	0.1 cm^2/s	this study
Q (activation energy)	60 kcal	this study
V (activation volume)	10 cm^3	this study

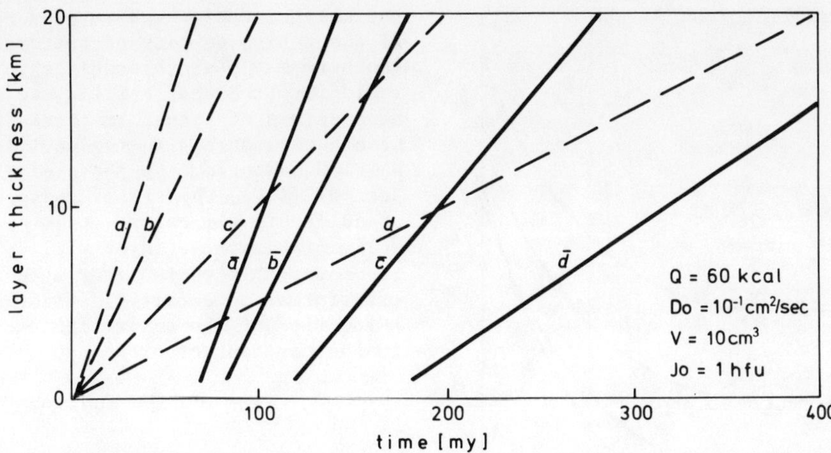

Fig. 8. Thickness of the sedimentary layer (dashed) and the stable phase layer (solid) as a function of time. Curve parameters are sedimentation rates: a) 300 m/Ma, b) 200 m/Ma, c) 100 m/Ma, d) 50 m/Ma.

ditions are costant temperature at the top of the model, continuity of heat flux across the layers and constant heat flux from below. The crust is initially completely composed of the metastable phase. As temperature increases with depth the transformation will start at the crust mantle boundary. The values of the parameters of the standard model may be ascertained from Table 1. Fig. 8 illustrates the results of the cases of sedimentation at a constant rate. The stable phase layer develops with a rate comparable with the sedimentation rate but lags behind the sedimentation with a certain delay. This first result charakterizes all models with constant sedimentation rate. The rate of thickening $\dot{x}_p$ of a stable phase layer is mainly a function of the sedimentation rates $\dot{s}$ and may be expressed by

$$\dot{x}_p = 3.367 \; \dot{s}^{0,761} \quad m/Ma \tag{3}$$

as long as the activation volume does not exceed 10 cm³/mole substantially (Spohn, 1979). The time delay Δt is a function of the sedimentation rate and all the other parameters. This can easily be made plausible if one considers, for the time being, the thermal and mechanical properties of the sediments to be equal to those of the crust. Then sedimentation, with isostatic compensation, means thickening of the crust and moving the lower crust into regions of higher pressure, temperature and then higher transformation rate. If the transformation has occurred at the crust mantle boundary, then the accumulation of a sedimentary layer of thickness s means moving the region of favourable transformation rate up within the crust that distance s. The increasing pressure will oppose the transformation but its efficiency to do so will be the lower, the lower the activation volume.

It remains to discuss the time delay between sedimentation and the development of the stable phase layer. The variation of the time delay with the activation volume is rather slight provided the latter does not exceed 10 cm³/mole (Fig. 9). The results of the numerical evaluation of the model are such that the variation of the time delay with the other cited parameters may be represented by the following equation:

$$t = 68064.71 \; \dot{s}^{-0,519} \cdot$$
$$\cdot \frac{Ks}{J_o} (0.276 \, Q + \log L - \log D_o - 11{,}552) + \frac{20 - X_m}{\dot{x}_p} \tag{4}$$

Thus the time delay (in Ma) is inversely proportional to nearly the square root of the sedimentation rate (in m/Ma) and to the initial heatflux out of the crust (in hfu). It is proportional to the heat conductivity (cal/cm² Ks), to the activation energy (kcal/mole), to the logarithm of the grain size (cm) and the diffusion constant (cm²/s). The calculations were done assuming an initial depth to the crust mantle boundary X_m (km) of 20 km below sea level. If the crust were initially of a different thickness the time delay would vary according to the additional term in eq. (4).

The funktion Δt (eq. (4)) has an infinite number of zeros. For example with an activation energy of 45.5 kcal/mole and the values of the other parameters equal to the standard model the time delay will diminish. The rate of thickening will in these cases, however, initially be smaller then those given by eq. (3). If the activation energy were even lower, there would be a stable phase layer in the lower crust even without sedimentation.

The case of a sedimentary layer of constant thickness deposited on top of the crust with different rates has been discussed by Spohn and Neugebauer (1978). Significantly new

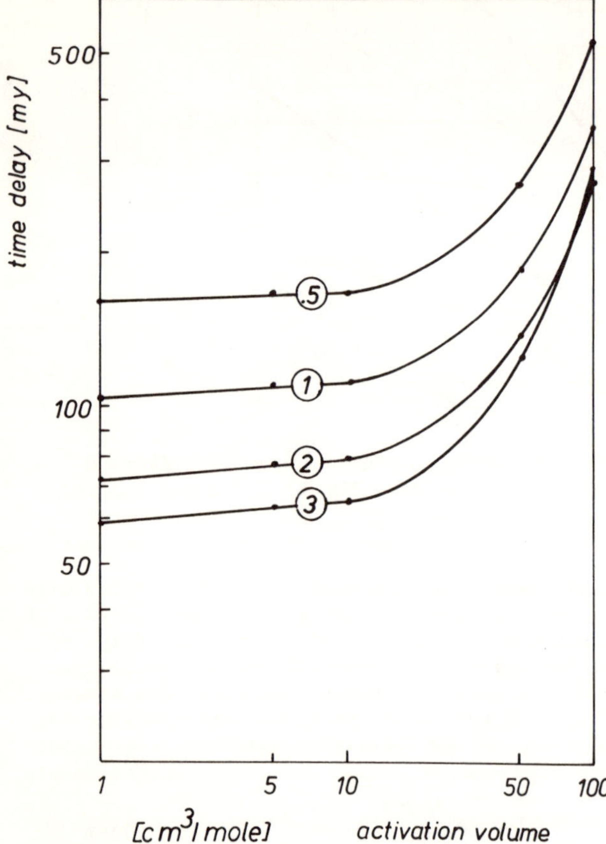

Fig. 9. Time delay between sedimentation
and thickening of the stable phase as a function of the activation volume. Curve parameters are sedimentation rates in 100 m/Ma.

results are at present not available. The standard
model of Spohn and Neugebauer (1978) used an activation energy of 80 kcal instead of 60 kcal and a
diffusion constant of 10^4 cm^2/s (L = 0.01 cm) instead of 0.1 cm^2/s. These changes of the parameters do nearly compensate (eq. (4)) and the results discussed in the earlier paper will be valid
with the modified model too.

Finally, we may turn to the question of what
assemblage the stable phase will be. Assuming the
Green and Ringwood (1967) phase diagram, as has
been done with the above model, the stable phase
would initially be an eclogite. But this will only
remain so if the geotherm stays in the eclogite
field during sedimentation. This would imply that
the thermal and mechanical properties of the sediments were close to those of crust or that the
sedimentation rate is so high (>300 m/my) that
pressure increases relatively fast compared to
temperature. Of course, pressure increases immediately during sedimentation while the temperature
lags behind because of the thermal inertia of the
system. Considering observed sedimentation rates
($\leq$ 50 m/Ma) the geotherm will move into the garnet
granulite field. Under some possible but rather

unlikely circumstances such as high initial heat
flow and high density of the sedimentary rock, the
geotherm may pass through the garnet granulite
field into the basalt field, so that there would
be a series of transformations from metastable
basalt to stable garnet granulite to stable basalt. Considering the phase diagram of Ito and
Kennedy (1971) the stable phase would be a garnet
granulite in all cases.

The stable phase layer will be separated from
the metastable phase layer by a zone where the
transformation occurs. The width of this zone is
a function of the activation energy and the diffusion constant according to

$$\Delta x = 0.1\ Q + 0.7\ \log D_o + 2.7$$

and of the activation volume. But as in the case
of the time delay and the rate of thickening its
influence is negligible unless it does not exceed
10 cm^3/mole substantially (Spohn, 1979). Fig. 10
gives the degree of transformation of the metastable rock at time,

$$t = \Delta t + f \cdot (\dot{x}_p)^{-1} \cdot 10^3\ Ma, \qquad (5)$$

where f is the curve parameter, in a 10 km thick
section of the crust above the crust mantle boundary. The degree of transformation is by definition 1 where the rock has been completely transformed to the stable phase assemblage while it is
zero where the assemblage is purely metastable.
After the time Δt the transformation zone will be
fully established and moves upward in front of the
stable phase layer keeping its width constant.
According to Whitten (1976) sedimentation rates at

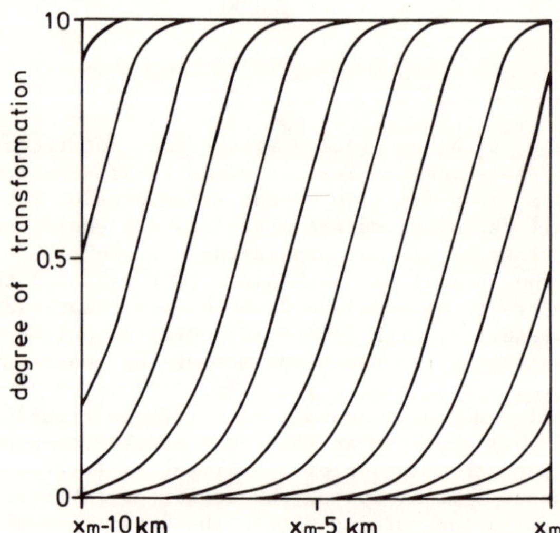

Fig. 10. Degree of transformation of the
crustal rock in a section of 10 km above the
crust-mantle boundary at times:

$$t = \Delta t + f \cdot (\dot{x}_p)^{-1} \cdot 10^3\ Ma.$$

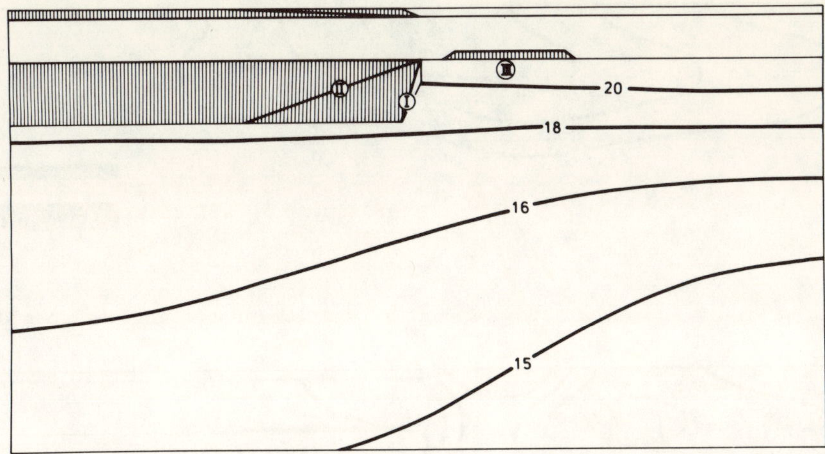

Fig. 11. Structural scheme of the dynamical models of a passive continental margin. Hatched areas indicate differential loading and buoyancy of the continental structure I, II and loading of the oceanic structure due to the transition of metastable phases, III. Contours represent strain rates per second for negative power index and a constant maximum shear stress of one bar. 20 reads: $10^{-20}s^{-1}$.

the Atlantic continental margin of Northern America may be as much as 200 m/Ma. But sedimentation at such high rates will last not more than about 10 Ma. Sedimentation averaged over some hundred million years occurs at rates that range up to 50 m/Ma. Thus the development of a stable phase layer will begin with the standard model at the earliest after about 160 Ma of sedimentation. At this time the transformation zone will be fully established and the geotherm will have crossed the eclogite – garnet granulite boundary according to Ringwood (1975) but still lie close to it. The increase in density of the partly transformed rock may be taken as the product of the degree of transformation times the density increase across the completed transformation which may amount to 0.5 g/cm if eclogite is the stable phase. As the density increases nearly linearly across the garnet granulite field the density increase across the complete transformation in the Ito and Kennedy garnet granulite case discussed above may be taken to be about 0.4 g/cm^3. On calculating the dynamic response of the continental margin structure to the phase change it may be convenient to replace the transformation zone by an equivalent stable phase layer of the same mass. The 7 km thick transformation zone illustrated in Fig.10 will in this sense be equivalent to a nearly 2.5 km thick stable phase layer. Stable phase layers of 1.25 km, 0.5 km and 0.18 km are equivalent to the transformation zones marked by −1, −2 and −3 in Fig. 10 respectively. These zones will develop in 143–314 Ma, 130–262 Ma and 113–210 Ma for sedimentation rates between 50 m/Ma and 10 m/Ma.

The development of a stable phase layer or an equivalent stable phase layer causes isostatic readjustment of the sedimentary basin. The force acting on a lithospheric column may be calculated from the excess mass of the uncompensated column

$$\Delta m = (1 - \varrho_b/\varrho_e) X_p \varrho_m \qquad (6)$$

with ϱ_b, ϱ_e, ϱ_m being the densities of the metastable phase, the stable phase, and the mantle respectively. Additional loads act as the deepening basin becomes filled with water and sediments.

2. Dynamics of Passive Margin Evolution

The structural features at a continent-ocean transition across a passive margin are not well known. While numerous seismic and other data reveal the tectonic and sedimentary characteristics at the top, much less is known about the course of the crust-mantle boundary across the passive continental margin. In addition to seismic wave velocities which usually find values intermediate between typical lower crustal and upper mantle velocities, gravity data and isostatic interpretation will provide constraints on the possible position of the crust-mantle boundary (Scrutton, 1976). In order to study the dynamic character of a passive margin structure by means of numerical experiments, we applied a ratio of standard oceanic to continental crustal thickness of about 2.8 after Worzel (1974). The transition between oceanic and continental crust has been modelled in a very simplified way, as shown in Fig. 11. The two-dimensional model structure corresponds to an overall depth of 200 km and a total length across the continental margin of 380 km. The numerical calculations demonstrate that this is a sufficient size für the model.

The numerical results on the dynamics of the established structure will be controlled by the applied deformation mechanisms and the imposed boundary conditions including the applied loads. Thermal contraction and sedimentary loads will not be considered since the phase reaction is most

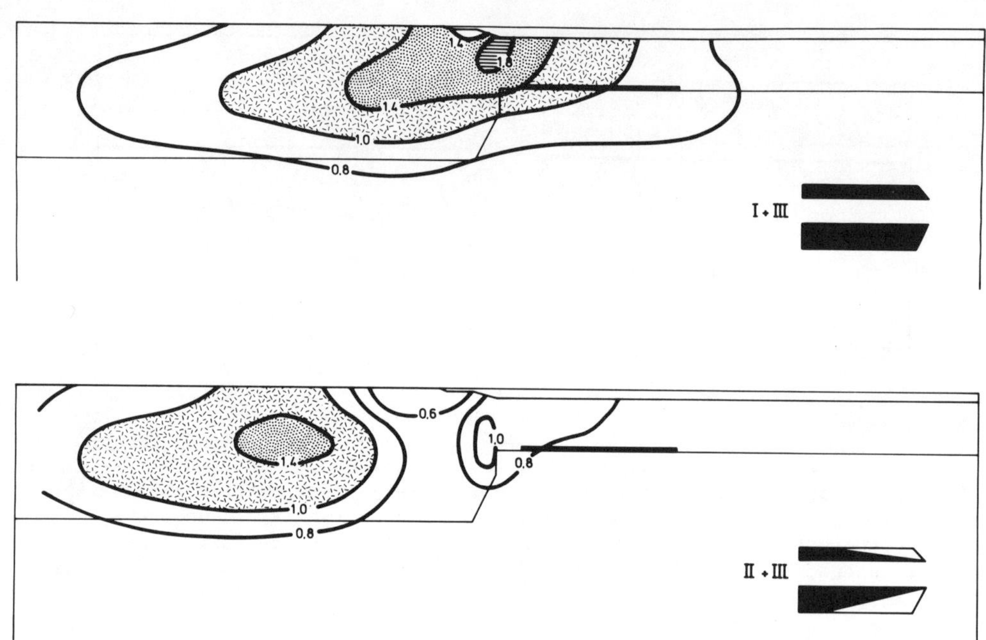

Fig. 12. Maximum shear stress according to the entire loads described in the text. Contours in 10^2 bars. Case I and II correspond to the different decrease of balanced loads, in Fig. 11.

efficient after the cooling effect has ceased and for the gravitational energy isostatic equilibrium was assumed.

The rheology of the modelled cross-section will be represented appropriately by linear diffusional flow and additional nonlinear dislocation creep, which both depend on the thermal regime of the structure in the same way (Stocker and Ashby, 1973). The transition between the dominance of the two mechanisms can be taken at about 1 bar shear stress. Measurements published by Kohlstedt and Goetze (1974) suggest a shear stress power index for nonlinear creep equal to three. A more detailed description of the rheology applied is given by Neugebauer and Spohn (1978). Fig. 11 shows strain rate contours per second which are based on the above mechanisms at a constant stress level of one bar. Thus the contours express essentially the influence of the assumed absolute temperature, while induced shear stresses above one bar will increase the shown rate according to the mentioned power index of the shear stress. In our model the crust will be represented by rates of 10^{-22} to 10^{-23} per second in the context of the mentioned conditions. This means for instance actual crustal strain rates of 10^{-16} s^{-1} at onehundred bar shear stress.

In this study we distinguished two load cases at the passive margins generating a local stress system which gives rise to substantial amounts of structural deformation including the subsidence of the modelled shelf and rise regions.

The first one is the dominant stress system which arises from body forces associated with a variation in thickness of the low density crust.

Hence, the lateral change of density from the ocean ρ_o to that of crustal rocks ρ_c involves differential gravitational loading of the continental crust according to $\Delta\rho_1 = (\rho_c - \rho_o)$, while at the level of the lower continental crust the change from crustal to the upper mantle density ρ_m will cause a buoyancy force proportional to $\Delta\rho_b = (\rho_c - \rho_m)$. Provided that the structure is in isostatic equilibirium of the Airy type, these forces, affecting the continental crust, will balance each other, imprinting a typical stress system on the lithosphere structure. Such a load system acting on an elastic structure has been investigated numerically by Bott and Dean (1972) while Neugebauer (1979) investigated the nonlinear creep behaviour of such a model. He demonstrated that the shape and the topology of the stress system, arising from differential gravitational loading and upthrusting at the margin, will distinctly be affected by the slope of the loads decreasing towards the oceanic crust: There are two reasons for caring about the gradual decline in the amount of loads balancing each other near the juxtaposition of continental and oceanic crust. First the more or less continuously accumulating sediments at the shelf will replace crustal rock densities by those of sediments and will thus imply a reduction of gravitational loading in this area. On the other hand margins of this type are assumed to be fairly near to isostatic equilibrium. If this holds generally during the marginal development, as was concluded by Worzel (1974), we would also expect a decrease of the buoyancy associated with the area of sedimentation. So one will have significant differential load and buoyancy at the continental edge and shelf if only few sediments have accumulated there, but a progressive decrease in these effects for pronounced sedimentary wedges. In this study we assumed two different stages, case I and II in Fig. 11, indicating a decrease of the forces to zero over a horizontal distance of 10 and 80 km respectively. Beside this fundamental load system which will exist at any marginal transition zone, we have investigated as a second load case the possible consequences on the dynamics of the structure of crustal loading along the continental rise. According to the derived net density increase by metastable phase transition in the oceanic crust we will exhibit numerical calculations based on a density increase of 0.1 g/cm taken for 1 km thick layer and a horizontal distance up to 60 km, as indicated in Fig. 11. These values fit into the sequence of models studied by Neugebauer and Spohn (1978). In this paper, however, we relate the dynamic models more to the discussed transition zone and the average amount of observed values of subsidence, thus the values are on the low side.

Thus the boundary conditions of the model structure will be related to the above load system, if there are no other volume or body forces applied to the structure. As the loads due to phase transitions at the oceanic crust will give rise to crustal bending, the crustal structure at the upper left has been fixed. The right end of the oceanic crust is horizontally fixed and vertically free. The model structure beneath the crust can be seen as a representation of a half-space with the rheological character of the upper mantle.

The numerical calculations have been carried out by means of the finite element technique. The principles of this method are outlined elsewhere and are not discussed here. For the numerical representation of creep the so called initial strain method was used. After each solution of the system of equations, which represents the structure, the internal load system will be corrected due to creep deformations. This will be done until steady state deformation rates are reached. Thus the procedure requires 40 to 50 solutions of the system of equations to obtain the steady state approach associated with a given load system. For further explanation of the model stated see Neugebauer and Spohn (1978).

The State of Stress

Observed phenomena like steep faulting related to basin formation, substantial subsidence and in some cases seismic activity give evidence for stress systems at passive continental marginswhich may vary in their local position and with time. The sedimentary record, as for instance compiled by Sheridan (1974), reveals tectonic activities like normal faulting of shelf basins and the shelf edge in the very beginning of their development, while subsidence seems to be a lasting process with laterally variable rates. From a comprehensive study of intraplate tectonism by Sykes (1978) it can be recognized that a so called passive continental margin is not necessarily free of earthquake activity although seismicity is much lower than at active margins.

We will now discuss whether the tectonic activity can possibly be associated with the progressive release of gravitational energy. This will be done on the basis of the prediction of the dynamics by the numerical experiments and in conjunction with observed data.

A load system like the balance of gravitational loads and upthrust induces vertical directed maximum principal stresses of 700 bars at the maximum in the continental crust for the chosen density model. Approaching the transition from continental to oceanic crust the orientation of the maximum principal stress turns towards horizontal orientation and to an amount of about 120 bars in the part of the model which represents the oceanic crust. Clearly, the amount of the horizontal maximum principal stress in the oceanic crust is affected by the existence of the horizontally fixed boundary at the left. It seems, however, that this is the most acceptable condition in the context of the dynamics of a continuous plate.

According to the stresses which could have an effect upon the tectonic development of the margin, the distribution of maximum shear stress throughout the structure would be of great interest. Fig. 12 shows two typical patterns of maximum shear stress as a result of the numerical calculations for steady state creep conditions. The stresses are contoured in 10^2 bars. The difference between the two states is a steep decrease of standard loading towards the juxtaposition of the two crusts, I, and a gradual one over a horizontal distance of 80 kilometers, II, in accord with Fig. 11.

In both models the loads due to lower crustal metastable phase transition beneath the continental rise, III, area are included. The share of maximum shear stress which is contributed by the latter loads (III) is not greater than 30 bars and mainly confined to the zone of loading. Thus the stresses shown are dominated by the differential load system.

From the distribution of maximum shear stress in Fig. 12 we can conclude that there exists a distinct concentration of shear stresses across the marginal structure which is associated with the gravitational energy involved. The maximum of 150 - 200 bars (above) is confined to the narrow transition zone between the two crustal types. While there is a steep gradient of shear stresses towards the oceanic structure, we observe a rather high level within the upper half of the continental structure, just where the development of the continental shelf is expected, and a gradual decrease towards the interior of the continent.

A second conclusion that results from the numerical experiments is that of a possible displacement of the shear stress concentration into the continental structure as a function of the gradual decrease of the standard loads of the margin. The

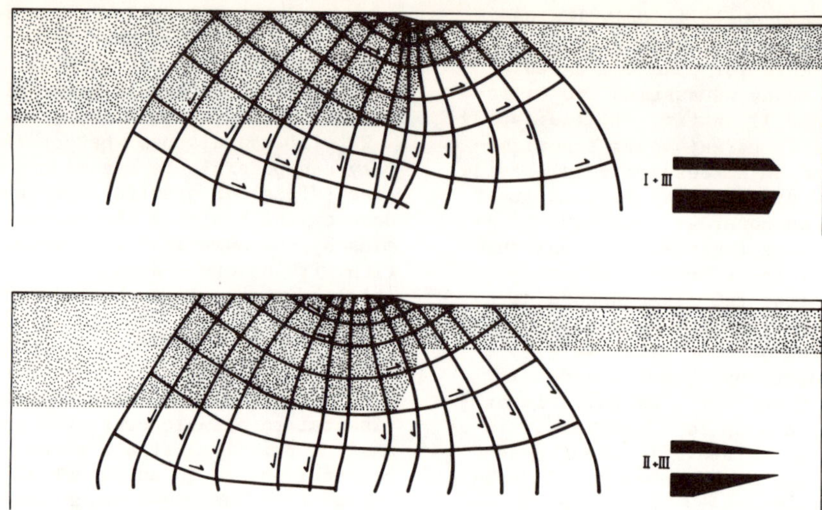

Fig. 13. Maximum shear stress slip lines at the modelled transition corresponding to model
I + III and II + III in Fig. 12.

principal maximum is situated where the linear decline of the load system starts, while a second
maximum remains where the loads attain zero,
Fig. 12 (below). Examining the tectonic activities across the margin in the context of the predicted shear stresses, one should place emphasis
on the following observations.

The existence of local concentrations of maximum
shear stress favours the deformation by creep
there, keeping in the mind that the rate depends
on the third power of the maximu shear stress. The
higher shear stress areas appear to coincide with
observed zones of substantial subsidence very
well.

Furthermore the variable position of high shear
stresses due to the decline of gravitational energy gives rise to the assumption that differential
gravitational loading and equalizing upthrust will
be able to influence the tectonic development locally along a broad horizontal distance over the
shelf and slope of the margin. This view will be
supported by the course of the maximum shear
stress slip-lines, Fig. 13, corresponding to
Fig. 12. Here high angle faulting will be favou-
red either around the transition between the two
crustal types or landwards from this position. It
will be obvious that the predicted amount of
shear stress is not sufficient to cause faulting,
however, it might be high enough to reach the level to reactivate the preexisting tectonic framework in a specific sense. This could be important
for the tectonically controlled subsidence across
margins.

Finally Neugebauer (1979) found from a compilation of earthquake epicenters, that seismicity at
old margin structures corresponds to the shear
stress pattern of the type illustrated in the lower diagram of Fig. 12, while young margins seismicity fits the distribution the upper diagram
of Fig. 12. This supports the view that the re-

lease of gravitational energy at the margin is a
long lasting process with a gradual decrease, and
the transition between continental and oceanic
crust is subject to a progressive decay. The
structural changes as predicted by the numerical
experiments in relation to the discussed load systems will be presented and discussed in the
light of the corresponding structural data in the
following section.

Subsidence across Marginal Structures

The variable state of stress predicted by the
numerical experiments provides a suitable tool to
evaluate the influence of the specific load system
on the tectonic activity along a marginal structure. However, the numerical model is a continuous
representation of the natural marginal structure,
so processes like discontinuous deformations along
faults or similar behaviour are not covered by the
calculations presented. Thus the mechanical response of the modelled structure will be a continuous, steady state creep deformation. In order to
quantify the progressive decay of the modelled
marginal structure, we focus our attention on the
lateral change of the rate of subsidence. Again we
distinguish two different local systems: the differential loading and upthrust governing the overall flow pattern, and the loads associated with
the metastable phase transition which have a substantial effect on the calculated rates. The
creep deformations of the structure indicate horizontally oriented flow for the continental
structure, turning into a predominant vertical
direction approaching the transition to the oceanic structure. The maximum rate is reached at the
oceanic part of the structure, where the flow
turns to the horizontal direction for deeper
parts of the modelled oceanic upper mantle.

According to the loads, Fig. 11, case I and II,

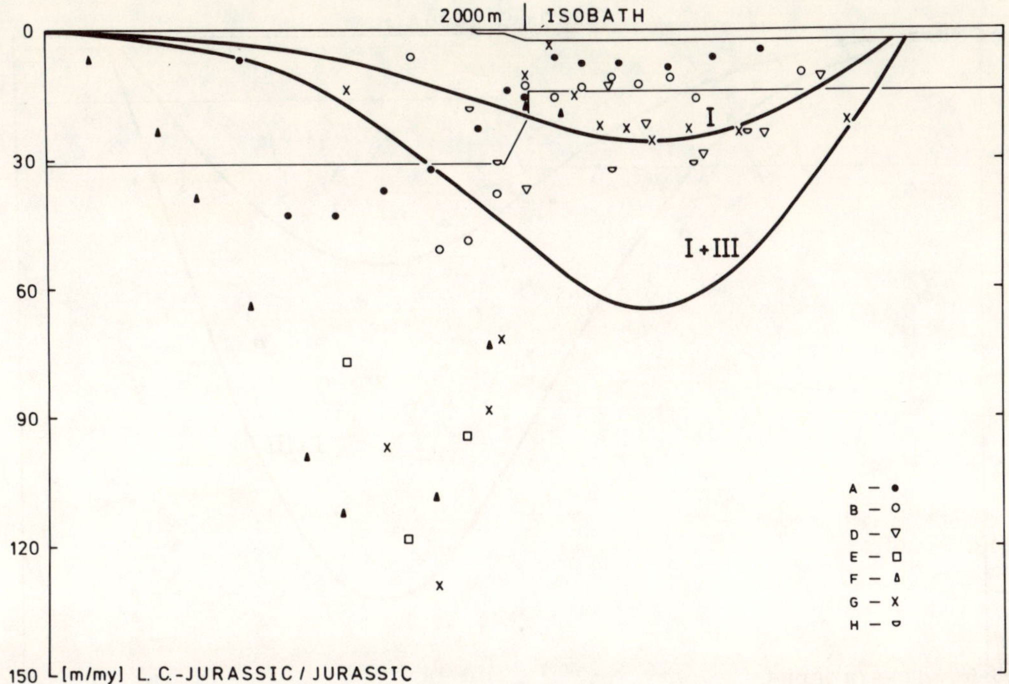

Fig. 14. Average Jurassic or Upper Cretaceous to Jurassic rates of sedimentation after Sheridan (1974). Cross section: A - Labrador, B - Newfoundland, D - Laurentian channel, E - Nova Scotia, F - Gulf of Maine-Georges Bank, G - New Jersey, H - Cape Hatteras. The horizontal distance has been normalized to that of the model. As a common point for all profiles the position of the 2000 m Isobath was chosen. Curve I represents modelled rate distribution for differential loading and upthrust, Fig. 11 case I, curve I + III for additional loads at the oceanic structure due to metastable phase transition.

the numerical models yield corresponding characteristic distributions of rates of subsidence for the top of the structure. These rates, which represent the progressive decay of the structure associated with isostatic conditions only, are shown in Fig. 14 and 15, I and Fig. 16, II. Thus the steep decline of balanced loads causes a continuous increase of the rate of subsidence from zero at the left boundary to 20 m/Ma at the continental margin proper and a maximum rate of about 26 m/Ma at the oceanic structure adjacent to the margin. In contrast to model I the gradual decline of loads in model II provides rates of 15 m/Ma and 19 m/Ma at the same positions. This systematic difference agrees with the concept of progressive decrease of gravitational energy across a passive continental margin. As a consequence, model I could be associated with a young, model II with a mature marginal structure. The observed decline of the rate of subsidence along the cross section suggests that the progressive decay will be a lasting process which affects the development of a continental margin substantially. This will be demonstrated using the subsidence history of a marginal structure which is documented by its sedimentary deposits. For this purpose data from the North American Atlantic continental margin have been chosen for reference because

their structure is one of the best known and one of the oldest (about 210 - 170 Ma). Hence it is suitable to study the characteristic features of a long term development.

If one whishes to discuss the history of the entire margin comprising shelf, slope and rise, the most detailed possible dating of the development is the subdivision into Jurassic, Cretaceous and Tertiary times. This yields a rather rough picture of the tectonic activity compared with the high resolution of observed subsidence rates as demonstrated for single wells by Whitten (1976). However, the procedure chosen is sufficient to emphasise the essential features of the subsidence history across a continental margin.

Sheridan (1974) presented a comprehensive compilation of the knowledge of stratigraphic data across the above continental margin. These data provide average rates of Jurassic, Cretaceous and Tertiary sedimentation along the marginal cross section, which are taken to be equal to the mean rate of subsidence for the subsequent consideration.

Jurassic Subsidence. According to the geophysical and stratigraphic data the major onset of drastic subsidence along the Atlantic continental margin of North America can be dated as Jurassic.

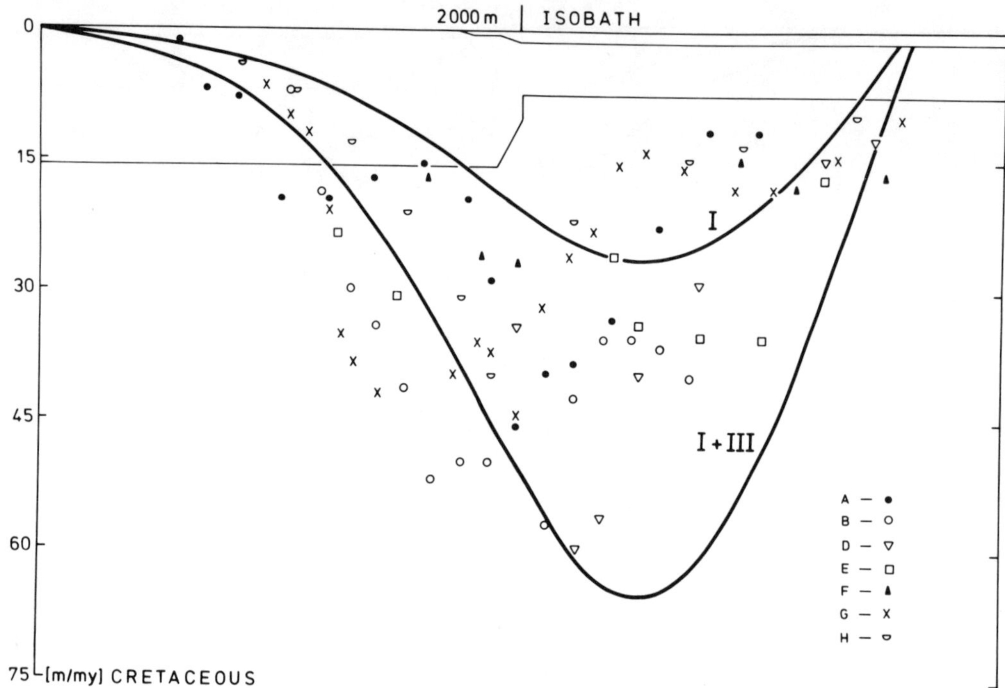

Fig. 15. Average Cretaceous rate distribution, for explanation see Fig. 14. The modelled rates are the same as in Fig. 14.

The pre-Jurassic basement is interpreted to be of a blockfaulted, rifted style with the accumulation of great thicknesses of shallow water sediments in subsided basins along the entire margin. Such faulting is assumed to be initiated by the rift forming process and it seems to control basin development at the young margin. The distinguishable blocks, broken by faults, cause different stratigraphic units to thicken or thin in response to block movement.

Fig. 14 summarizes the average Jurassic and in some cases Jurassic to lower Cretaceous rates along seven distinct sections across the Atlantic continental margin of North America after Sheridan (1974). The most outstanding characteristics are the probable fault controlled rather high rates on the shelf which correspond to basin formation, while the rates on the slope and rise are apparently at a very low level. In terms of the lateral change of rates as predicted by the numerical models, curve I, the progressive decay of the modelled structure appears to be sufficient to fit the trend of the data at the rise and the slope. The additional effect of loads due to metastable phase transition on the calculated rates, curve I and III, is not required to improve the correspondence shown. Comparing the calculated rates and the observed data one would suggest that the fault controlled basin subsidence is superimposed upon a continuous marginal subsidence. The latter might be adequately explained by modelled rates in responce to the gravitational load system associated with the marginal structure.

Cretaceous Subsidence. The Cretaceous rates, shown in Fig. 15, indicate a distinct change in the spatial distribution of subsidence. While during the Jurassic, fault controlled subsidence is concentrated at the shelf, the maximum of Cretaceous rates is evidently concentrated around the outer shelf and continental slope. Obviously the fault controlled contribution to the subsidence ceases with time and the continuous flexure of the structure gradually becomes predominant, because both the continental as well as the oceanic crust are forced to subside with rates of the same order of magnitude.

Again the type of modelled subsidence rate curves as in Fig. 14 have been used to interpret the characteristic features of Cretaceous data in terms of structural loading, curve I, and additional phase transition loads at the rise, curve I + III. The latter load case causes an increase of the maximum rate from 26 to 65 m/Ma overall and from 20 to 51 m/Ma at the modelled slope. While the low values are due to the squeezing effect in isostatic equilibrium, the increase in the rates corresponds to the loads in the sense of bending. There is a gross similarity between the predicted and derived rates, while the spatial trend of the two appears to be displaced in such a way that the maximum of the numerical results is more distant from the continental edge. It should be emphasised that the rate curve according to progressive decay of the structure due to the release of gravitational energy presents a lower bound for the derived average rates, at least for the shelf. This re-

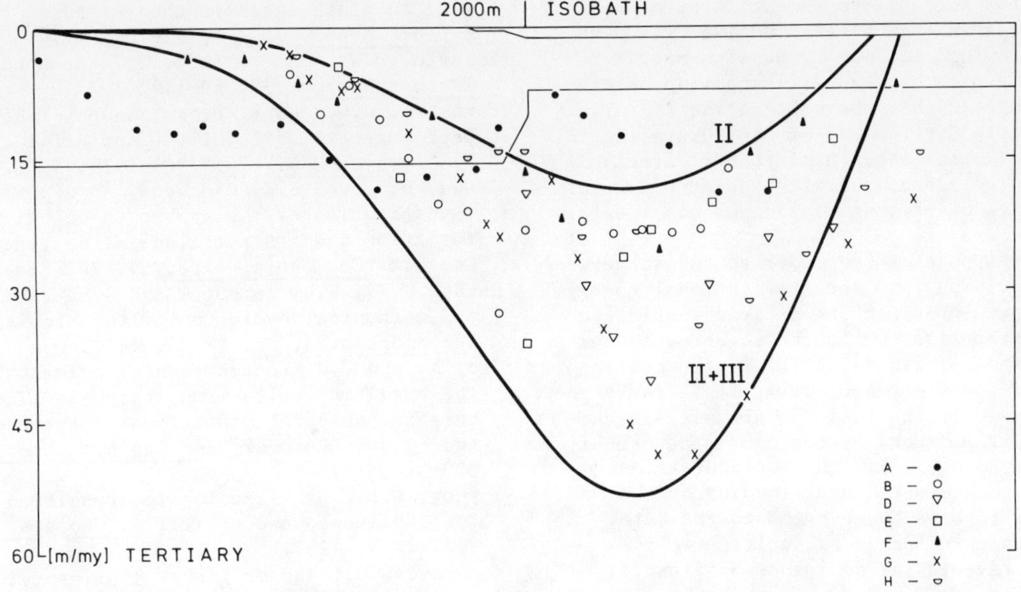

Fig. 16. Average Tertiary rate distribution for the cross sections in Fig. 14. Here the model-led rates, curve II and II + III, correspond to the decrease of differential loads and upthrust shown in case II, Fig. 11 and additional loads at the oceanic structure due to metastable phase transition, III.

quires, in terms of the chosen model, an additional driving mechanism, like the proposed metastable phase transition. The resulting increase of rates along the structure implied by additional loads at the rise appears to be of geodynamic relevance.

Tertiary Subsidence. Considering the third, Tertiary, compilation of derived rates, Fig. 16 we recognise a generally lower rate of subsidence, and the maximum rates are evidently shifted from the slope area oceanward to the continental rise.

As with the previous sediment accumulations we applied the progressive decay load case, model II, Fig. 11, as a basis for the numerical rates, and superimposed rise loading, curve II + III, in Fig. 16. Now the distribution of derived data agrees reasonably well with the area bounded by the numerical predictions. Again, as for the Cretaceous case, additional load is required which exceeds the effect of the progressive decay of the structure. Here the rates rise from 19 to 53 m/Ma at the maximum and from 16 to 44 m/Ma at the slope. The obvious coincidence of the data shown suggest that the typical features of Tertiary subsidence might be interpreted by the modelled rates. This means the average Tertiary rates might be associated with the effect of the structural loads and a regionally varying amount of loading of the continental rise due to gradual transformation of metastable rocks to a stable phase.

Tectonic Development. The tectonic development of a continental margin may now be discussed from the point of view of the average sedimentation ra-

tes across the margin, and in the context of the spatial and temporal efficiency of proposed mechanisms for subsidence. The spatial distribution of tectonic activity as inferred for the stratigrafic periods discussed above demonstrated that there are, more or less, typical styles of tectonic activity that change during time. Furthermore regional activities, for example at the shelf or rise, are not necessarily representative of the entire marginal structure. Thus it appears that a superposition of primary causes such as thermal contraction, progressive decay of the transition in crustal thickness and the transition of metastable phases is probably required in addition to the secondary effect of sedimentary loading.

On the basis of the studies of the kinetics of metastable phase transitions and the modelled dynamics of passive continental margin we suggest that the tectonic development of a passive margin is generally a superposition of a fault controlled high subsidence rate in the early stages and the response of a crust of varying thickness to different loads. The body forces associated with variation in thickness of the crust will affect the entire development of the margin.

The investigations have shown that the transition from metastable basalt to a stable phase might occur in the lower oceanic crust. The corresponding increase in density may reach a sufficient amount within geologically relevant times to affect the tectonic development of the margin substantially. Together with the progressive decay of the margin it serves to explain the tectonic development of the mature margin.

In connection with the rough estimates of sedimentation rates that have been made we will suggest the following sequence of passive margin development. The time constants for cooling of crustal rocks suggest that thermal contraction might affect the early development of the future continental shelf. On the other hand the gravitational potential energy associated with the structure of the margin will be present as soon as continental separation occurs.

The creep response of the model to the structural loads resulting from the gravitational energy yields an upper bound for the derived subsidence rates across the rise, while the stresses favour the reactivation of faults in the continental crust. During the Cretaceous large scale fault movement ceases while the tectonic activity is characterized more and more by the continuous subsidence curves. However, now the horizontal change of rates due to the structural loading of the crust appears to be a lower bound to the data. The disappearing influence of faulting in the shelf region reveals the dominance of flexural response during the Tertiary. Here the data fi the bandwidth and shape of modelled rate variations bounded by the response to differential gravitational loading and buoyancy on one side and additional metastable phase transition loading on the other side. Thus the numerical calculations suggest in accord with the average subsidence rates a progressive decrease of the decay of the margin but a relative increase of the loads resulting from metastable phase transition.

Acknowledgement. This work was supported by the Deutsche Forschungsgemeinschaft.

References

Ahrens, T.J. and G. Schubert, Gabbro-Eclogite reaction rate and its geophysical significance, Rev. Geophys., 13, 383-400, 1975.

Bott, M.H.P. and D.S. Dean, Stress systems at young continental margins, Nature, 235, 23-25, 1972.

Clark, S.P. and A.E. Ringwood, Density distribution and constitution of the mantle, Rev. Geophys., 2, 35-88, 1964.

Cohen, L.H., K. Ito and G.C. Kennedy, Melting and phase relations in an hydrous basalt to 40 kbars, Am. J. Sci., 265, 475-518, 1967.

Dewey, J.F. and J.M. Bird, Mountain belts and the new global tectonics, J. Geophys. Res., 75, 2625-2647, 1970.

Drake, C.L., M. Ewing and G.H. Sutton, Continental margins and geosynclines: the east coast of North America north of Cape Hatteras, Phys. Chem. Earth, 5, 110-198, 1959.

Engel, A.E.J., G.C. Engel and R.G. Havens, Chemical characteristics of oceanic basalts and the upper mantle, Geol. Soc. Am. Bull., 76, 719-734, 1965.

Falvey, D.A., The development of continental margins in plate tectonic theory. Aust. Petr. Explor. Assoc. J., 14, 95-106, 1974.

Forsyth, D.W., The evolution of the upper mantle, Tectonophysics, 38, 89-119, 1977.

Forsysth, D.W. and F. Press, Geophysical tests of petrological models of the spreading lithosphere, J. Geophys. Res., 76, 7963-7979, 1971.

Green, D.H. and A.E. Ringwood, An experimental investigation of the gabbro to eclogite transformation and its petrological applications, Geochim. Cosmochim. Acta, 31, 767-833, 1967.

Haxby, W.F., D.L. Turcotte and J.M. Bird, Thermal and mechanical evolution of the Michigan Basin, Tectonophysics, 36, 57-75, 1976.

Ito, K. and G.C. Kennedy, An experimental study of the basalt-eclogite transition, in: The structure and physical properties of the crust, edited by J.C. Heacock, AGU Geophys. Monogr. 14, 303-314, 1971.

Joyner, W.B., Basalt-eclogite transition as a cause for subsidence and uplift, J. Geophys. Res., 72, 4977-4998, 1967.

Kennedy, G.C., The origin of continents, mountain ranges and ocean basins, Am. J. Sci., 47, 491-504, 1959.

Kittel, C., Introduction to solid state physics, 5th ed., Wiley New York, 599 p., 1976.

Lovering, J.F., The nature of the Mohorovicic Discontinuity, Trans. Am. Geophys. Un., 39, 947-955, 1958.

Manning, J.R., Diffusion kinetics of atoms in solids, D.v.Nostrand, Princeton, 276 p., 1968.

McDonald, G.J.F. and N.F. Ness, Stability of phase transitions within the earth, J. Geophys. Res., 65, 2173-2190, 1960.

Mitchell, A.H. and H.G. Reading, Continental margins, geosynclines, and ocean floor spreading, J. Geol., 77, 629-646, 1969.

Neugebauer, H.J., Passive continental margins: development and state of stress (abstract), XVII IUGG Canberra, 1979.

Neugebauer, H.J. and T. Spohn, Late stage development of mature Atlantic-Type continental margins, Tectonophysics, 50, 275-305, 1978.

O'Connell, R.J. and G.H. Wasserburg, Dynamics of the motion of a phase change boundary to changes in pressure, Rev. Geophys., 5, 329-410, 1967.

Rao, C.N.R. and K.J. Rao, Phase transitions in solids, McGraw Hill, New York, 330 p., 1978.

Ringwood, A.E., Composition and Petrology of the Earth's mantle, McGraw Hill, New York, 618, p., 1975.

Salmang, H. and H. Scholze, Die physikalischen und chemischen Grundlagen der Keramik, 5th ed., Springer, Berlin, 450 p., 1968.

Scrutton, R.A., Crustal structure at the continental margin south of South Africa, Geophys. J. R. Astr. Soc., 41, 601-623, 1976.

Sheridan, R.E., Atlantic Continental Margin of North America, in: The Geology of continental margins, edited by C.A. Burk and D.L. Drake, p. 391-407, Springer, Berlin, 1974.

Spohn, T., Transformation metastabilen Basalts als mögliche Ursache epirogenetischer Bewegungen der

Kruste, _Reports Inst. Meteorology and Geophysics_, Frankfurt, 39, 155 p., 1979.

Spohn, T. and H.J. Neugebauer, Metastable phase transition models and their bearing on the development of Atlantic-type geosynclines, _Tectonophysics_, 50, 387-412, 1978.

Stocker, R.L. and M.F. Ashby, On the rheology of the upper mantle, _Rev. Geophys. Space Phys._, 11, 391-426, 1973.

Sykes, L.R., Intraplate seismicity, reactivation of preexisting zones of weakness, alkaline magmatism, and other tectonism postdating continental fragmentation, _Rev. Geophys. Space Phys._, 16, 621-688, 1978.

van de Lindt, W.J., Movement of the Mohorovicic discontinuity under isostatic conditions, _J. Geophys. Res._, 72, 1289-1297, 1967.

Whitten, E.H.T., Cretaceous phases of rapid sediment accumulation, continental shelf, eastern USA, _Geology_, 4, 237-240, 1976.

Wetherill, G.W., Steady-state calculations bearing on geological implications of a phase transition Mohorovicic discontinuity, _J. Geophys. Res._, 66, 2983-2993, 1961.

Worzel, J.L., Standard oceanic and continental structure, in: _The Geology of continental margins_, edited by C.A. Burk and C.L. Drake, p. 59-66, Springer, Berlin, 1974.

SUBSIDENCE HISTORY AND TECTONIC EVOLUTION OF
ATLANTIC–TYPE CONTINENTAL MARGINS

M. S. Steckler[1] and A. B. Watts

Lamont–Doherty Geological Observatory and Department of Geological Sciences
of Columbia University Palisades, New York 10964

Introduction

The continental margins which border the
Atlantic, Indian and Arctic oceans are characterized by substantial thicknesses of seaward dipping sediments. The relatively old margins off
North America and Africa, which are characterized
by thicknesses in excess of >7 km, appear to
overlie a faulted basement. The relatively young
margins of the Red Sea and western Mediterranean,
on the other hand, are characterized by thinner
sediments (<4 km) which overlie a faulted continental basement.

Dietz (1963) suggested that the loading of
sediments on the continental slope and rise
would produce a regional downwarp or flexure of
the adjacent continental crust. Gunn (1944),
Walcott (1972), Cochran (1973), and Watts and
Ryan (1976) have quantitatively shown that substantial thicknesses of shallow and deep water
sediments may accumulate at a margin simply as
a result of sedimentary loading. They used
simple flexural models with the lithosphere
responding to loads as an elastic plate overlying
a weak fluid.

Biostratigraphic data from deep commercial
boreholes in the outer continental shelf off
eastern North America (Scholle, 1977; Gradstein
et al., 1975; Jansa and Wade, 1975), northwest
Africa (Von Rad and Arthur, 1979) and the western
Mediterranean (Cravatte et al., 1974) show that
much of these sediments were deposited in continental or neritic to inner shelf marine environments. While large thicknesses of relatively
shallow water sediments cannot be produced by
sedimentary loading alone, all hypotheses for the
origin of the subsidence of continental margins
are in agreement that sedimentary loading must
contribute to a major portion of the subsidence.

A useful approach to the problem, therefore,
is to quantitatively account for the effects of
sedimentary loading during margin evolution
(Watts and Ryan, 1976; Steckler and Watts, 1978;

[1]Present Address: Bullard Laboratories, Univ. of
Cambridge, Madingley Rise, Madingley Road, Eng.

Keen, 1979). The resulting tectonic subsidence
of the margin is that part of the subsidence not
caused by the weight of sediment and water loads
through time.

Most studies now consider that the dominant
mechanism affecting the tectonic subsidence of
continental margins is thermal contraction following heating and thinning of the crust at the time
of initial rifting (Sleep, 1971; McKenzie, 1978).
Sleep (1971) suggested the subsidence was caused
by crustal thinning following uplift and erosion
at the time of initial rifting. This model cannot, however, explain the substantial thicknesses
of sediments observed off eastern North America
(Steckler and Watts, 1978). Another difficulty
is that there is little stratigraphic evidence at
margins for large amounts of uplift and erosion
at the time of initial rifting (Kent, 1976). In
fact, initial rifting appears to be characterized
by normal faulting (de Charpal et al., 1978).
McKenzie (1978) proposed a model, which addressed
the uplift problem, in which the lithosphere
undergoes a passive uniform extension or stretching at the time of rifting. Extension causes
necking or thinning of the crust which subsequently subsides with time. This model appears
to be in agreement with the evidence for extension (listric faults and rotated fault blocks)
during the early rifting history of margins.

The purpose of this paper is to review the
principal factors that contribute to the subsidence of Atlantic-type continental margins. We
will briefly discuss the effects of sediment
compaction, paleobathymetry, and sea-level
changes. The effects of sedimentary loading and
thermal contraction will be discussed using simple
thermal and mechanical models for the development
of a margin. The overall objective of the paper
is to better understand the origin and evolution
of Atlantic-type continental margins.

Structure

The development of seismic techniques which
use the multi-channel array along with commercial
and DSDP/IPOD drilling have greatly improved

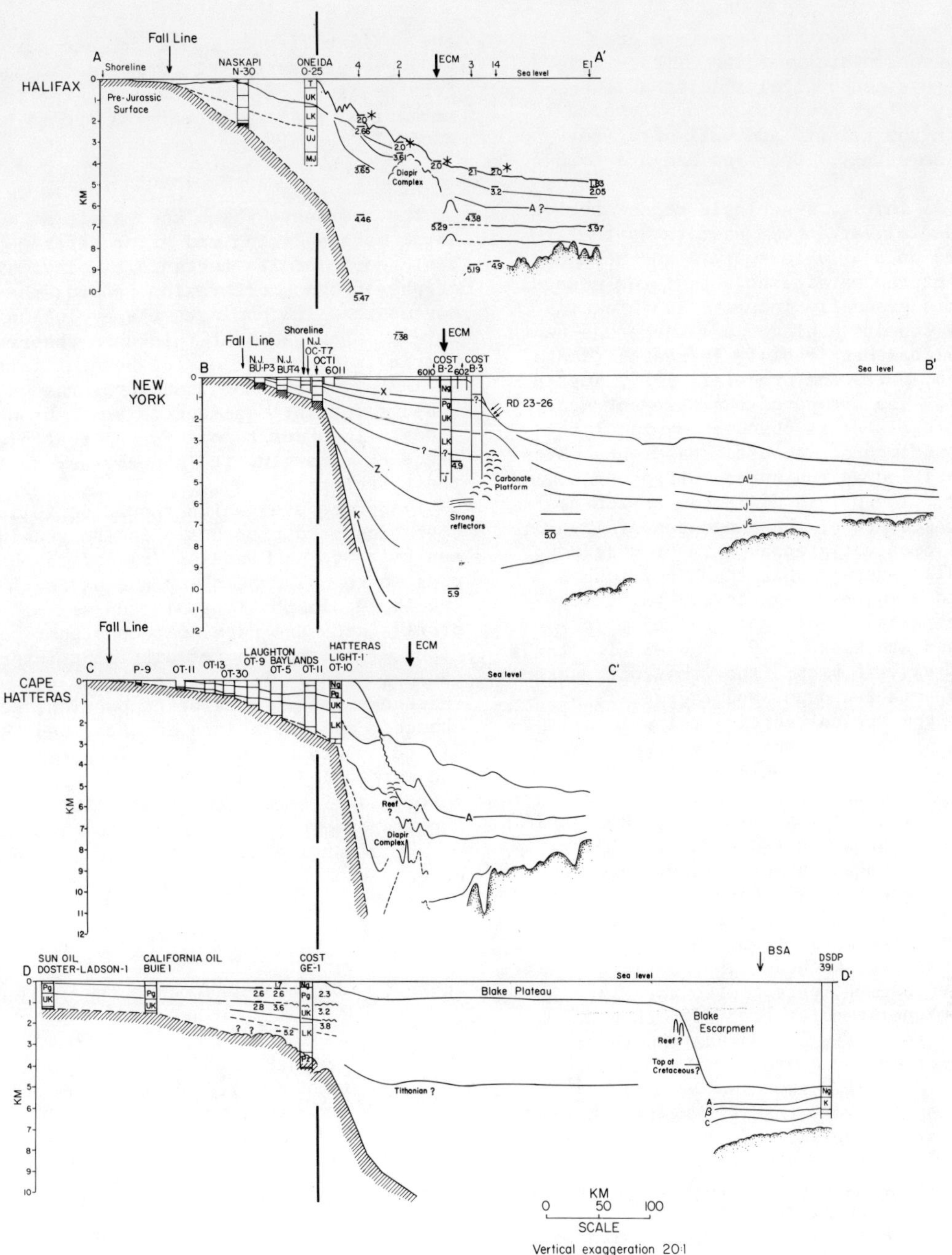

Figure 1. Summary geologic cross sections along the continental margin off North America aligned along the hinge zone (heavy dashed line). The stratigraphy of the land boreholes are from Brown et al. (1972). The stratigraphy of the wells offshore from Nova Scotia are from Williams (1975), the COST B-2 well from Scholle (1977) the COST GE-1 well from Amato and Bebout (1978) and DSDP site 391 from Scientific Party (1976). The solid lines in the four sections indicate prominent seismic reflectors identified on nearby multi-channel seismic profiles (Given, 1977; Grow et al., 1979; Grow and Markl, 1977; Buffler et al., 1979; Dillon et al., 1979; Sheridan, 1976). The arrow labeled ECM refers to the position of the East Coast Magnetic Anomaly and BSA to the Blake Spur Anomaly.

knowledge of the sedimentary structure of
Atlantic-type continental margins. There have
now been numerous geophysical studies of these
margins. Only a few, however, have attempted to
integrate both the seismic and well data (for
example, Schlee et al., 1976; Von Rad and Arthur,
1979).

We summarize in Fig. 1 geologic cross-sections
of the continental margin off eastern North
America, based on available seismic and well data.
Each profile of the margin shows that the coastal
plain sediments gradually increase in thickness
from the fall line to a hinge zone where the depth
to continental basement rapidly increases (Jansa
and Wade, 1975; Watts and Steckler, 1979; Austin
et al., 1980). The nature of the basement seaward of the hinge zone is obscured by the large
thickness of sediments. Acoustic basement, which
has been identified as the upper surface of
oceanic layer 2, occurs at distances as far as
100 to 300 km seaward of the hinge zone (Fig. 1).

Studies of biostratigraphic data from deep
commercial wells suggest that the hinge zone
corresponds to a major change in crustal thickness across margins. Geoid and gravity modelling (Watts and Steckler, 1979; Grow et al., 1979;
Hutchinson et al., in prep.) show that off eastern
North America both the crust and thermal lithosphere thins from typical continental values landward of the hinge zone to oceanic values seaward
of the hinge zone. In the Gulf of Lion, the
hinge zone is indicated by an abrupt increase in
the depth to continental basement. Steckler and
Watts (1980) have shown that a significant lateral heat flow is required across the hinge zone
in order to explain the well data. In the Gulf
of Aden, there is a 3 to 4 km basement offset
between the uplifted margins and the magnetic
quiet zone offshore. Cochran (in press), who has
correlated the basement offset with the hinge
zone and the magnetic quiet zone with thinned
continental lithosphere, found that the thinning
across the hinge zone was greater than a factor
of 2.

Fig. 1 shows, however, that off eastern North
America there are large variations in the overall
distribution of sediments along the margin. Off
New York and Florida the shelf break in slope
extends seaward of the hinge zone while off
Halifax and Cape Hatteras it is slightly landward
of the hinge zone. The larger distance to the
shelf break off New York may be due in part to
the supply of terrigeneous and bioclastic sediments to the margin in the Early Jurassic, prior
to the deposition of Horizon β (Tucholke and
Mountain, 1979). In fact, the shelf break off
New York appears to have extended even further
seaward than at present and may have been eroded
by 20–30 km by counter-currents which began in
the middle Tertiary (Grow et al., 1977).

These considerations suggest that while the
overall tectonic evolution of margins may be
controlled by the hinge zone, the position of the
shelf break and the width of the shelf depend on
the availability of sediments and surficial
geologic processes. These effects should therefore be isolated and removed if the thermal and
mechanical history of margins are to be understood.

Sediments

The sediments which accumulate at a continental margin during and after rifting provide the
best record of its tectonic history. The stratigraphy of the shelf region records the vertical
movements which dominate the evolution of the
margin. The sediment thickness observed at a
margin is a consequence of several factors,
including compaction, sea-level changes and the
response of the basement to sediment and water
loads. In order to obtain the tectonic subsidence at a margin, it is necessary to account for
these factors.

The procedures which should be followed have
been discussed previously in the geohistory
analysis of Van Hinte (1978) and the backstripping approach of Steckler and Watts (1978).
Available biostratigraphic and seismic reflection
profile data are used to reconstruct the stratigraphy of the margin for different intervals of
geologic time. In reconstructing the sedimentary
thickness the effects of compaction, sea-level
changes, and variations in water depths of deposition should be included. Then, the sediment
and water loads are backstripped for the various
intervals of geological time using a model for
the response of the basement to these loads.

The general equation for backstripping can
be written

$$Y = \Phi * \left[S^* \frac{(\rho_m - \rho_s)}{(\rho_m - \rho_w)} - \Delta_{SL} \frac{\rho_w}{(\rho_m - \rho_w)} \right] + W_d - \Delta_{SL} \quad (1)$$

where Y = depth of basement without the effects
      of sediment and water loads

  S^* = sediment thickness corrected for
      compaction

  ρ_s = mean density of sediments

  ρ_m = mean density of mantle

  ρ_w = mean density of water

Δ_{SL} = sea level relative to the present day

  Φ = basement response function

The first terms in brackets in equation (1) therefore represents the effect of the response of the
basement to the weight of sedimentary infill and
changes in sea-level, while the last term represents the water depth relative to the present day
sea level. The last term is that part of the
tectonic subsidence which has not been filled in
and loaded by sediments. In order to backstrip
both the sediment and water loads, the basement
response must be estimated during the development
of a margin.

a) Compaction Correction.

As the weight of overburden increases, sediments will expel pore fluids and compact. Thus

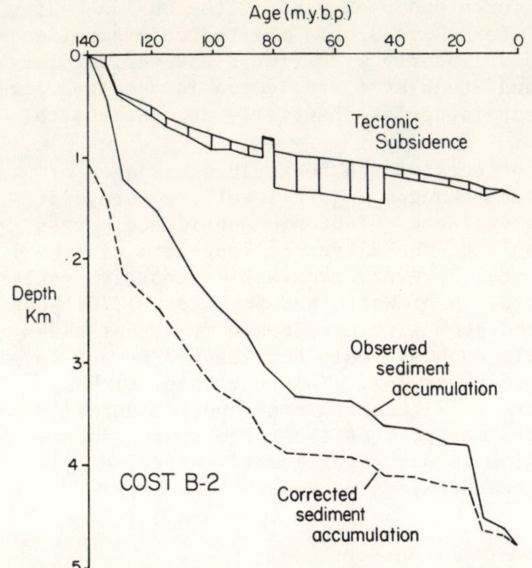

Figure 2.   Tectonic subsidence and sediment
accumulation at the COST B-2 well through geologic time. The tectonic subsidence has been
computed assuming an Airy model (Fig. 2) for
the response of the basement to sediment loads.
The range in the tectonic subsidence indicate
variations in paleobathymetry through time.
The solid line indicates the observed sediment
accumulation and the dashed line indicates the
observed sediment accumulation corrected for
the effects of compaction.

the present day thicknesses of older strata at a
margin do not reflect the actual amount of
subsidence during those time intervals. In order
to correct for compaction it is therefore necessary to know how the porosity varies with depth
during the evolution of the margin. Lithologic
logs (sonic, density, porosity) only provide
information, however, on the present day variation of porosity with depth.

The actual procedures which should be used to
correct for compaction are complicated because
the processes by which sediments compact are not
fully understood. The compaction of sediments
(for example, Engelhardt, 1977) is not only a
mechanical process which depends on the depth of
burial of the sediments. The chemical processes
of mineral solution and recrystallization play an
increasingly larger role with depth. These
processes, like the mechanical process, gradually
cause a reduction in the pore volume of sediments.

Steckler and Watts (1978) assumed the simple
model of a constant porosity versus depth profile
through geologic time. Sclater and Christie
(1980) used a similar approach but allowed each
lithology to compact along separate exponential
porosity versus depth curves.

We illustrate in Fig. 2 the magnitude of the
compaction correction at the COST B-2 well off
New York (Fig. 1). Fig. 2 shows the total sedi-

ment accumulation in the well (solid line) along
with the sediment thickness corrected for compaction (dashed line). The compaction correction
was calculated using a porosity versus depth
profile based on Smith et al. (1976) and
Rhodehammel (1977). The size of the compaction
correction is substantial, reaching nearly 1 km
for sediments which formed 140 m.y.B.P.

The calculation of the compaction correction
in Fig. 2 assumes, however, the porosity remained
constant beneath the base of the well and the
porosity versus depth profile is independent of
lithology. The lack of knowledge of the porosity
between the base of the well and the basement
reduces the accuracy of the compaction correction.
Sediments such as sandstones and shales compact
with increasing depth in a similar way but because of a number of factors, including the
depositional history and thickness of the overburden, this compaction may vary. We believe,
however, that the form of the corrected sediment
accumulation curve shown in Fig. 2 would not be
altered significantly if a different porosity
versus depth profile had been used for each
lithology in the well.

b) Paleobathymetry

The tectonic subsidence of a margin provides
a depression in which sediments infill. The
sediments do not necessarily, however, fill the
depression to sea-level. The water depth which
remains, relative to present day sea level, is
therefore, an important part of the tectonic
subsidence of the margin. Unfortunately, estimates of the water depth through time (paleobathymetry) are difficult to obtain and constitute a major uncertainty in obtaining the
tectonic subsidence of a margin.

Although the distribution of some faunal
assemblages seem to be related directly to depth
(for example, Van Hinte, 1978) the availability
of direct water depth indicators is limited.
Estimates are made either by direct comparisons
to present day occurrences of certain species or
assemblages, or by quantitatively determining the
relative abundance of, for example, benthonic/
planktonic forams and radiolarians or ostracods.
In general, estimates of water depths are most
accurate in regions where recent faunal assemblages are well known (for example, Gulf Coast,
U.S., eastern Mediterranean) and in sediments
formed in shallow-water environments (neritic
and/or shelf facies). Estimates are less precise
in older sediments and in sediments formed in
deep-water environments.

The paleobathymetry at the COST B-2 well is
shown in Fig. 2. This well was characterized by
relatively shallow water depths early in its
evolution. Water depths increased significantly
during the Late Cretaceous to Early Tertiary.
Relatively shallow water depths characterize the
Middle and Late Tertiary history at the well.
Fig. 2 shows that paleobathymetry modifies the

shape and amplitude of the tectonic subsidence at
the well. The effects of paleobathymetry at
this well are, however, still small in comparison
to the overall shape and amplitude of the tectonic subsidence.

c) Sea Level

The variations in sea-level which occur
through time (for example, Vail et al., 1977;
Pitman, 1978) contribute in two main ways to the
tectonic subsidence at a margin. The height of
sea-level is the reference surface for paleobathymetry and sediment thickness estimates.
Thus changes in sea-level with respect to the
present day are required in order to provide a
reference surface for the tectonic subsidence.
In addition, the excess water associated with a
highstand in sea-level acts as a load on the
basement and depresses it. Steckler and Watts
(1978) and Watts and Steckler (1979) modelled
this effect by assuming the basement responds to
the water load as an Airy-type crust. This is
justified in the case of the Late Cretaceous sea-
level highstand because of its large areal extent.
However, near the shoreline flexural effects are
important (Chappell, 1974) and should be included.
The magnitude of the long-term sea-level
changes through time is a subject of controversy
at the present time (Vail et al., 1977; Pitman,
1978; Watts and Steckler, 1978; Bond, 1978).
Pitman (1978) estimated that sea-level has fallen
by about 350 meters since the Late Cretaceous
using changes in the volume of mid-ocean ridge
crests through time. This estimate appears to
agree with that of Sleep (1976), based on the
present elevation of Cretaceous sediments in a
tectonically undisturbed region of the continental interior. Watts and Steckler (1979) estimated that sea-level has fallen by less than
about 200 meters since the Late Cretaceous using
well data off eastern North America. Their values
yield a minimum estimate, but are in better agreement with the magnitudes estimated from continental flooding (Wise, 1974; Bond, 1978).
Vail et al. (1977) have proposed that superimposed on the long-term changes in sea-level are
short-term changes. Based on the recognition of
sedimentary sequences on seismic reflection
profiles they have recognized slow rises in sea-
level followed by abrupt falls. The major periods
of sea-level rise were grouped by Vail et al.
(1977) into supercycles, which range from about 5
to 100 m.y. in duration. There is considerable
debate at present, however, whether supercycles
identified by Vail et al. (1977) represent sea-
level changes or whether they represent tectonic
changes (for example, Bally, 1980; Watts et al.,
in press).
The use of the different curves for long-term
changes in sea-level significantly affects the
tectonic subsidence at a margin (Watts and
Steckler, 1979). For example, if the Pitman

(1978) curve had been used at the Oneida O-25 well
off Halifax, Nova Scotia the tectonic subsidence
increases linearly with time. However, if the
Watts and Steckler (1979) curve is used the tectonic subsidence exponentially decreases with
time.
The effects on the tectonic subsidence of
short-term changes in sea-level are more difficult to evaluate. Tectonic subsidence curves
based only on the different long-term sea-level
curves are, however, remarkably smooth (Steckler
and Watts, 1978; Watts and Steckler, 1979; Keen,
1979) and show little evidence of abrupt rises
and falls such as would be expected if sea-level
was rapidly varying. This suggests, although
there are still uncertainties on the detailed
paleobathymetry, that the tectonic subsidence
at margins is not significantly affected by
short-term sea-level changes.

d) Basement Response

The response of the basement to sediment and
water loads at a continental margin have been
modelled either as Airy-type or as flexure of a
thin elastic plate overlying a weak fluid (Gunn,
1944; Walcott, 1972; Cochran, 1973; Chappell,
1974; Watts and Ryan, 1976; Turcotte et al.,
1977; Steckler and Watts, 1978; Watts and
Steckler, 1979, 1981; Keen, 1979; Keen et al.,
1981). Seismic reflection profiling indicates
that active faulting accompanies the early stages
of rifting and that the basement beneath the
margin often consists of horsts and grabens or
half-grabens (Boeuf and Doust, 1975; Ponte and
Asmus, 1976; Given, 1977; de Charpal et al.,
1978; Steckler and Watts, 1980). This suggests
that an Airy type model, in which the sediment
and water loads are locally compensated, is most
applicable early in the rifting history. The
presence of gently dipping post-rift sediments
and a wide coastal plain, however, suggest that
flexure is active later in margin evolution. In
the case of Airy loading the basement response
function (equation 1) is unity. If flexure is
active, then the deflection due to sediment and
water loads depends not only on the loads directly above it but also on the lateral strength of
the lithosphere and the load distribution. The
basement response function in equation (1) represents a weighting function relating the load to
the tectonic subsidence.
We show in Fig. 2 the tectonic subsidence at
the COST B-2 well off New York based on an Airy
model for the response of the basement to sediment and water loads. Fig. 2 shows that the
tectonic subsidence comprises a significant part
of the total sediment accumulation at the well.
The tectonic subsidence is generally smoother
than the sediment accumulation curves, indicating
that sedimentary processes such as compaction and
paleobathymetry have been satisfactorily accounted for by backstripping.

Flexure

The effective flexural rigidity of the oceanic lithosphere is a strong function of the age of the sea-floor at the time of loading (Watts, 1978; Caldwell and Turcotte, 1979; Forsyth, 1979; Watts et al., 1980). Rheological models of the oceanic lithosphere (Goetze and Evans, 1979; Bodine et al., 1981) indicate that the temperature dependence of the strength of rocks and hence, the equivalent elastic thickness T_e, can be modelled as the depth to the 500° isotherm of the oceanic lithosphere (Watts, 1978; Watts et al., 1980; Bodine et al., 1981). Watts et al. (1980) have shown that free-air gravity anomalies over loads which form on or near a ridge crest can be adequately explained by $T_e = 5$ km while loads formed off-ridge can be explained by $T_e = 25$ km.

In most current models for the evolution of continental margins (Sleep, 1971; Kinsman, 1975; Steckler and Watts, 1978; McKenzie, 1978; Falvey and Middleton, 1981) the post-rift tectonic subsidence is caused by thermal contraction following heating and thinning of the lithosphere at the time of rifting. Thus we would expect

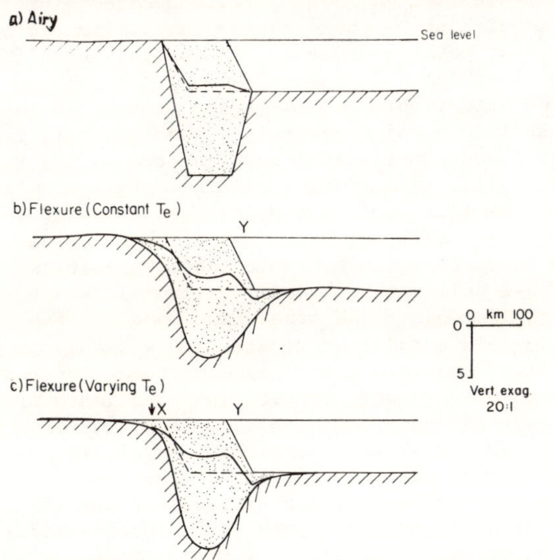

Figure 3. Theoretical cross-sections across a continental margin due to different models of sediment loading. The assumed tectonic subsidence in each model is delineated by the dashed line and reaches a maximum value of 5 km beneath the shelf. a)-Airy model, b)-Flexure model with constant elastic thickness, c)-Flexure model with varying elastic thickness. The solid lines indicate the positions of a theoretical stratigraphic horizon in which 2/3 of the tectonic subsidence lies below and 1/3 above. This corresponds approximately to the depth of the Jurassic/Cretaceous boundary off eastern North America (Fig. 1). The basement response for the three models is discussed in the text.

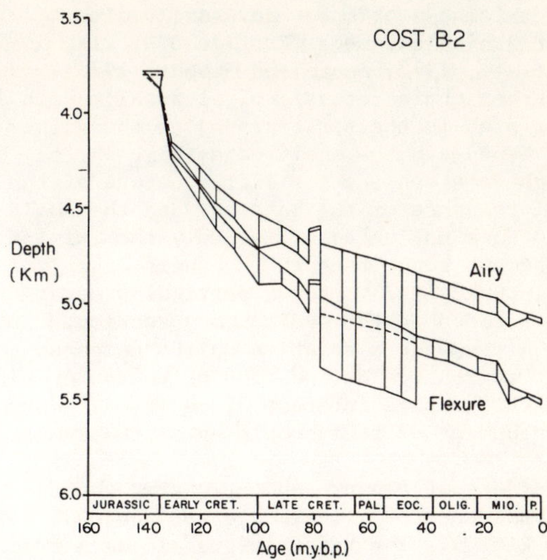

Figure 4. Tectonic subsidence at the COST B-2 well based on different isostatic models for backstripping sediments through time. The upper curve is based on the Airy model and the lower curve is based on the flexure model with an effective flexural rigidity that increases with time.

that T_e at a margin would increase with time as the basement cools.

We illustrate in Fig. 3 the response of the basement at a margin to sedimentary loads using both Airy and flexure models. In the models it is assumed that the tectonic subsidence (shown by a dashed line in Fig. 3) increases steeply seaward of the original shore-line to a constant value of 5 km beneath the continental shelf and slope. In Fig. 3a it was assumed that the lithosphere responds in an Airy manner and the shape of the resulting sedimentary basin exactly matches that of the tectonic subsidence. Fig. 3c assumed a flexure model in which the first 2/3 of the sediments at a continental margin loads a 5 km thick elastic plate and the remaining 1/3 loads a 25 km thick plate. The main effect of using a model in which the basement rigidity increases with time is that younger sediments progressively overstep the older sediments. This is most easily seen (Fig. 3) at the edges of the basin and should be recognized in seismic reflection profiles and well data from margins.

The proportions of the sediments assumed in Fig. 3 correspond closely to that of the Jurassic and Cretaceous/Tertiary sections off eastern North America. Thus Fig. 3 suggests that the Jurassic sediments pinch out beneath the Cretaceous/Tertiary sediments, forming a subcrop near the hinge zone. This subcrop is, in fact, observed along most of the margin off eastern North America (Fig. 1). For example, neither the Island Beach well nor the COST GE-1 well (Profiles BB' and DD', Fig. 1) sampled Jurassic sediments and the Hatteras Light-1 only penetrated a small thickness of Upper Jurassic sediment before reaching basement (Profile CC', Fig. 1; Brown et al., 1972; Amato and Bebout, 1978).

A further characteristic of flexural models is an outer high in the stratigraphy at the point labeled Y beneath the shelf edge (Fig. 3b, c). This high is a consequence of the nature of the flexural response of the basement and the shelf edge and does not necessarily imply the existence of a basement ridge beneath the shelf edge (for example, Burk, 1967). For a particular stratigraphic horizon in the sedimentary section, however, it suggests relative uplift compared to adjacent regions. The location of the outer high is of stratigraphic interest since it may control the development of reef complexes at the shelf edge.

The effect of flexure on estimating the tectonic subsidence is illustrated for the COST B-2 well in Fig. 4. The biostratigraphic data from the well was corrected for compaction, paleo-bathymetry and sea level. The upper curve was backstripped using an Airy response to the sedimentary load. For the lower curve the sedimentary section across the continental margin at the COST B-2 well was divided into six time sections, each of which was flexurally backstripped using successively higher flexural rigidities. The elastic thickness used in the calculations was the average depth to the 500° isotherm for the extension model of rifting (McKenzie, 1978) with β = 3.7. This value satisfactorily explains the overall shape of the tectonic subsidence and the total sediment thickness at the well (Watts, 1981).

The tectonic subsidence predicted by flexural backstripping is 10% larger than the estimated using an Airy response. In a flexural model, the deflection due to the load is distributed over a larger area and the deflection directly beneath the load is smaller. The tectonic subsidence needed to account for sediment accumulation in the center of a basin, therefore, is larger. For other locations across the margin, the errors induced by not including flexure can be substantially greater. For example, most of the subsidence of the coastal plain may be due to flexure.

Thermal and Mechanical Models

The result of backstripping sediment and water loads through time is to obtain the tectonic subsidence at a margin (for example, Fig. 2). The tectonic subsidence can then be compared to different geophysical models for the origin of margins. Most recent studies indicate that the post-rifting subsidence of margins is thermal in origin (Sleep, 1971; Kinsman, 1975; Steckler and Watts, 1978; McKenzie, 1978; Royden et al., 1980; Royden and Keen, 1980; Falvey and Middleton, 1981; LePichon and Sibuet, 1981; Beaumont et al., in press). These studies suggest that following heating and crustal thinning at the time of

rifting, the lithosphere cools and subsides slowly with time due to thermal contraction.

The actual geological processes that are active during rifting, however, are poorly understood. Among the processes that have been proposed are thermal uplift and erosion (Sleep, 1971; Kinsman, 1975), subcrustal flow (Bott, 1971, 1973), metamorphic phase changes (Falvey, 1974; Falvey and Middleton, 1981), necking or extension of the lithosphere (Artemjev and Artyushkov, 1971; McKenzie, 1978), pervasive dike intrusion (Royden et al., 1980), listric faulting (de Charpal, 1978), and depth-dependent extension (Royden and Keen, 1980). The relative importance of these factors, some of which are not mutually exclusive, is still uncertain.

The simple extension model proposed by McKenzie (1978) and its variations (Royden et al., 1980; Royden and Keen, 1980) have emphasized the syn-rift subsidence, produced by a balance between isostatic subsidence from crustal thinning and uplift due to thermal expansion. The simple extension model is a passive model of rifting in that all of the lithospheric heating is a result of the extension of the lithosphere, and predicts a large proportion of syn-rift to post-rift subsidence. The ratio of syn-rift to post-rift subsidence may, in fact, provide a constraint in distinguishing whether continental rifting is an active or passive process (Sclater and Christie, 1980; Steckler and Watts, 1980).

The main problem at margins, however, is that the sediments which accumulate soon after rifting often obscure the stratigraphy of the sediments formed prior to or during rifting. This complicates the interpretation of the tectonic subsidence. For example, Watts and Steckler (1981) have shown that the tectonic subsidence at the COST B-2 well can be equally well explained by the cooling plate and stretching models. The cooling plate model attributes the maximum amount of subsidence to thermal contraction and at the COST B-2 well results in an estimate of the tectonic subsidence that exceeds that of a mid-ocean ridge. The stretching model, in contrast, attributes the smallest amount of the subsidence to thermal contraction. A significant proportion of the subsidence in this model occurs prior to or during rifting. There is presently too little seismic and lithologic data beneath the COST B-2 well to satisfactorily distinguish between these models.

There is now general agreement, however, that sediment loading and thermal contraction are the main factors affecting the subsidence of continental margins. Flexure studies suggest that the response of the Earth to sedimentary loads is a strong function of temperature. We can expect therefore, a coupling between thermal and flexural properties of the lithosphere during the development of a margin.

We show in Fig. 5 a simple thermal and mechanical model for the development of a continental margin. The model is based on the uniform

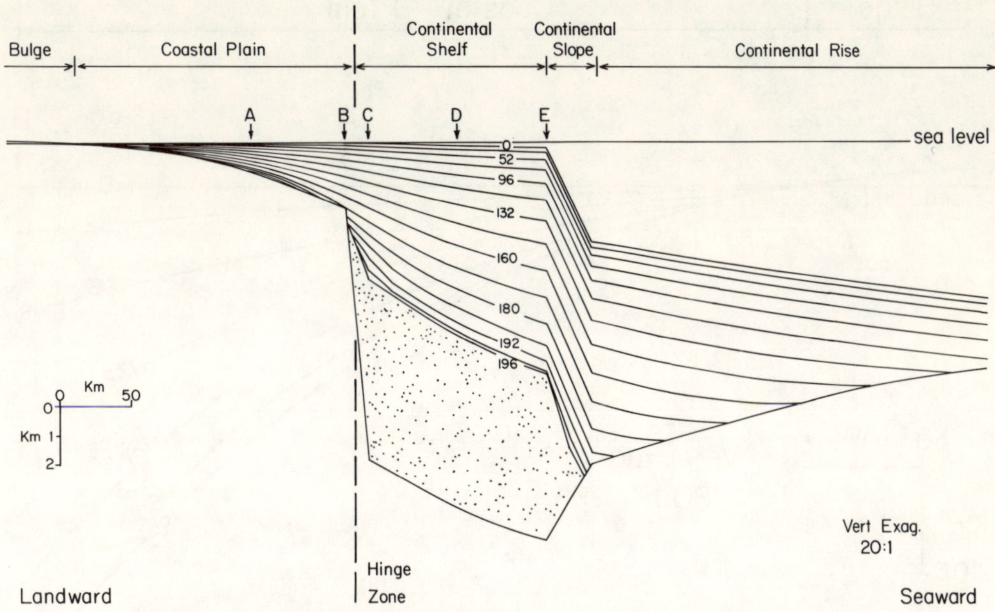

Figure 5.   Stratigraphy of a theoretical Atlantic-type continental margin based on a simple thermal and mechanical model. The thermal model is based on a stretching model (McKenzie, 1978) in which the lithosphere undergoes a passive extension at the time of rifting. Landward of the hinge zone is unstretched continental crust and lithosphere. From the hinge zone to the mid-point of the continental slope, the lithosphere has been stretched by $\beta = 3$. Seaward of the mid-point of the slope is oceanic lithosphere. The mechanical model is based on the flexure model of isostasy. The effective elastic thickness, which is assumed to be given by the depth of the 500° isotherm, varies across the margin. The margin is shown 196 m.y. after rifting with stratigraphic units every $\sqrt{t}$ time interval for the post-rift sediments and stippling for the syn-rift sediments.

stretching model (McKenzie, 1978) and a flexural rigidity of the lithosphere that increases with age during margin evolution. Similar models have recently been discussed by Keen et al. (1981) and Beaumont et al. (in press). The subsidence, temperature and heat flow in the model in Fig. 5, however, are computed using a modified form of the stretching model. In the modified model, a mid-ocean ridge includes a 5 km thick crust and still subsides correctly. A mid-ocean ridge, in the modified model, corresponds to $\delta = 6.24$ and $\beta = \infty$ (see Royden and Keen, 1980). An initial crustal thickness of 31.2 km and a coefficient of thermal expansion of $3.4 \times 10^{-5}°C^{-1}$ were assumed. The flexural rigidity is estimated from the depth to the 500°C isotherm (with Young's modulus of 6.5×10^{11} N-m and Poisson's ratio of 0.25). The model includes the effects of horizontal as well as lateral heat conduction.

The model (Fig. 5) assumes a simple margin in which there is no crustal thinning landward of the hinge, thinning by a constant factor of 3 in the transition region extending to the middle of the continental slope, and oceanic lithosphere seaward of the slope. Sediments of density 2.5 g/cm³ are assumed to progressively infill the subsiding

margin and to maintain a constant bathymetry profile through time. The stipple in Fig. 5 shows the syn-rift sediments (which are assumed to load an Airy-type crust) and the solid lines indicate individual stratigraphic horizons every $\sqrt{t}$ m.y. interval within the post-rift sedimentary sequence.

The simple model predicts a number of stratigraphic features at a margin including a well developed hinge zone, a seaward tilting of sediments beneath the outer continental shelf and, an onlap of stratigraphic horizons within the coastal plain (Figs. 5 and 6). The seaward tilting of the sediments and basement beneath the outer shelf and the lack of an outer high (Fig. 2) is due to loading in continental slope and rise regions (the model in Fig. 3, for example, does not include a slope and rise region) while the onlap of stratigraphic horizons within the coastal plain is mainly due to the increase in flexural rigidity with time.

Although the rifting history assumed in the model in Fig. 5 is simplified, the model is generally similar to the stratigraphic cross-section off New York (Fig. 1). We have not, however, included the effects of compaction, paleobathy-

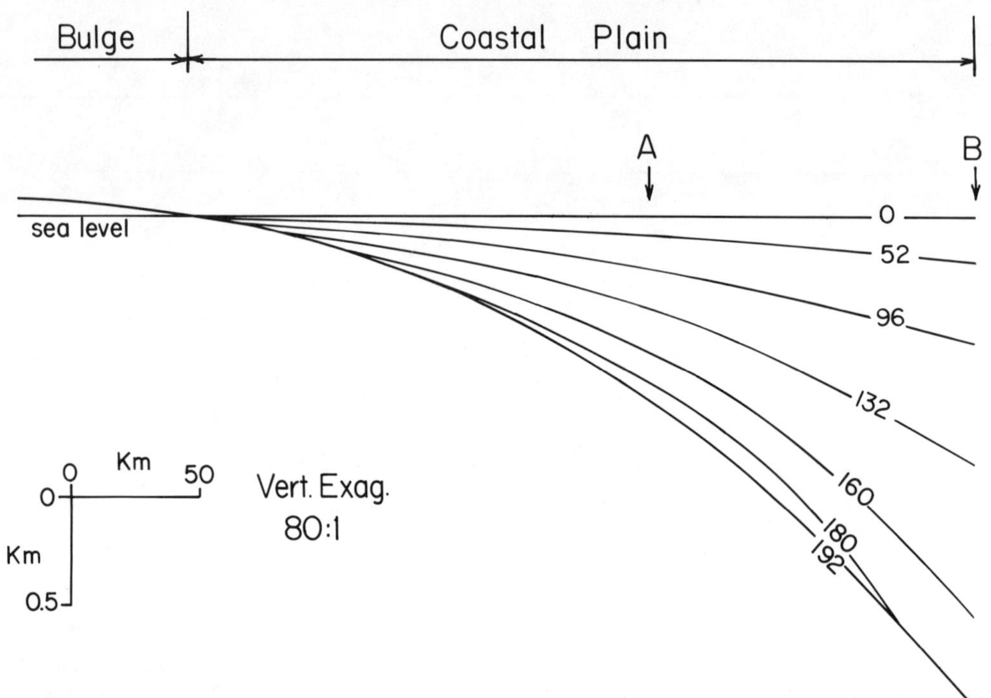

Figure 6.   Stratigraphy of the coastal plain region of the thermal and mechanical model in Fig. 5.   The stratigraphic units are shown every other $\sqrt{t}$ time interval. The coastal plain is characterized by a progressive onlap of younger sediments on older sediments.

metry and sea-level changes during the development of the margin.   The comparison of the model results with observations suggests, therefore, that sedimentary loading and thermal contraction are the major factors controlling the development of the margin off New York.   The model results in Figs. 5 and 6 are important since, in addition, they enable the effects of flexure and lateral heat flow during development of a margin to be examined.

The effect of flexure and lateral heat flow on the sediment accumulation and heat flow across the model of the margin in Fig. 5 is illustrated in Figs. 7 and 8.   The solid curves indicate the sediment accumulation (Fig. 7) and heat flow (Fig. 8) calculated at different locations A to E across the margin.   The dashed curves in Figs. 7 and 8 indicate the sediment accumulation and heat flow at C, D and E (shelf regions) that would be expected if lateral effects of flexure and heat flow had not been included.   Comparison of these curves show that flexure and heat flow have contrasting effects during the development of a margin.   Flexural effects are most significant later in margin evolution while lateral heat flow is most important early in margin evolution.   The sediment accumulation at C is less than expected because flexure becomes important near the hinge zone.   E, on the other hand, increases its sediment accumulation mainly because of loading in continental rise regions.

The heat flow at C is less than expected because of the effects of lateral flow of heat from the shelf to the region of unstretched crust landward of the hinge zone.   Heat flow at E, however, is larger because of lateral heat flow from oceanic lithosphere to the region of stretched crust.   The solid horizontal line in Figs. 7 and 8 indicate the sediment accumulation and heat flow at A and B (coastal plain) that would be expected if lateral effects of flexure and heat flow had not been included.   The sediment accumulation at A is greater than expected because of flexure.   B, on the other hand, actually is uplifted because of the additional heat flow into the region of unstretched crust. Flexure becomes increasingly important after about 10-16 m.y. and B subsides at similar rates to C.   Near the hinge zone, where lateral flexure and heat flow are largest, stresses may cause faulting and decoupling across the hinge zone for the first 10-15 m.y. after rifting, increasing the tendency for the coastal plain to remain emergent.

The model studies (Fig. 5 and 6) therefore suggest that lateral effects due to flexure and heat flow are important in the development of Atlantic-type continental margins.   These effects interact with each other during the development of a margin.   Early in margin development the effects of lateral heat flow are most significant and could cause uplift and

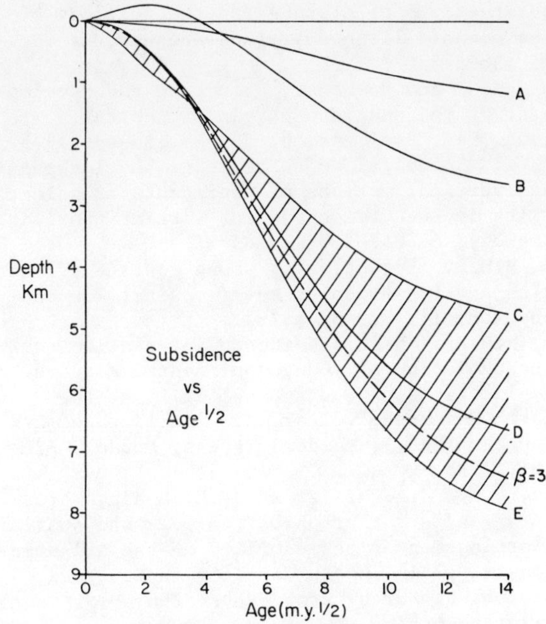

Figure 7. Subsidence vs. $\sqrt{t}$ time for the 5 locations (A to E) across the continental margin in Fig. 5 (solid curves). The dashed curve is the expected subsidence based on the stretching model with β = 3. The shaded region indicates the effect of lateral heat flow and flexure on the subsidence at points C, D, E. The two effects compete early in margin evolution. Later in margin evolution (>16 m.y.) flexural effects dominate.

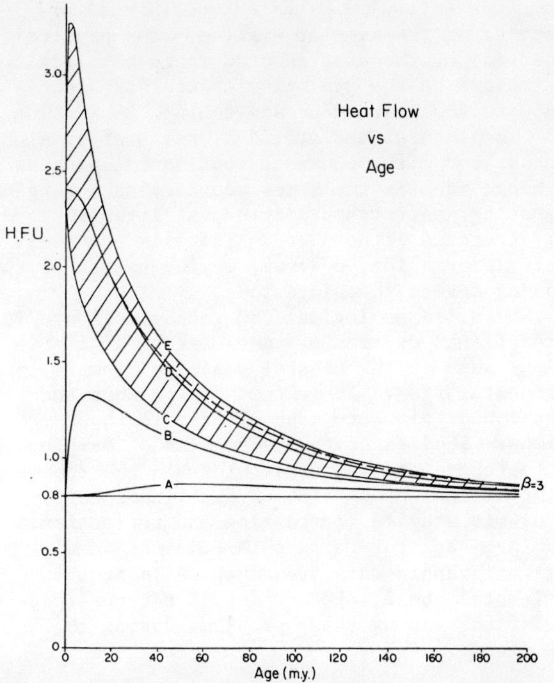

Figure 8. Heat flow vs. time for the 5 locations (A to E) across the margin in Fig. 6 (solid curves). The dashed curve is for the one-dimensional stretching model with β = 3. The shaded region indicates the effect of lateral heat flow (heat flow is unaffected by flexure) on the heat flow at points C, D, E. The effect is largest early in margin evolution. The heat flow gradually approaches the background heat flow (0.8 H.F.U.) of unstretched lithosphere later in margin evolution.

erosional unconformities in flanking regions of the newly formed ocean basin. Flexural effects overcome the effects of lateral heat flow later in margin development and cause broad downwarping of the crust and seaward tilting of strata on both sides of the hinge zone. Flexural effects are, therefore, significant enough that they should be taken into account in backstripping studies particularly for well or seismic data in the region of the hinge zone. In addition, lateral heat flow effects should be considered in thermal history studies of sediments at a margin. Although the present day heat flow at a margin is similar to that expected on the basis of the simple one-dimensional model (Fig. 8), the heat flow varies significantly during the history of the margin. Therefore, methods which estimate the maturation history of sediments by integrating the temperature-depth history at a margin should include lateral heat flow effects.

Summary

This paper has reviewed some of the factors that control the subsidence of Atlantic-type continental margins. The most important are sedimentary loading and thermal contraction of the lithosphere following heating and thinning at the time of rifting. Thermal and mechanical models have been constructed in which loading and thermal effects are coupled. These models satisfactorily explain the main stratigraphic features of margins including a well developed hinge zone, coastal plain onlap, outer highs and the transition from a relatively narrow basin at the time of rifting to a broad basin following rifting. Other factors controlling the subsidence include sea-level, paleobathymetry and compaction but these are small compared to sediment loading and thermal contraction.

There are a number of outstanding problems at continental margins. These include:

1. Detailed measurements of the subsidence and uplift history of continental margins. An important problem is to document the subsidence and uplift episodes that occur at widely separated margins. Backstripping techniques should be used in an attempt to separate local tectonic effects from widespread tectonic or eustatic effects.

2. Geological and geophysical studies of the crust and upper mantle at continental margins. Most mechanisms for the origin of the post-rift subsidence of margins assume the crust and litho-

sphere was extensively heated and thinned or stretched at the time of rifting. The physical properties of these attenuated rocks may provide constraints on the geological processes associated with the heating or stretching.

3. The lateral and vertical extent of stretched crust and lithosphere at continental margins. The hinge zone is the major boundary at a margin separating unstretched continental lithosphere from stretched lithosphere. There is presently little information, however, on the nature of the thinning across the hinge zone.

4. Detailed geological and geophysical studies of the flanks or rims of continental margins in regions such as the coastal plain and lower continental rise. The effects of flexure and lateral heat flow are most evident in these regions. Studies of flanks or rims of margins may therefore provide constraints on the thermal and mechanical properties of the lithosphere.

Seismic studies (refraction and reflection) with large aperture arrays in combination with biostratigraphic data from deep wells in the continental shelf, appear to hold the most promise of addressing these problems during the next decade.

Acknowledgments

This work has been supported by a Phillips Petroleum Fellowship to M.S.S. and National Science Foundation grant OCE 77-10647 to A.B.W. We thank R. Scrutton, J. Cochran and G. Karner for helpful comments on the manuscript and A. Hales for allowing us to revise and update the original manuscript. Lamont-Doherty Geological Observatory Contribution Number 2999.

References

Amato R.V. and J.W. Bebout, (Eds.), Geographical and operational summary COST No. GE-1 well, Southeast Georgia embayment area, South Atlantic OCS, U.S.G.S. Open File Report 78-668, 122pp., 1978.

Artemjev, M.E. and E.V. Artyushkov, Structure and isostasy of the Baikal rift and the mechanism of rifting, J. Geophys. Res., 76, 1197-1211, 1971.

Austin, J.A., E. Uchupi, D.R. Shaugnessy, and R.D. Ballard, Geology of New England Passive margin, Am. Assoc. Pet. Geol. Bull., 64, 501-526, 1980.

Bally, A.W., Basins and subsidence - A summary; in A.W. Bally, P.L. Bender, T.R. McGretchin and R.I. Walcott, (Eds.), Dynamics of Plate Interiors, Geodynamics Series, Vol. I, 5-20, 1980.

Beaumont, C., C.E. Keen, and R. Boutillier, On the evolution of rifted continental margins: Comparison of models and observations for the Nova Scotian Margin; Geophys. J. R. Astr. Soc., 1982, in press.

Bodine, J.H., M.S. Steckler, and A.B. Watts, Observations of flexure and the rheology of the oceanic lithosphere; J. Geophys. Res., 86, 3695-3707, 1981.

Boeuf, M.G. and H. Doust, Structure and development of the southern margin of Australia, Aust. Pet. Expl. Assoc. J., 15, 33-43, 1975.

Bond, G., Speculations on real sea-level changes and vertical motions of continents at selected times in the Cretaceous and Tertiary periods, Geology, 6, 247-250, 1978.

Bott, M.H.P., Evolution of young continental margins and formation of shelf basins; Tectonophysics, 11, 319-327, 1971.

Bott, M.H.P., Shelf subsidence in relation to the evolution of young continental margins, in Implications of continental drift to the Earth Sciences, vol. 2, D.H. Tarling and S.K. Runcorn (Eds.), Academic Press, London, 675-683, 1973.

Brown, P.M., J.A. Miller and F.M. Swain, Structural and stratigraphic framework and spatial distribution of permeability of the Atlantic coastal plain, North Carolina to New York, Geological Survey Prof. Paper 796, U.S. Govt. Printing Office, Washington, D.C.

Buffler, R.T., J.S. Watkins and W.P. Dillon, Geology of the offshore southeast Georgia embayment, U.S. Atlantic continental margin, based on multichannel seismic reflection profiles, in J.S. Watkins, L. Montadert, P.W. Dickerson (Eds.), Geological and Geophysical Investigations of Continental Margins, AAPG Memoir 29, 11-26, 1979.

Burk, C.A., Buried ridges within continental margins, Trans. New York Acad. Sci., 397-409, 1967.

Caldwell, J.G. and D.L. Turcotte, Dependence of the thickness of the elastic oceanic lithosphere on age; J. Geophys. Res., 84, 7572-7576, 1979.

Chappell, J., Late quaternary glacio- and hydro-isostasy on a layered Earth, Quaternary Res., 4, 405-428, 1974.

Cochran, J.R., Gravity and magnetic investigations in the Guinea Basin, Western Equatorial Atlantic, GSA Bull., 84, 3249-3268, 1973.

Cochran, J.R., The magnetic quiet zone in the eastern Gulf of Aden: implications for the early development of the continental margin; Geophys. J. R. Astr. Soc., in press.

Cravatte, J., P.H. Dufaure, M. Prim and S. Rouaix, Les sondages du golfe du Lion: stratigraphie, sédimentologie, in Notes et Mémoires No. 11, Compagnie Francoise Pétroles, Paris, 209-274, 1974.

de Charpal, O., P. Guennoc, L. Montadert, and D.G. Roberts, Rifting, crustal attenuation and subsidence in the Bay of Biscay, Nature, 275, 706-711, 1978.

Dietz, R.A., Collapsing continental rises: An actualistic concept of geosynclines and mountain building, J. Geol., 71, 314-333, 1963.

Dillon, W.P., C.K. Paull, R.T. Buffler, and J-P. Fail, Structure and development of the southeast Georgia embayment and northern Blake Plateau: Preliminary analysis, in J.S. Watkins, L. Montadert and P.W. Dickerson, (Eds.), Geological and geophysical investigations of continental margins, AAPG Memoir 29, 27-42, 1979.

Englehardt, W. von, The origin of sediments and sedimentary rocks: E. Schweixerbert'sche Verlagsbuchhandlang, Stuttgart 2nd edition, trans. by W.D. Johns, 359p., 1977.

Falvey, D.A., The development of continental margins in plate tectonic theory, Aust. Pet. Explor. Assoc. J., 14, 95-106, 1974.

Falvey, D.A. and M.F. Middleton, Passive continental margins: Evidence for a prebreakup deep crustal metamorphic subsidence mechanism; Oceanologica Acta, Proceedings of 26th International Congr., Geology of Continental Margins Symposium, Paris, 1980, 143-153, 1981.

Forsyth, D.W., Lithospheric flexure; Rev. Geophys. Space Phys. 17, 1109-1114, 1979.

Given, M.M., Mesozoic and early Cenozoic geology, Bull. Can. Petr. Geol., 25, 63-91, 1977.

Goetze, C. and B. Evans, Stress and temperature in the bending lithosphere as constrained by experimental rock mechanics, Geophys. J. R. Astr. Soc., 59, 463-478, 1979.

Gradstein, F.M., G.L. Williams, W.A.M. Jenkins, and P. Ascoli, Mesozoic and Cenozoic stratigraphy of the American continental margin, eastern Canada, in Canada's continental margin and offshore petroleum exploration, C.J. Yorath, E.R. Parker, and D.J. Glass (Eds.), Can. Soc. Petr. Geol. Memoir 4, 103-130, 1975.

Grow, J.A. and R.G. Markl, IPOD-USGS multichannel seismic reflection profile from Cape Hatteras to the Mid-Atlantic Ridge, Geology, 5, 625-630, 1977.

Grow, J.A., R.E. Mattick, and J.S. Schlee, Multichannel seismic depth sections and interval velocities over outer continental shelf and upper slope between Cape Hatteras and Cape Cod, in Geological investigations of continental margins, J.S. Watkins, L. Montadert, and P.W. Dickerson (Eds.), A.A.P.G. Memoir 29, 65-83, 1979.

Gunn, R., A quantitative study of the lithosphere and gravity anomalies along the Atlantic coast, Franklin Inst. J., 237, 139-154, 1944.

Jansa, L.F. and J.A. Wade, Geology of the continental margin off Nova Scotia and Newfoundland, in Offshore geology of eastern Canada, Geological Surv. of Canada paper 74-30, v.2, 51-105, 1975.

Karner, G.D. and A.B. Watts, On isostasy at Atlantic-type continental margins, J. Geophys. Res., in press.

Keen, C.E., Thermal history and subsidence of rifted continental margins. Evidence from wells on the Nova Scotia and Labrador shelves, Can. J. Earth Sci., 16, 505-522, 1979.

Keen, C.E., C. Beaumont, and R. Boutillier, Preliminary results from a thermo-mechanical model for the evolution of Atlantic-type continental margins, Oceanological Acta, Proceed-

ings of 26th International Geol. Congr., Geology of Continental Margins Symposium, Paris, 1980, 123-128, 1981.

Kent, P., Major synchronous events in continental shelves, Tectonophysics, 36, 87-92, 1976.

Kinsman, D.J.J., Rift valley basins and sedimentary history of trailing continental margins, in A.G. Fisher and S. Judson (Eds.), Petroleum and global tectonics, Princeton U. Press, 83-128, 1975.

Le Pichon, X. and J.C. Sibuet, Passive Margins: A model of formation, J. Geophys. Res., 86, 3708-3720, 1981.

McKenzie, D.P., Some remarks on the development of sedimentary basins, Earth Planet Sci. Lett., 40, 25-32, 1978.

Parsons, B. and J.G.Sclater, An analysis of the variation of ocean floor bathymetry and heat flow with age, J. Geophys. Res., 82, 803-827, 1977.

Pitman, W.C.,III, The relationship between eustacy and stratigraphic sequences of passive margins, Bull. Geol. Soc. Am., 89, 1389-1403, 1978.

Ponte, F.C. and H.E. Asmus, The Brazilian marginal basins: Current state of knowledge, in de Almeida, F.F.M. (Ed.), Symposium on Continental Margins of Atlantic-type, São Paulo, Brazil, 1976.

Rhodehammel, E.C., Sandstone porosities, in P.A. Scholle (Ed.), Geological studies on the COST B-2 well, U.S. Mid-Atlantic outer continental shelf, U.S. Geol. Surv. Circ. 750, 23, 1977.

Royden, L., J.G. Sclater, R.P. von Herzen, Continental margin subsidence and heat flow: important parameters in formation of petroleum hydrocarbons, Amer. Assoc. Petr. Geol. Bull., 64, 173-187, 1980.

Royden, L. and C.E. Keen, Rifting processes and thermal evolution of the continental margin of eastern Canada determined from subsidence curves, Earth Planet Sci. Lett., 51, 343-361, 1980.

Schlee, J., J.C. Behrendt, J.A. Grow, J.M. Robb, R.E. Mattick, P.T. Taylor, and B.A. Lawson, Regional geologic framework off northeastern United States, Amer. Assoc. Petr. Geol. Bull., 60, 926-951, 1976.

Scholle, P.A., Geological Studies on the COST B-2 well, United States Mid-Atlantic outer continental shelf area, in geological studies of the COST B-2 well U.S. Mid-Atlantic outer continental shelf area, P.A. Scholle, (Ed.), U.S. Geol. Surv. Circ. 750, 1-3, 1977.

Sclater, J.G. and P.A. Christie, Continental stretching: An explanation of the post Mid-Cretaceous subsidence of the Central North Sea basin, J. Geophys. Res., 85, 3711-3739, 1980.

Sheridan, R.E., Sedimentary basins of the Atlantic margin of North America, Tectonophysics, 36, 113-132, 1976.

Sleep, N.H., Thermal effects of the formation

of Atlantic continental margins by continental breakup, Geophys. J. R. Astr. Soc., 24, 325-350, 1971.

Sleep, N.H., Platform subsidence mechanisms and "eustatic" sea-level changes; Tectonophysics, 36, 45-56, 1976.

Smith, M.A., R.V. Amato, M.A. Furbush, D.M. Pert, M.E. Nelson, J.S. Hendrix, L.C. Tamm, G. Wood, Jr., and D.R. Shaw, Geological and operational summary, COST no. B-2 well, Baltimore Canyon Trough area, Mid-Atlantic outer continental shelf, U.S. Geol. Surv., open-file rept. 76-744, 79pp, 1976.

Steckler, M.S., The thermal and mechanical evolution of Atlantic-type continental margins, Ph.D. Thesis, Columbia Univ., 261p. 1981.

Steckler, M.S. and A.B. Watts, Subsidence of the Atlantic-type continental margin off New York, Earth Planet Sci. Lett., 41, 1-13, 1978.

Steckler, M.S. and A.B. Watts, The Gulf of Lion: Subsidence of a Young Continental Margin, Nature, 287, 425-429, 1980.

Turcotte, D.L., J.L. Ahern, and J.M. Bird, The state of stress at continental margins, Tectonophysics 42, 1, 1977.

Tucholke, B.E. and G.S. Mountain, Lithostratigraphy and Paleosedimentation patterns in the North American basin, Maurice Ewing Symposium, Series 3, AGU Washington, 58-86,1979.

Vail, P.R., R.M. Mitchum, Jr. and S. Thompson, III, Part Four: Global cycles of relative changes of sea level, Amer. Assoc. Pet. Geol. Memoir 26, 83-98, 1977.

Van Hinte, J.E., Geohistory analysis - application of micropaleontology in exploration geology, Am. Ass. Petr. Geol. Bull., 62, 201-222, 1978.

Von Rad, U. and Arthur, M.A., Geodynamic, sedimentary and volcanic evolution of the Cape Bojador continental margin (NW Africa), Maurice Ewing Symposium series 3, AGU Washington, D.C., 187-204, 1979.

Walcott, R.I., Gravity, flexure and the growth of sedimentary basins at a continental edge, Geol. Soc. Amer. Bull, 83, 1845-1848, 1972.

Watts, A.B., An analysis of isostasy in the world's oceans, 1. Hawaiian-Emperor seamount chain, J. Geophys. Res., 83, 5989-6004, 1978.

Watts, A.B., The U.S. Atlantic continental margin: subsidence history, crustal structure and thermal evolution; Amer. Assn. Petrol. Geol. Education Course Note Series # 19, 75pp., 1981.

Watts, A.B. and W.B.F. Ryan, Flexure of the lithosphere and continental margin basins, Tectonophysics, 36, 25-44, 1976.

Watts, A.B. and M.S. Steckler, Subsidence and eustasy at the continental margin of eastern North America, Maurice Ewing Symposium Series 3, A.G.U., Washington, D.C., 218-234, 1979.

Watts, A.B., J.H. Bodine, and N.M. Ribe, Observations of flexure and the geological evolution of the Pacific ocean basin, Nature, 283, 532-537, 1980.

Watts, A.B. and M.S. Steckler, Subsidence and tectonics of Atlantic-type continental margins, Oceanologica Acta, Geology of Continental margins Symp., Proceedings of 26th Intern. Congr., Paris, 4, 143-153, 1981.

Watts, A.B., G.D. Karner, and M.S. Steckler, Crustal flexure and the driving mechanism of sedimentary basin formation, Proc. Royal Society, London, in press.

Williams, G.L., Dinoflagellate and spore stratigraphy of the Mesozoic-Cenozoic, offshore eastern Canada, in Offshore geology of Eastern Canada, Geological Survey of Canada paper 74-30, 107-161, 1975.

Wise, D.U., Continental margins, freeboard and the volumes of continents and oceans through time, in C.A. Burke and C.L. Drake (Eds.), The geology of continental margins, 45-58, 1974.

SUBSIDENCE OF CONTINENTAL MARGINS: THE CASE FOR ALTERNATIVES TO THERMAL CONTRACTION

L.L. Sloss

Department of Geological Sciences, Northwestern University, Evanston, Illinois 60201

Abstract. Initial heating and consequent
uplift of trailing continental margins by the
creation of new ridge systems, followed by
cooling and subsidence of such margins is a
powerful and attractive concept quantitatively
supported by elegant modeling. However, basins
of cratonic interiors document subsidence
histories that are coincident in mode and in
timing with passive margins although no evidence
exists for localized uplift at basinal sites nor
for localized and synchronous heating events such
as would be required to precede thermal
contraction and subsidence. Investigators are
thus faced with a choice between two proposi-
tions: 1) thermal contraction is a subsidence
mechanism unique to continental margins and
another mechanism is required for the synchronous
downwarp of craton-interior basins; or, 2) a
single, globally effective mechanism exists to
explain the history and geometry of subsidence of
interior and continent-margin basins.

The proposition involving multiple unique
solutions is rejected; instead, the attention of
the geodynamics community is directed to a
previously enunciated, but unsubstantiated,
theory that invokes variation in the rate of flow
of melt from continental to oceanic asthenosphere. Such variation is postulated to result
from alternate trapping of melt beneath
continents, causing uplift, and draining of melt
to oceanic asthenosphere circulation, causing
continental deflation and buckling accompanied
by accelerated sea-floor spreading. The theory
aids in unifying mechanisms responsible for the
subsidence of cratonic and passive-margin basins,
eustatic sea-levels, convergent-margin orogenesis, and other concomitants of plate
tectonics.

Introduction

Entia non sunt multiplicanda praeter
necessitatum - William of Occam ca. 1300.

Mechanisms responsible for the subsidence of
continental crust to accommodate significant
thicknesses of sediments at passive margins have
been the subject of continuing speculation over
the past decade. As reviewed by Bott [1979], the
more acceptable hypotheses include gravity
loading, thermal contraction, and crustal creep.
Watts and Ryan [1976], with amplification by
Watts and Steckler [1979], have shown that
sediment loading is inadequate to account for the
amplitude of subsidence; these workers appeal to
thermal contraction as the driving force in
subsidence. Thermal contraction as a subsidence
mechanism has been investigated with rigor by
Sleep [1971 and subsequent papers including this
issue]; variants involving thermally-induced
phase changes [e.g., Falvey, 1974] have been
suggested. Oceanward creep of lower continental
crust [Bott and Dean, 1972] is conceived as
resulting from topographically-created gravity
potentials at continent-ocean margins.

Thus, among currently fashionable subsidence
models, sediment loading and concomitants in
terms of eustatic sea levels [Pitman, 1978; Vail
and others, 1977] are demonstrably inadequate
unless amplified by other "driving forces",
leaving thermal contraction and crustal creep as
apparently viable hypotheses. Of these, thermal
contraction is by far the more popular; after
all, the mechanism is susceptible to elegant
modeling, the existence of a heat source at
spreading axes is widely acknowledged, and the
predicted exponential decay of contraction rates
appears to be consonant with the time scale and
decline of subsidence rates. Stratigraphers
(this writer is one such), perhaps because they
cannot see the light under a bushel of burden-
some detail, perhaps because of a natural
distrust in any model requiring partial-
differential expressions, or, more charitably,
because of a wider experience in sedimentary
basins and their tectonic evolutions, tend to
search for other solutions than those accepted,
somewhat myopically, by geophysicists. This
paper reviews one stratigrapher's reservations
with respect to the thermal contraction
hypothesis and reiterates a potential alternative.

The Data

Stratigraphic data derived from individual
exposures or individual boreholes are inherently

variable as a consequence of the vagaries of
erosion, shifting depocenters, local deformations and other factors, both random and
systematic, that affect observed sedimentary
thicknesses. Such variations, ignored or
passed over commonly as irrelevant noise by the
geophysics community, constitute essential
signals to the stratigrapher concerned with
environmental reconstruction or the migration and
entrapment of pore fluids. In response to
stratigraphic concerns, many syntheses and
compilations in the form of cross sections, and
other displays, which constitute source materials
for geophysical modelers, tend to emphasize
differences from datum point to datum point,
making it difficult for the modeler to choose
the "typical" or "average" set of sedimentary
thicknesses for a given basin or shelf region.
This circumstance, in turn, works to obscure
similarities from region to region and from
continent to continent.

Many regions of sedimentary accumulation have
been penetrated by sufficient numbers of boreholes and crossed by enough reflection-seismo-
graph profiles to permit the construction of
isopach maps of closely successive stratigraphic
units. Such maps are useful integrations of
sedimentary thicknesses and, therefore, of
subsidence histories. Digitization of isopach
maps yields two parameters: total volume and
mean thickness of each stratigraphic unit.
Volume is a measure of the degree to which
crust or subcrust has been displaced, deformed,
or increased in density to accommodate the
sedimentary mass. Mean thickness is a measure
of basinal subsidence independent of basin size.
Both parameters have been applied to analysis of
the subsidence histories of sedimentary basins,
producing results that are somewhat at variance
with those derived from individual boreholes or
from interpretation of seismic-reflection
profiles [Sloss, 1979].

The accompanying figure illustrates the
subsidence history of the Michigan Basin from
Middle Ordovician through Early Carboniferous.
The data derive from digitization of isopach
maps and show the net subsidence of the basin
area in terms of the volume of sediment deposited
and preserved. Note that part of Early
Ordovician and much of Early Devonian time are
not represented by preserved sediment, indicating
non-deposition and/or erosion. A similar major
stratigraphic lacuna intervenes between Early
and Late Carboniferous.

The Dilemma

Investigation of my admittedly small sample
of craton-interior basins and craton-margin
depositional sites reveals four salient
characteristics:

1) Where data permit analysis covering long
spans of time (typically > 100 million years),
subsidence history cannot be fit to a simple

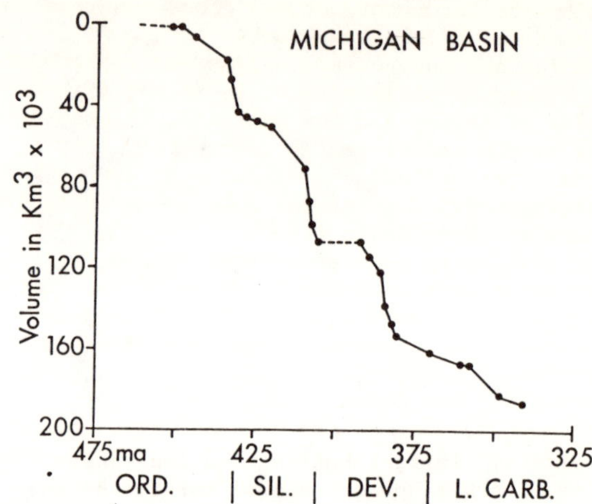

Figure 1. Middle Paleozoic subsidence curve of
the Michigan Basin expressed as net volume of
sediment preserved. Data sources as in Sloss
[1979].

exponential decline in rates of subsidence.

2) Distant basins exhibit a surprising degree
of synchrony, sharing episodes of rapid and slow
subsidence and of relative stability.

3) Basins of cratonic interiors and basins of
passive continental margins are broadly (sometimes closely) synchronous in terms of subsidence
events.

4) Basinal subsidence, both in the interior
and at margins, tends to exhibit distinct and
globally concurrent episodes dominated by either
relatively slow flexural downwarp or by more
rapid subsidence accompanying fracture and
faulting [Kent, 1977].

If repeated episodes of interior- and marginal-
basin subsidence reflect thermal contraction,
then episodic local heating events are called
for. No evidence exists for a repetition of
such events in the thermal history of basins,
nor is there any record of localized uplift and
erosion at basinal sites such as would precede
cooling and downwarp. Instead, although basin
interiors are commonly exposed to subaerial
erosion between periods of active subsidence,
both the elevation above baselevel and the
duration of non-deposition in basin interiors
are markedly lower than on surrounding areas.
These observations indicate that basins maintain
tendencies of relative subsidence even during
times of broad cratonic emergence. When it is
further noted that many subsidence events start
slowly, build to a climax and decline in a
pattern more sinusoidal than exponential, it
becomes difficult to accept a thermal-contraction
model. An equal strain on thermal concepts is
imposed by the apparent synchrony among distant
basins, since simultaneous subsidence would
require simultaneous precursor heating; as noted,
evidence of suitable cratonic hot spots cannot
be found, let alone coincident and repeated

heating/cooling events at positions hundreds to thousands of kilometers apart.

Heating and subsequent cooling and depression of a divergent plate margin, as at the Atlantic margins of Africa and the Americas, is beautifully simple and elegantly documented. But why do interior basins and basins at nondivergent margins share the subsidence histories of margins of Atlantic type?

The similarity of craton-interior and continental-margin basins is not confined to the timing of subsidence episodes but extends to coincidence in changes of mode of subsidence from broadly flexural to fault controlled and vice versa. The later Jurassic – Early Cretaceous opening of the Atlantic, was, to be sure, accompanied by related rifting of adjacent cratonic blocks and the development of fault-bounded basins confined within a few hundred km of the newly formed plate boundaries; these troughs are readily accommodated by prevailing concepts. Cratonic history is marked, however, by other periods characterized by rapidly subsiding, fault-related basins that are not clearly accompaniments of continental rifting or other obvious manifestations of plate divergence or convergence. From late in Early Carboniferous (Namurian) time to Late Triassic and Early Jurassic all cratons were occupied by basins of rapid subsidence and rapid sediment accumulation. Many were plainly bounded by high-angle faults (some adjacent to complementary uplifted blocks that became detrital sediment sources); others are asymmetric in a manner suggestive of control by deep-seated basement faulting. In North America (as in South America and southern Africa) such basins, although reduced in amplitude of subsidence as the center of the craton is approached, extend from margin to margin and appear to be independent of the nature (convergent, divergent, or transcurrent) of inter-plate motions at the margins.

It is popular to dismiss the faulted Late Triassic – Early Jurassic basins of the eastern U.S. as precursors of Atlantic rifting. Late Paleozoic basins of similar style in adjacent areas of New England and the Maritimes are not so judged, however, and these have synchronous counterparts from Illinois to Utah and from Michigan to Texas (and from the Yukon to the Arctic Islands). It is more reasonable to consider all these late Paleozoic – early Mesozoic basins (and the Gondwana basins of the southern hemisphere and certain of the Hercynian basins of Europe) as representative of a widely-sensed episode during which a special tectonic mode pervaded all cratons. Similar globally effective episodes may be identifiable in the late Precambrian and Cenozoic; none is capable of being viewed with comfort and equanimity as a simple corollary of parochial thermal events.

In sum, then, although thermal contraction provides a delightful explanation for some tens of millions of years of subsidence at passive margins, it fails to satisfy the demands of concurrent interior and margin subsidence or the much longer, interrupted time spans of downwarp of interior basins; nor does the theory aid in understanding significant, globally effective, and synchronous shifts in the mode and tempo of subsidence. Thus, in view of the demonstrable inadequacy of eustatic sea-level change and concomitant loading as a first-order factor in subsidence and with recognition of the apparently insuperable problems raised by thermal contraction, students of passive continental margins are faced with the same dilemma as that which confronts colleagues concerned with cratonic interiors.

A Possible Alternative

It is clear that a unified field theory capable of accommodating observations from both continental interiors and continental margins is not at hand, at least not in a form readily acceptable with equal enthusiasm by stratigraphers and geodynamicists. The bare outlines of an hypothesis were set forth a few years ago [Sloss and Speed, 1974] but, in the absence of certain elements of stratigraphic data and without further elaboration of geophysical theory, these thoughts have, understandably, had little impact. Continuing analysis of the evolution of sedimentary basins, in both the rate and geometry of subsidence, reinforce earlier observations and support the conclusions, if not the mechanism proposed. The significant observations are those (numbered 1 – 4) noted in previous paragraphs; other points requiring consideration are based on acceptance of the spacing of dated magnetic-anomaly stripes as indicative of the rate of spreading at divergent oceanic plate margins and, therefore of the magma discharge rate at mid-ocean ridges:

1) It follows that episodes of rapid spreading are matched by times of rapid plate consumption at convergent margins, with all the consequences of large-scale subduction.

2) As noted by other workers [e.g., Johnson, 1971], the rate of tectonic subsidence of cratonic basins rises and falls in harmony with the apparent intensity (= rate of convergence) or orogenesis at continental margins.

3) An obvious corollary is that some episodes of slow sea-floor spreading may be equated with times of relative tectonic stability and high freeboard of cratons and their margins.

4) Other episodes of relatively slow spreading at ridge axes, as today, are not times of quiescent, stable cratons, but are characterized by marked vertical movements within cratons, commonly associated with high-angle faulting.

5) The sedimentary-tectonic history of cratons prior to the preserved oceanic record of spreading rates is an indication of oceanic tectonic history over the past one-to-two billion years.

Sloss and Speed [1974] assume that a constant

radial heat flow continuously generates partial
melting of continental asthenospheres. When
such melt is trapped below cratons the latter
are relatively buoyant and are elevated with
respect to sea level. Upon release of subcontinental melt (or melt + solid) to convective
flow of the oceanic asthenosphere, extrusion at
mid-ocean ridges is increased (initiating the
sequellae of accelerated plate motions) and
continents deflate with consequent buckling at
pre-existing sites of upwarp (arches) and downwarp (basins and passive margins). Pre-Late
Carboniferous Phanerozoic history suggests three
cycles of alternating cratonic inflation and
deflation. These cycles are termed "submergent"
and "emergent" episodes by Sloss and Speed [1974]
and are represented in the Paleozoic by the first
three cratonic stratigraphic sequences [super-
sequences of Mitchum, Vail, and Thompson, 1977]
and their bounding unconformities.

It would appear that sealing off asthenospheric
flow from continents to oceans does not always
result in simple continental emergence followed
by slow deflation and buckling as continental-
oceanic flow patterns are re-established. The
postulate here discussed suggests that periods
of continental elevation accompanied by abrupt
vertical movements, volcanism and plutonism of
cratonic interiors, and spreading axes of
back-arc basins ["oscillatory" episodes of Sloss
and Speed, 1974] are caused by rapid and
pulsatory release of asthenospheric melt along
short pathways within cratons and at their
margins.

The Sloss/Speed proposition accommodates much
of the body of stratigraphically oriented data
familiar to students of cratonal tectonics but
it suffers a weakness in lacking an explanation
for the synchronous trapping and release of melt
from separated cratons. This deficiency could
be overcome if there were a theoretical basis
for episodic variation in the rate of global
radial heat flow. More important is the
suggestion that the "pulse of the earth" is
detectable and that a controlling global
mechanism exists under which sea-floor spreading,
sea-level change, continent-margin orogenesis,
and the subsidence of basins of craton interiors
and passive margins are all second-order
manifestations.

Acknowledgments. Although many of the thoughts
expressed in this paper are antithetical to those
of Norman Sleep, he has provided valued advice
and counsel which are deeply appreciated, as are
the efforts of Dorrel Edstrand in data analysis
and in preparation of manuscript and illustra-
tion. Much of the research reported here has
been supported by National Science Foundation
grants EAR-7622499 and EAR-7813644.

References

Bott, M.H.P., Subsidence mechanisms at passive
continental margins, _In_: J.S. Watkins, L.
Montadert, and P.W. Dickerson (eds.),
Geological and Geophysical Investigations of
Continental Margins, American Association of
Petroleum Geologists, Memoir 29, pp. 3-9, 1979.

Bott, M.H.P., and D.S. Dean, Stress systems at
young continental margins, _Nature (Phys. Sci.)_,
_235_, 23-25, 1972.

Falvey, D.A., The development of continental
margins in plate tectonic theory, _J. Australian
Petroleum Exploration Assoc._, _14_, 95-106, 1974.

Johnson, J.G., Timing and coordination of orogenic, epeirogenic and eustatic events, _Geol.
Soc. America Bull._, _82_, 3263-3298, 1971.

Kent, P.E., The Mesozoic development of aseismic
continental margins, _J. of the Geological Soc._,
_134(1)_, 1-18, 1977.

Mitchum, R.M., Jr., P.R. Vail, and S. Thompson,
III, The depositional sequence as a basic unit
for stratigraphic analysis, _In_: Seismic
Stratigraphy -- Applications to Hydrocarbon
Exploration, American Association of Petroleum
Geologists, Memoir 26, pp. 53-62, 1977.

Pitman, W.C., III, The relationship between
eustacy and stratigraphic sequences of passive
margins, _Geol. Soc. America Bull._, _89(9)_,
1389-1403, 1978.

Sleep, N.H., Thermal effects of the formation of
Atlantic continental margins by continental
break-up, _Geophysical J. of the Royal Astron.
Soc._, _24_, 325-350, 1971.

Sloss, L.L., Global sea level change: A view
from the craton, _In_: J.S. Watkins, L.
Montadert, and P.W. Dickerson (eds.), Geological and Geophysical Investigations of Continental Margins, American Association of Petroleum
Geologists, Memoir 29, pp. 461-477, 1979.

Sloss, L.L., and R.C. Speed, Relationships of
cratonic and continental margin tectonic
episodes, _In_: W.R. Dickinson (ed.), Tectonics
and Sedimentation, Society of Economic
Paleontologists and Mineralogists Special
Publication No. 22, pp. 98-119, 1974.

Vail, P.R., R.M. Mitchum, Jr., and S. Thompson,
III, Global cycles of relative changes of sea
level, _In_: Seismic Stratigraphy -- Applications of Hydrocarbon Exploration, American
Association of Petroleum Geologists, Memoir 26,
pp. 83-97, 1977.

Watts, A.B., and W.B.F. Ryan, Flexure of
lithosphere and continental margin basins,
Tectonophysics, _36_, 25-44, 1976.

Watts, A.B., and M.S. Steckler, Subsidence and
eustasy at the continental margin of eastern
North America, _In_: Implications of Deep
Drilling Results in the Atlantic Ocean,
American Geophysical Union, Ewing Volume,
(in press), 1979.